Climate–Vegetation

PLANT AND VEGETATION

Volume 4

*Series Editor:* M.J.A. Werger

For further volumes:
http://www.springer.com/series/7549

# Climate–Vegetation

## Afro-Asian Mediterranean and Red Sea Coastal Lands

by

**M.A. Zahran**

*Deptartment of Botany*
*Faculty of Science*
*Mansoura University*
*Mansoura 35516*
*Egypt*

Edited by

**F. Gilbert**

*School of Biology*
*Nottingham University*
*Nottingham NG72RD UK*

Springer

M.A. Zahran
Department of Botany
Faculty of Science
Mansoura University
Mansoura 35516
Egypt
zahrancabi2001@yahoo.com

ISBN 978-94-007-3181-3 e-ISBN 978-90-481-8595-5
DOI 10.1007/978-90-481-8595-5
Springer Dordrecht Heidelberg London New York



Softcover reprint of the hardcover 1st edition 2010



Printed on acid-free paper

Springer is part of Springer Science+Business Media (www.springer.com)

*To Egyptian – British collaboration*
*To Prof M. Kassas, Cairo University, who gave me much to learn*
*To the soul of the late Prof. A.J. Willis, Sheffield University (UK), co-author of three of my books*
*To Dr. Francis Gilbert, Nottingham University (UK)*
*To my colleagues and students*
*To my family: Ekbal, Amal, Eman and Ahmed*

# Foreword

This is a treatise that surveys and interprets features of plant life in two maritime basins: the Mediterranean (the southern sector) and the Red Sea.

The two enclosed water bodies have very different geological histories: the former is what remains of a great ocean (the Tethys Sea), and the latter is part of the less-ancient Great Rift Valley. Their geographical alignments are very different. The Mediterranean extends east–west, spanning ca. 40° longitude and ca. 10° latitude. The Red Sea extends north–south, spanning about 20° latitude. These geographical alignments are associated with different climate patterns. The Mediterranean has its characteristic climate with sub-divisions based on the volume of rainfall. In the Red Sea basin, a tropical (summer rainfall) climate prevails in the southern part and winter rainfall in the northern part, but aridity is rampant. The Red Sea is a corridor that links tropical seas (the Indian Ocean) and the temperate (Palaearctic) world to the north. The opening of the Suez Canal in 1869 made the link real, and allowed a number of tropical marine species to enter the Mediterranean.

The two seas have a common feature: the both have a hydrological deficit because evaporation exceeds water income (rainfall and river discharges). These deficits are compensated for by inflows from the Atlantic through the Straits of Gibraltar, and from the Indian Ocean through Bab-al-Mandeb.

The cultural histories of the two basins are also different. The Mediterranean claims justifiably to be the cradle of civilization. The birth of agriculture occurred on the southeastern edges of its basin (the Fertile Crescent). Human settlements have a very long history all over this basin. Human impacts on the natural vegetation are ecological factors producing extant forms of plant growth that are in most instances different – or very different – from the pristine forms. Sustained human occupation was not a special feature of the history of the Red Sea basin. It may have been the route for the northward spread of early man from his inception in East Africa towards the Mediterranean basin and beyond: the Red Sea remains one of the principal trade routes of the world.

To write a book dealing with this great ecological diversity is a Herculean venture. The author, Professor M.A. Zahran, depends on his first-hand experience with several parts of the North African Mediterranean basin and the Red Sea coastal lands in Egypt and Saudi Arabia, and his familiarity with the extensive literature on the ecology of the two basins. He is a well-seasoned fieldworker in surveys of the vegetation. The text of this welcome book displays these attributes: surveys of plant growth, ecological analyses (climate–landforms–soils–human impacts), and the utility of the vegetation and use of individual species.

As with earlier books authored by Professor Zahran, the completion of this book was the commendable product of collaboration between British and Egyptian colleagues. We are grateful to Dr. Francis Gilbert of the University of Nottingham, for his invaluable contribution.

Cairo, Egypt
October 2009

M. Kassas

# Preface

Global climatic changes are likely to be dominated by the influence of the greenhouse effect caused by increasing concentrations of carbon dioxide ($CO_2$) methane ($CH_4$), nitrous oxide ($N_2O_2$), ozone ($O_3$) and halocarbons (CFCs etc.). These greenhouse gases, individually and collectively, change the radiative balance of the atmosphere, trapping more heat near the Earth's surface and causing a rise in the global mean surface air temperature. Large environmental changes will occur as a consequence (Jeftice et al., 1996).

Coastal and continental deserts often contain within them a complex mosaic of climates and microclimates. There are also great contrasts in the climate and surface characteristics from one desert to the next. There are "cold deserts" and "hot deserts", deserts with winter precipitation, deserts with summer precipitation, deserts with virtually no precipitation, perpetually foggy coastal deserts, continental deserts with near the maximum possible sunshine, barren deserts, heavily vegetated deserts, rocky deserts, salt deserts and deserts with recognizable sand formations (Warner, 2004). The existence and distribution of desert climates have profoundly shaped the historical pattern of human activities including travel, settlement, communication and economic development. Most of the world's great civilizations developed at the margins of deserts. Virtually, all of the world's great contemporary religions (Judaism, Christianity, Islam, Hinduism and Buddhism) were born in desert regions. In fact, deserts should get more attention from scientists and decision-makers because they are the only alternative for a better future for the peoples of arid countries.

The Afro-Asian Mediterranean and Red Sea coastal deserts are affected by different climates associated with various habitats and biota (flora and fauna). The climate of the Afro-Asian Mediterranean coast is subdivided according to the volume of rainfall, which mainly occurs in winter. On the other hand, the climate of the Red Sea is tropical, with summer rainfall in its southern section and winter rainfall in its northern section. Aridity is the main climatic feature of the coastal lands of both seas, already under stress from population pressure and conflicting uses. The apparent effects of these threats include loss of biodiversity, rapid deterioration in land cover and depletion of water availability through destruction of catchments and aquifers. Changes of climate will interact with these environmental changes, causing further stresses. A sustained increase in mean ambient temperature beyond 1°C would cause significant changes in these ecosystems (Anonymous, 1998a). Many organisms are already at or near their tolerance limits, and some may not be able to adapt further under hotter climates.

This book is an attempt to compile, integrate and discuss the ecological information about the climatic features, habitat types and vegetation forms of the coastal deserts of the Afro-Asian Mediterranean and Red Sea basins. The countries involved are Morocco, Algeria, Tunisia, Libya, Egypt and Palestine–Israel (along the Afro-Asian Mediterranean coast); and Egypt, Sudan, Eritrea, Djibouti, Saudi Arabia and Yemen (along the Afro-Asian Red Sea coast). It is the first regional book of its kind in the Middle East: other similar books (e.g. Zohary, 1962; Zahran, 1982a; Danin, 1983; Al-Eisawi, 1996; Zahran and Willis, 1992, 2008) deal with the ecology of the vegetation of individual countries.

The subject of this book is described under four main chapters: the climate–vegetation relationships of the Afro-Asian Mediterranean coastal deserts; the Afro-Asian Red Sea coastal deserts; a short account of the distribution pattern of the major elements of the flora; and a comprehensive report on the role of representative economically valuable plants in the sustainable development of the coastal deserts.

Mansoura, Egypt
October 2009

M.A. Zahran

# Acknowledgments

I am indebted to Prof M. Kassas, Cairo University, for his help and encouragement; his library in the Faculty of Science of Cairo University was the main source of ecological literature documented in this book. My sincere thanks to Dr. Francis Gilbert of Nottingham University (UK) who edited this book. The assistance of my son Ahmed, a 5th-year student in the Faculty of Medicine, Ain Shams University, Cairo, during the preparation of the book is highly appreciated.

Mansoura, Egypt M.A. Zahran
October 2009

# Contents

# About the Author

Professor Mahmoud Abdel Kawy Zahran was born in Samalut (Minia Province, Upper Egypt) on 15 January 1938. He graduated in 1959 from the Faculty of Science, Cairo University, from where he also obtained his MSc (1962) and PhD (1965) in the field of plant ecology.

Professor Zahran worked as research assistant and researcher in the National Research Center (1959–1963) and Desert Institute (1963–1972). In October 1972 he was appointed Assistant Professor in the Faculty of Science, Mansoura University, and promoted to Professor of Plant Ecology in November 1976. He was Head of the Department of Botany from 1983 to 1989.

Professor Zahran joined the Faculty of Meteorology & Environmental Studies of King Abdul Aziz University, Jeddah, Saudi Arabia between 1977 and 1983, Sana'a University in Yemen between 1992 and 1994, and UAE University in the United Arab Emirates between 1995 and 1997.

Through a scholarship from the British Council, Professor Zahran spent one academic year (1966–1967) in Lancaster University (UK), and with a fellowship from the European Commission, another academic year (1994–1995) as a visiting professor in Sheffield University (UK). He has also been a Visiting Professor in various other universities of Europe (e.g. Norway, Sweden, Poland, etc.).

For his scientific achievements, Professor Zahran received the State Prize of Egypt from the Academy of Scientific Researches and Technology (1983), the first class Gold Medal of the Egyptian President (1985), the Diploma of the International Cultural Council of Mexico (1987) and the Major Prize of Mansoura University in Basic Sciences (1991).

Professor Zahran is the author of 15 books published worldwide, three of them co-authored with the late Professor A.J. Willis of Sheffield University. He has contributed to 14 other books, and edited and translated others. All deal with plant ecology and applied plant ecology of the River Nile and deserts of Egypt and the Arab World.

# Introduction

a. *Climatic changes*

During the long geological history of the earth its climate has been continually changing influenced by both terrestrial and extraterrestrial forces (Jeftic et al., 1996). The trends of these changes were gradual imperceptible measured against the length of human life. However, relatively sudden changes could be expected by catastrophic events, e.g. volcanic eruptions. Man's perceptible impacts on the climate are of recent origin started by the advent of the industrial revolution when increasing quantities of gases were emitted into the atmosphere creating the "Green House Effect". The accumulation of gases in the atmosphere will inevitably lead to changes in global climate. Realizing the potential threat of such changes, in 1995 the United Nation Organizations, and particularly UNEP and UNESCO, established task forces to assess the environmental, social and economic problems of seas and coastal lands, likely to follow on the heels of predicted climatic changes and to identify suitable measures and policy in response that might mitigate or avoid the negative impacts of these changes.

Global climatic changes are affected mainly by increasing the concentrations of the atmospheric green house gases, namely: carbon dioxide ($CO_2$), methane ($CH_4$), nitrous oxide ($N_2O$), ozone ($O_3$) and halocarbons ($CFC_{11}$) (Wigley, 1992). Between the pre-industrial period (1750–1800) and 1990 the concentrations of all green house gases have increased substantially: $CO_2$ from 279 to 354 ppm, $CH_4$ from 790 to 1,720 ppbv, $N_2O$ from 285 to 310 ppbv, $CFC_{11}$ from 0 to 280 pptv, $CFC_{12}$ from 0 to 440 pptv. These gases individually and/or collectively change the radiative balance of the atmosphere, trapping more heat near the earth's surface and causing a rise in global-mean air temperature.

Substantial global warming may be virtually certain, but the attendant changes in climate at the regional level is highly uncertain Jones et al. (1986a, b, c). Future changes in climate can be set in context by a brief description of changes that have occurred during the past century. In fact, fluctuations of climatic conditions from year to year, decade to decade and even long periods are noticeable both regionally and in terms of global mean values. For example, the near-surface air temperature averaged over the globe has increased by about 0.5°C since the late nineteenth century: there have been noticeable changes in rainfall during the twentieth century on all spatial scales-average levels show an upward trend from 1920 to the present in mid-to-high latitudes (35–70°N) (Bradely et al., 1987). Most if not all other climatic variables show evidence of a continually changing climate. At the regional level, and perhaps even at the global mean scale, during the past 100 years these climatic changes have been attributed to natural variability. Our understanding is that climatic variation undoubtedly occurs due to: (1) changes in the reflectance (or albido) of the earth-atmosphere system influenced by explosive volcanic eruptions (which produce dust and aerosol layers in the stratosphere); (2) changes in cloudness; (3) changes in surface characteristics (vegetations, snow and ice cover etc.); (4) changes in solar irradiance and (5) changes in the concentrations of green house gases ($CO_2$ etc.).

Climatic changes will cause many troubles (Broom, 2008). Heat waves, storms and floods will kill many people and harm many others. Tropical diseases which will increase their range as the climate warms and will exact a toll in human lives. The changing pattern of rainfall will lead to local shortages of food and safe drinking water. Large-scale human migrations in response to rising sea levels and other climate-induced stresses will impoverish many people. According to World Health Organization (WHO) estimates, as long ago as 2000 the animal death toll from climate change had already reached more than 150,000. Also, Broom (2008) stated "the European heat wave of 2003 is estimated to have killed 35,000 people. In 1998 floods in China adversely affected 240 millions". According to Anonymous (2008a), 160 square miles of Antarctica's Wilkens ice shelf disintegrated on February 2008 and the present carbon dioxide ($CO_2$) level in the atmosphere, 385 ppm, is already too high. This means that there is a substantial risk that within just a few centuries the earth could lose its great Antarctic and Green land ice sheets causing catastrophic rises in sea level, unless $CO_2$ levels are quickly brought back to 350 ppm. The Intergovernmental Panel on Climate Change (IPCC) assumes that by the end of this century (21st) atmospheric greenhouse gases will reach the warming equivalent of 600 ppm $CO_2$. At that level the IPCC suggests there is 67% chance that by 2100 the earth will have warmed by between 2 and 3°C (3.6–5.4°F) since pre-industrial times. The IPPC puts the chances at 5% that even if $CO_2$-eq levels stabilize at 550 ppm, global temperature will eventually rise by 8°C (14.4°F).

Climatic changes associated with land degradation, vegetation overuse, desertification and global food crises are actually the major issues worldwide, the problem is even worse in arid and semi-arid regions (Anonymous, 1998a). Their natural vegetation is under great threat due to the high population increase and the systems of land use. The apparent effects of these threats include loss of biodiversity, rapid deterioration in land cover (e.g. vegetation) and depletion of water availability through destruction of catchments and aquifers. Change of climate will interact with these underlying changes in the environment adding further stresses to a deteriorating situation. A sustained increase in mean ambient temperatures beyond 1°C would cause a significant change in forest, rangeland and desert cover, species distributions, composition and migration patterns and biome distribution. In fact, many desert organisms are already near their tolerance limits and some may not be able to adapt further under hotter conditions. Arid and semi-arid areas currently under land degradation and desertification are particularly vulnerable. However, correct management, conservation and sustainable utilization of the natural vegetation and other renewable resource, of the deserts and semi-deserts could play significant role in their development for the welfare of their people.

b. *Coastal deserts*

Since the major parts of the coastal lands of the Afro-Asian Red Sea and those of the African and SW Asian Mediterranean Sea (the subject of the present bock) are actually coastal deserts, let me consider the coastal desert on a world scale. Geographically, deserts are defined as arid lands where evaporation exceeds rainfall: they are classified, according to the amount of annual rainfall, into 3 types: extreme deserts, deserts and semi-deserts. Extreme deserts are area with rainfall below 70 mm/year, deserts having rainfall less than 120 mm/year whereas semi-deserts are receiving rainfall between 150 and 300 mm/year (Smith, 1996). Raheja (1966) stated that, the arid lands of the world are characterized by the minimum of annual precipitation and the maximum temperature, evaporation and aridity. Evaporation in deserts and semi-deserts is very high and ranges between 7 to more than 100 times as much as precipitation. Also, deserts are characterized by a wide daily range in temperature from hot by day to cool by night. Low humidity allow up to 90% of the solar isolation to penetrate the atmosphere and heat the ground. At night, however, the desert yields the accumulated heat of the day back to the atmosphere.

Deserts occur in two district belts between latitudes 15° and 35° in both north and south hemispheres, i.e. around the Tropics of Cancer and Caprican. Their approximate areas are as follows (Raheja, 1966): 2.24, 8.42 and 8.20 million square miles for the extreme arid, arid and semi-arid deserts respectively, i.e. about 36% of the area of the world. According to this classification, the area under extreme arid climate is far lower than that under arid or semi-arid climates.

The world's deserts are cold and hot, Allan and Warren (1993). Cold deserts have average winter temperature below 0°C (32°F) and are covered with snow most of the year. They do not receive much precipitation and whatever does fall, remains as snow. Cold deserts are commonly referred to as *tundra* if a short season of above-freezing temperatures occurs or as an *ice cap* if the temperature remains below freezing all year long, rending the land almost completely lifeless. Cold deserts are, therefore the result of very low temperatures, prohibiting life for must living organisms except those adapted morphologically, physiologically and ecologically to live and reproduce under such harsh cold climate. Hot deserts, on the other hand, have little rain with average summer temperature above 30°C (86°F) accompanied by a wide daily range of about 20°C. As in cold deserts, most living organisms cannot establish themselves in hot deserts, only drought tolerant and/or drought resistant plants (xerophytes) as well as salt tolerant plants (halophytes) could grow and reproduce. In the deserts, vegetation cover is usually thin supported by poor soil.

Hot deserts represent challenge for life on the earth (Oldfield, 2004). Though lack of water and extreme temperature make survival difficult for most living organisms, yet deserts are rich in wildlife. Also, Allan and Warren (1993) reported "though considered inhospitable, yet they (deserts) are the home of about 13% of the world's population and together (cold and hot deserts) cover almost 40% of the earth's surface. After oceans, deserts are among the most important elements in the global climate system".

Generally, hot deserts can be classified under two main types according to their proximity to the influence of the seas and oceans: coastal deserts and inland (continental) deserts. According to Meigs (1953), hot deserts, both coastal and inland, are world dry regions that receive little precipitation and could be classified under three categories:

1. Hyper-arid (extremely dry) lands having at least 12 consecutive months without rainfall. However, as there are virtually no places where rainfall is entirely unknown and according to Allan and Warren (1993), hyper-arid areas receive less than 25 mm (1 in.) annual rainfall,
2. Arid (dry) lands having a mean annual rainfall less than 250 mm, and
3. Semiarid lands having a mean annual rainfall between 250 and 500 mm and are generally referred to as grasslands or steppes rather than deserts. Meigs (1953) estimated that the total areas of the dry lands (deserts) to cover about one third of the Earth's land surface.

Shortage of water is the main factor hindering the development projects in the deserts. However, although coastal deserts are extending along the coastal lands of the seas and oceans of the tropical regions, the advent of cheaper methods of extracting fresh water from the seas has meant they have received increasing attention and interest. Since early 1970s, most of the coastal deserts could has desalinated water by solar radiation, electroanalysis, steam power, nuclear energy etc. Securing supplies of fresh water will expedite implementation of the development plans for coastal deserts which are, actually, rich in renewable natural resources including: solar and wind energies, different plants and animals adapted to live under aridity and/or salinity stress of the deserts.

Great environmental differences exist among the coastal deserts of the world. Such differences have significant effects on the potentiality for forms of ecological and economic developments. Climatically, the coastal deserts of the world could be: hot, warm, cool and cold (Meigs, 1973). Such classification is based on the mean monthly temperature. The hot deserts are tropical and subtropical with the mean monthly temperatures for the warmest and coldest months as follow: >30°C (86°F) and >22°C (72°F) and >30°C (86°F) and 10–22°C (50–72°F), respectively. For the other coastal deserts, the corresponding figures are as follow: 22–30°C (72–86°F) and 10–22°C (50–72°F) for the warm coastal deserts, <22°C (72°F) and 10–22°C (50–72°F) for the cool coastal deserts and <22°C (72°F) and <10°C (50°F) for the cold coastal deserts. Hot coastal deserts are the most extensive in the world having approximate length of about 16,000 km (10,000 miles) compared with 7,700 km (4,800 miles), 7,000 km (4,400 miles) and 1,300 km (800 miles) for warm, cool and cold coastal deserts, respectively.

c. *Climate–soil–vegetation relationship*

The relationships between climate, soil, vegetation and other components of desert ecosystems including those of the Mediterranean and Red Sea coastal lands, have been described by many authors: Shreve (1942), Hassib (1951), Tadros (1953), Vernet (1955), Kassas (1955), Chapman (1960), Zohary (1962), Batanouny (1973), Younnes et al. (1983), Ayyad and El-Ghareeb (1984), Evenari et al. (1985), Zahran and Willis (2008) etc. Such relationships are so interconnected that, in an ecological perspective, they can hardly be considered as separate entities. In arid lands, the interactions are of high significance, so that slight regularities in one component (biotic or abiotic) are likely to lead to substantial variations in others.

Climatically, induced processes of weathering, erosion and deposition are continuously at work, dissecting the desert landscape into a variety of landforms and fragmenting the physical environment into a complex mosaic of microenvironments. The impact of rainfall is unmistakable. A decade or even a century may pass before a desert ecosystem experiences heavy precipitation, but when rain does fall, it results in a great deal of erosion and deposition owing to the sparseness of vegetation which offers little or no protection to the soil. Major erosion forms now present in the deserts generally results from fluvial action (Hills, 1966). Some, such as wadis and their outflows, are undoubtedly relict features derived from past periods of heavy rainfall, but many are attributed to present day occasional heavy rainfall. Because of the scarcity of rainfall, the high evaporation rate and the sparseness of vegetation in arid lands, salt accumulation close to the soil surface is a common phenomenon. This is obvious in the coastal deserts affected by marine ingression and in the inland depressions and oases where the water-table is very shallow or exposed.

Vegetation influences soil through the effects of plant cover on soil temperature and water flux (intake, run-off and erosion) and through the production of organic matter and its incorporation into the soil, thus influencing soil structure, macro- and micro-flora and fauna, water intake and storage, and the balance of water and nutrient elements (Le Houerou, 1985). Reduction of plant cover and biomas leads to the dynamic processes of "steppization" and "desertization" (Le Houerou, 1969a). The organic content of soil is directly related to plant cover, species composition and biomass; it decreases from more than 5% to less than 0.5% when plant cover is reduced from 50 to 5%, with a reduction in above-ground phytomass from over 10,000 kg $ha^{-1}$ to less than 500 kg $ha^{-1}$. The nature of the organic compounds in the soil also changes: the C/N ratio decreases from 15–20 to 8–10; and the proportion of humic acids in the organic matter varies inversely with plant cover and biomass and increases with aridity. Such decrease in total organic content associated with a decrease in plant cover and biomass result in a reduction in the structural stability of soil aggregates. This leads to the destruction of soil structure in the upper horizons and hence to a decrease in water-holding capacity and increased run-off. In silty to clayey soils this can seal the soil surface which is an important factor is desertification.

As revealed by the studies carried by Le Houerou (1972, 1985), the primary productivity of the natural vegetation of the arid land is mainly affected by the soil-water balance. In some cases certain features of plant nutrition may be limiting especially nitrogen availability. Higher productivity was obtained from vegetation growing on deep sandy soils having some thin loam or silt layers at depth. Silty to loamy soil have high productivity during rainy years but productivity is greatly reduced in years of drought. Sandy soils, on the other hand, have much more even productivity even in years when rainfall is deficient. Le Houerou and Hoste (1977) estimated that range land productivity in deep sandy soils is of the order of 4–5 kg $ha^{-1}$ of total above-ground dry matter for each millimeter of rainfall whereas in fine-textured soil (silty or clayey) the corresponding figure is 1.5–2 kg $ha^{-1}$ and only 0.5–1 kg $ha^{-1}$ in shallow soils.

d. *Aim of the book*

The present book is an ecological study of the vegetation types of the different habitats of four Afro-Asian coastal lands extending along the Mediterranean and Red Seas together with an overview of the humid sections of the North African and SW Asian Mediterranean Coastal lands. The four areas are: the North African Mediterranean coast; the southern part of the Asian (western) Mediterranean coast; the African (western) Red Sea coast and the Asian (eastern) Red Sea coast. As climatic aridity prevails along the whole stretches of the Afro-Asian Red Sea coastal lands as well as along most of the African and Middle Eastern Mediterranean coastal lands, the book is concerned mainly with the natural vegetation of the arid coastal habitats, associated with both seas. The relationships between the climatic features of these four coastal lands, the main habitats, the geographical distribution of the different vegetation types, their communities and floristic elements will be discussed.

I hope that the bulk of the complied ecological information in the book constitute strong base for the agro-industrial and socio-economic development of these coastal deserts. No doubt the natural wealth of the flora, being the renewable resources for food, fodder, wood and raw materials for various strategic industries, e.g. drugs, paper, textiles, oils, perfumes etc., will be the backbone of such development plans. Apart from the introduction, the book contains four main chapters as follow:

Chapter 1 Afro-Asian Mediterranean Coastal Lands (AAMCL)
Chapter 2 Afro-Asian Red Sea Coastal Lands (AARSCL)
Chapter 3 Climate–Vegetation Relationships: Perspectives
Chapter 4 Climate–Vegetation and Human Welfare in the Coastal Deserts

# Chapter 1
# Afro-Asian Mediterranean Coastal Lands

## 1.1 Introduction

The Mediterranean gave birth to the whole world: all the world is her debt (Semple, 1932). Modern civilization traces back to seeds of culture-matured in the circle of the Mediterranean lands and transplanted thence to other countries whence they have been disseminated over the world. The Mediterranean Sea with its bordering lands has been a melting-pot for the peoples and civilization, which have rushed into it from continental hinterlands. It has been a catchment basin, and also a disturbutionary's center for its composite cultural achievements. This double role in history is an outgrowth of its geographical location and its relation to the neighboring continents.

Because of its geography and history, the Mediterranean Basin is an unique original region: the sea itself (Mediterranean means "in the midst of the land"), the complex and tortuose of landscape that surrounds it. Its unique climate, have all influenced the extraordinary development of the prehistoric civilization along its shores. Such development has deeply marked an irreversible-fragile environment with limited resources. Branigan and Jarrett (1975) stated that a significant role in the evolution of a succession of civilization have been played by the Mediterranean Sea and its surrounding lands. To Mesopotamia, Babylon, Syria, Palestine and Egypt, the Mediterranean was the Great Sea, their knowledge being confined to the Levant. To the ancient Greeks it became *Mare Interim*, "the interior sea" and later the Italians named it *Mare nostrum* (our sea). The name "Mediterranean" generally used at the beginning of the Christian era. It describes this almost landlocked sea and gives some prominence to the terra, the land that surrounds it. We may quote what is written by Semple (1932): Mediterranean civilization has given the world standards. These are embodied in classical culture and Christianity, and still represent ideals of achievement and conduct. The new faith born of a narrow tribal beliefs from the hill country of Judea and Galilee, emerged as a principle of universal brother-hood from the gradual calming of local religious incidents to an active intercourse between all parts of the Mediterranean Basin under the *Pax Romana*.

The Mediterranean Basin with its fantastic variety of landscapes, people, plants and animals, is one of the most complex regions on Earth in terms of geological,

M.A. Zahran, *Climate–Vegetation*, Plant and Vegetation 4,
DOI 10.1007/978-90-481-8595-5_1, © Springer Science+Business Media B.V. 2010

history, geographic, morphology and natural history. It is characterized by various habitats: high mountains, seashores, coastal wetlands, desert wadis, coastal dunes, small islets, dry scrublands, moist fir forests, grasslands etc. Diversity of living organisms between these habitats is obvious (Blondel and Aronson, 1999).

The Mediterranean climate is characterized by prolonged and intense summer drought of at least 2 or 3 and up to 11 months. Rains occur during the winter season. According to the respective length of the dry and rainy season and the total amount of rainfall, the Mediterranean climate can be differentiated, into six types: hyper-arid, arid, semi-arid, sub-humid, humid and hyper-humid. Mean annual rainfall can vary from 20–25 mm in the Mediterranean desert (arid) to 2,000–2,500 mm on mountain slopes or maritime areas exposed to rain-bearing winds. The temperature varies widely with latitudes, altitudes and continentally: the mean annual can drop below 10°C and rise above 25°C; the mean daily January minimum can vary from –10 to 15°C; and the mean daily maximum of July/August from 25 to 45°C (Le Houerou, 1992). That the arid lands of the Mediterranean Basin in North African Sahara to the west Asian desert spans a set of climatic systems with a rainless core and winter rainfall (Kassas, 1998).

The Mediterranean Basin in general has had a long history of human occupation: sites of old rain-fed agriculture settlements has been recognized in the East Mediterranean Basin (Fertile Crescent): irrigation farming prevailed in Mesopotamia and Egypt throughout recorded history; nomadic tribes with their flocks of camels, sheep and goats have remained features of human life in this region since millennia. The imacts of human occupations are evident everywhere. Accordingly, the Mediterranean Basin was and is still the focus of attraction of various scientists in the different fields of specializations (ecology, environmental sciences, biodiversity, climatology, geology, land use, agriculture, economy, history, sociology etc.). Endless number of publications has been produced, about the Mediterranean Sea and the land around it.

## 1.2 Geological History

The geological history of the Mediterranean Sea and its Basin has been described by Hus (1971) and Cohen (1980). They noted that the giant ancestor the Mediterranean Sea was the Tethys Ocean. Some 200 millions years before present, from the beginning of the Mesozoic era which formed a vast wedge-shaped unbroken seaway subdividing the super-continent Pangaea into the Laurasian proto-continent to the north and the Gondwanian proto-continent in the south. The physical geography of the future Mediterranean area changed continuously during this epoch as a result of several periods of continental convergence and collisions of tectonic plates, while seafloor spreading repeatedly rearranged the relative position of the continental plates and oceans in the ancient Tethys. Between the middle Jurassic and the Upper Cretaceous (165–180 million years BP), an eastward motion of Africa relative to Europe, which was then still joined to North America permitted the formation of the Atlantic Ocean. Later, the movement of Africa westwards began the separation from North Africa in the Lower Tertiary. Again, Africa moved eastward relative to Europe from the Upper Eocene (40 million years BP) enlarging the Atlantic. This led to the closing of the gap between Europe and Africa and the elimination of much of the original Tethys. At the same time, a northward movement of Africa initiated the creation of the mountain ranges that encircle the basin. All these complicated movement, of continental drift contributed to the intricate puzzle-like geography of the Mediterranean area. These movements generated the rotation of micro-continents or "micro-tectonic" plates that separated from the main continental blocks. During the Eocene, after more than 100 million years of transcontinental and reshuffling, the Tethys seafloor began to buckle between the colliding tectonic blocks. The Greco-Italian micro-continent, which was joined to the African plate during the Lower Cretaceous (120 million years BP) moved with it and collided with the Eurasian plate in the lower Eocene. In the Upper Oligocene (28 million years BP) the south-eastward motion of Africa relative to Europe caused the rotation of the Iberian plats which included all the large islands of the western Mediterranean and several crystalline blocks subsequently connected either to Africa or to Europe. In the Upper Oligocene (25 million years BP) the African plate, including Arabia, first made contact with south-west Asia, thus, dividing the Tethys Sea into two parts: the southern part, ancestor of the modern Mediterranean Sea, whilst further north and east, which was a shallow, brackish sea that geologist call the Paratethys.

The northward drift of Africa continued during the Miocene together with a general westerly movement up to the present. About 6 million BP present Africa collided with south-western Europe. The newly joined blocks then drifted along together from that point raising the Pyrenees to great heights and, for the first time in history closing the Mediterranean Sea at its western end. Thus, came into being the Maere *Medi-Terraneum* or "sea among the lands", some times also called the "inland sea".

## 1.3 Geomorphologic Features

The Mediterranean Sea lies between latitudes 30 and 45°N and longitudes 10 and 40°E covering a basin between southern Europe and northern Africa, opens to the Atlantic Ocean by the very narrow Strait of Gibraltar and entirely blocked at its eastern end by part of south-west Asia (Fig. 1.1). More than 3,600 km from Gibraltar to Lebanon with an areas of 1,146,000 2,978,000 km$^2$, it is all that is left of the great ancient ocean Tethys (Branigan and Jarrett, 1975). Blondel and Aronson (1999) stated that the Mediterranean Sea, the largest land sea in the world, extends over 2,969,000 km$^2$. With its satellites, the Black and the Azov Seas, it is a miniatures ocean. The sea is 740 km wide between Marseilles and Algeria, and 400 km wide between southern Greece and Tripolitania (Libya). Its coastline is 40,000 km in length and several "interior seas" are delimited by the shorelines of the larger archipelagos and northern peninsulas: the Balearic Sea between the Balearic Island

**Fig. 1.1.** Map of the Mediterranean basin

and continental Spain, the Tyrrhenian Sea between Corsica-Sardinia and the Italian Peninsula, the Adriatic and Lonian Seas between Italy and the Aegean Peninsula and the Aegean Sea between Greece and the Anatolian Peninsula.

The Mediterranean Sea is a great gulf of the Atlantic Ocean cutting back into the landmass of the Eastern Hemisphere. It carries the waters of the ocean 3,726,000 of km inland to the foot of the Lebanon Mountains and yet further through the Black Sea to the rugged coast of Caucasus. "It gives Asia an Atlantic seaboard, and hence rendered it possible for Asiatic navigators from Phoenicia to make the first recorded discovery of Britain. This is the maere *internium*, enclosed by three contrasted continents which it helps to divide, but whose difference in climate, flora, fauna and peoples and civilizations it helps to reconcile" (Semple, 1932).

The Mediterranean consists of a deep gash in the crust of the earth. This prolonged deep gash opens up a natural communication between the Atlantic and the Indian oceans. Also, by splitting the massive block of Europe and Asia, it acts like a link, as a hyphen, between East and West. We have here a geographical phenomenon for exceeding any local framework, and without exaggeration it is of worldwide consequence and proportions (Siegfried, 1948).

The Mediterranean Sea has an Asiatic, European and African coastlines and each continent possesses distinctive features of ecosphere, relief, hinterland connections and differentiated types of the prevailing Mediterranean climate. Europe has a long coastline of about 13,000 km measured from the Strait of Gibraltar to the mouth of Don River (Tanais), the ancient boundary of eastern Europe. The coastline of Africa measures only 5,000 km and is sharply divided into a long stretch of mountain coast to the west, and a longer stretch of plateau coast to the east, each is rather uniform until they combine in the Tunisian Peninsula and each is backed by the same desert hinterland. This part of Africa suffers from serious heat and drought, except where the coastal mountains receive a more or less ample rainfall in the winter. On the eastern edge of the African part lies Egypt, a striking contrast to the rest. The Asiatic Coastline runs from the Don River to the Isthmus of Suez, measures 6,000 km and therefore exceeds that of Africa. It forms a natural gateway between the Orient and Occident (Semple, 1932). A striking geomorphologic features of the Mediterranean Basin is the steepness of the coast, which almost everywhere sinks abruptly to the seafloor at 2,500–4,000 m. The average depth of the Mediterranean Sea is 1,400 m but deeper troughs occur in some places, for example at the southern tip of the Peloponnese (5,100 m) and southeast of Sicily (4,100 m). The average water salinity is 3.7–3.9%, varying little except in some southern regions, and to a lesser degree at the southern end of the Adriatic Sea.

The Mediterranean Basin is characterized by a wide topographic diversity which result in large part from the numerous mountain chains with heights ranging between 1,540 m (Mt Chambi in Tunisia) to 2,330 m (Mt Hodna-Muket in Algeria), and 3,090 m (in Lebanon). Higher peaks occur in Italy (3,260 m), Morocco (4,165 m) and Iran (5,600 m). Mountain chains are found throughout the length of the Mediterranean Basin along its European border and in the African side from the Atlantic to the Gulf of Gabes in Tunisia. East of these there are no encircling range neither in Libya nor in Egypt. The African shield as far as the Sinai Peninsula drops from the tableland by a series of faulted terraces to the coast, and this step-little formation is continued northwards in the floor of the sea. The eastern margin of the Mediterranean Basin is also lacking in mountain folds of any size. The Arabian Plateau, which in the Levant lands underlies sedimentary rocks of considerable thickness in some places, lies outside the region most affected by the mountain-building process, and although there are some folding of the sediments only is Lebanon was this sufficiently intense to produce high mountains; even there is only one peak that rises to more than 3,000 m (Querns es Suda, 3,057 m, Branigan and Jarrett, 1975).

The Mediterranean Basin is divided into two: western and eastern. The western basin stretches from the Strait of Gibraltar to the Sicilian Channel. At its western end, the Strait of Gibraltar, which is only 14.8 km across the narrow point, has a floor that rises to within 366 m of the sea surface and preventing the deeper waters of the Mediterranean flowing outwards to the Atlantic. The deepest point of the Western basins 4,265 m midway between the Islands of Capris and Sardinia. Most of the Eastern basin is occupied by the Ionian Basin, which stretches from the Sicilian Channel almost to the Levantine coast and its deepest point is 5,121 m in Agen Sea.

## 1.4 Climate: Main Features

The Mediterranean climate is found in five main regions of the world (California, Chile, South Africa. Australia and the Mediterranean Basin) covering about 2% of the total land area of the world. It is primarily coastal and is described as a maritime climate. The Mediterranean Basin is the largest region of land with the

Mediterranean climate (60%) followed by Australia (22%). California (10%), Chile (5%) and the Western Cape (South Africa, 3%). All these regions, located on the western or southwestern coasts of continents, and share the general Mediterranean climate features (Newbigen, 1924; Kendrew, 1961; Von Chi-Bonnardel, 1973; Le Houerou, 1992; Dallman, 1998). The Mediterranean Basin is the only climate region that includes parts of three continents (Europe, Africa and Asia), giving it a very rich flora particularly where continents meet.

The Mediterranean type of climate is most extensive along the edges of the Mediterranean Sea; its dry phase is most extensive on the south and south-east sides of the Sea towards the North African and Asian deserts. This climate border the North African coast everywhere except in the most humid parts of Tunisia, Algeria and Morocco and continues southwards into the Sahara until the rainfalls becomes too meagre and irregular in occurrence to support it. At the east end of the Mediterranean Sea, the dry climates with winter rain come down to the coast in Sinai, southern Israel and south-eastern Turkey, and extend inland throughout the Middle East including the plateau of Turkey (Erinc, 1949, 1950) as well as the northern half of Arabia (Meigs, 1962). The Mediterranean climate is classified by Emberger (1955) as a non-tropical with regular annual rainfall with summer as the dry season.

The Mediterranean climate, including the Mediterranean basin has been divided into four main zones (Meigs, 1962; UNESCO/FAQ, 1963), by the degree of drought of a given dry month, defined as the number of days in the month in which can be deemed dry from a biological point of view (denoted "$x$" below):

1. xerothermomediterranean, i.e. warm and dry when $150 < x < 200$;
2. thermomediterranean subdivided in turn into accentuated with a long dry season where $125 < x < 150$ and "attenuated" with a shorter dry season, where $100 < x < 125$;
3. mesomediterranean again subdivided into "accentuated" with a long dry season where $75 < x < 100$ and "attenuated" with a shorter dry season, when $40 < x < 75$;
4. sub-mediterranean, a transitional type when $0 < x < 40$. This type of climate is not regarded as Mediterranean climate proper (not eumediterranean).

An alternative classification, Le Houerou (1992), uses precipitation and evapotranspiration to define eight zones: hyper-arid, arid lower, air-upper, semiarid, sub-humid, humid, hyper-humid and wet. Each is characterized by its own vegetation types as follow: desert, desert bush, thorn woodland, very dry forest, dry forest, moist forest, wet forest and rain forest.

Yet a third classification based on the average duration of summer drought is given below. The Mediterranean Sea extends over 3,600 km from west to east and in that distance the total annual rainfall near sea level decreases from 891 mm at Gibraltar to 389 mm at Athens and to 84 mm at Port Said. The decrease in annual rainfall eastward is due mainly but not entirely to increasing distance from the Atlantic. Patterns of rainfall distribution, like those of temperature variations, are also affected by relief. Where mountain range lies across moisture bearing winds, such as in Morocco, their windward slopes may have rainfall totals greater than many wet areas of northern Europe, but they often also act as rainfall barriers and cast extensive rain shadows. In these enclosed and elevated regions, winter temperature are lower than the Mediterranean averages. Summers are also apt to be very hot away from the moderating influence of the sea.

According to Branigan and Jarrett (1975), because of the position and form of the Mediterranean Sea, the lands around it are one of the most clearly defined climatic units in the world. In summer, following the northward march of the sun, the Mediterranean region lies in the belt of north-east trade winds and of high pressure and it is therefore dry. In winter, with the retreat of the sun south of the equator, the region falls under the general influences of the westerlies and is therefore wet because of the depressions associated with them. In general the Mediterranean region has a temperate and changeable climate for the winter half of the year and a more uniform climate of the hot desert type during the summer. Long periods of sunshine and cloudless skies are experienced at all seasons.

The general climactic features of the Mediterranean Basin may be summarized as follow:

1. Rain falls in the winters half of the year with an annual average, of less than 100 mm to more than 2,000 mm. The total amount of rainfall is highly variable from year to year, with much coming in storms lasting only as few days.
2. Summer is the period of drought which may be partial or complete, and at least 2 up to 11 months are dry.
3. The annual average of temperature is small, because typically the winters are mild and summers are hot.
4. Mean annual rainfall varies from 20 to 25 mm in the Mediterranean arid lands (deserts) to 2,000–5,000 mm in the mountain slopes or locally in mountainous areas exposed to rain-bearing winds.
5. Temperature varies latitudinally, altitudinally and continentally: the annual mean can drop below 10°C or rise above 25°C, the mean daily minimum of January may vary from –10 to + 15°C and the mean daily maximum of July/August varies from 25 to 45°C.
6. Winter cold stress, an important biological attribute can be non-existent, light mild, intense or prolonged depending upon latitude elevation and continentally. Emberger (1930) stated "cold stress becomes a potent discriminating factor of vegetation distribution pattern, crop selection and land use".
7. Mediterranean climate areas, experience decreasing rainfall as the latitude approaches 30° towards the equator. Correspondingly, there is a gradual changes in vegetation towards xerophytes adapted to semi-arid and arid deserts. In contrast as latitude approaches 40–45° towards the poles, rainfall steadily increase and becomes distributed more evenly throughout the year.
8. Elevation is important in relation to temperature and rainfall. In the mountains, temperature decreases by roughly 5°C/1,000 m. Rainfall is typically greater with increased elevation. Vegetation, correspondingly changes with elevation. Orientation of mountain slopes has also a profound effects on vegetation.

9. The Mediterranean climate and adjacent areas can be described on the basis of the average duration of summer drought into six sub-climatic zones: (Koppen, 1931; Thornthwait, 1948; Emberger, 1955; Bagnouls and Gaussen, 1957; Meigs, 1962; Di Castri, 1991):
   1. Hyper-arid (11–12 dry months).
   2. Arid (9–10 dry months),
   3. Semi-arid (7–8 dry months),
   4. Sub-humid (5–6 dry months),
   5. Humid (3–4 dry months),
   6. Hyper-humid (1–2 dry months).

These categories correspond reasonably well with the variation in vegetation and particularly suitable for the diverse Mediterranean Basin itself, because its seasons of peak rainfall varies substantially from one region to another.

## 1.5 Mediterranean Basin Countries

The Mediterranean Basin countries (21) are; Albania, France, Greece, Italy, Portugal, Spain, Yugoslavia[1] (in Europe), Morocco, Algeria. Tunisia, Libya and Egypt (in North Africa), and Cyprus, Iran, Iraq, Egypt (Sinai), Palestine-Israel, Jordan, Lebanon, Syria, Saudi Arabia and Turkey (in Asia). The surface areas of the territories of these countries under the influence of the Mediterranean climate (shown in Table 1.1) reveal the following (Le Houerou, 1992) (Fig. 1.1):

1. The total surface areas of these 21 countries under the influence of the Mediterranean climate is 7,357,000 km$^2$ distributed in Europe (9%, 690,000 km$^2$), Africa (46%, 3,347,000 km$^2$) and Asia (45%, 3,320,000 km$^2$).
2. 100% of the surface areas of the territories of eight countries (Morocco, Tunisia, Cyprus, Iraq, Israel, Jordan, Lebanon, and Syria), 80% of Iran, 64% of Spain, 62% of each of Greece and Portugal and 50% of each of Libya, Egypt and Saudi Arabia territories are under the influence of the Mediterranean climate.
3. 40, 30, 18, 16 and 10% of the territories of Italy, Turkey, Albania, France and Yugoslavia,[1] respectively, are influenced by the Mediterranean climate.

The geographical distribution of the six bioclimatic zones of the Mediterranean climate within the territories of the 21 countries of the Mediterranean Basin (Table 1.2) shows that (Le Houerou, 1992):

1. Territories of all North African countries (Morocco, Algeria, Tunisia, Libya and Egypt), most of the Middle East countries (Iran, Iraq, Syria, Jordan, Palestine,

---

[1]Yugoslavia is no longer a country since February 2003 and officially is divided into three independent countries: Serbia, Montengro and Kosovo.

**Table 1.1** Surface areas of the territories of 21 Mediterranean basin countries of the European, African and Asian continents under the effect of the Mediterranean climate (after Le Houerou, 1992)

| Countries | $S$ ($10^3$ km$^2$) | $P$ (%) | Countries | $S$ ($10^3$ km$^2$) | $P$ (%) |
|---|---|---|---|---|---|
| *A- Europe* | | | *C- Asia* | | |
| 1- Albania | 5 | 18 | 1- Cyprus | 9 | 100 |
| 2- France | 87 | 16 | 2- Iran | 1,300 | 80 |
| 3- Greece | 81 | 62 | 3- Iraq | 434 | 100 |
| 4- Italy | 118 | 40 | 4- Israel–Palestine | 20 | 100 |
| 5- Portugal | 57 | 62 | 5- Jordan | 97 | 100 |
| 6- Spain | 317 | 64 | 6- Lebanon | 10 | 100 |
| 7- Yugoslavia (former) | 25 | 10 | 7- Syria | 144 | 100 |
| | | | 8- Saudi Arabia | 1,075 | 50 |
| | | | 9- Turkey | 231 | 30 |
| *Total Europe* | 690 | 9 | *Total Asia* | 3,320 | 45 |
| B- Africa | | 100 | *Grand total* | 7,356 | 100 |
| 1- Morocco | 320 | 50 | | | |
| 2- Algeria | 191 | 100 | | | |
| 3- Tunisia | 156 | 50 | | | |
| 4- Libya | 880 | 50 | | | |
| 5- Egypt | 500 | 46 | | | |
| *Total Africa* | 3,347 | 46 | | | |

$S$ = Surface area, $P$ (%) = percentage from the total surface area of the country.

Israel and Turkey) are under the influence of the hyper-arid and arid bioclimatic zones together with small parts of Spain and Italy.

2. All North African (except Egypt), Asian and European countries, have territories influenced by the semi-arid bioclimatic zone.
3. Mediterranean Basin countries with territories in the humid bioclirnatic zone include Morocco. Algeria and Tunisia (in North Africa), all Asian and European countries.
4. The hyper-humid bioclimatic zone prevails in the Mediterranean territories of all seven European and six of the Middle East (Turkey, Lebanon. Jordan, Iran, Iraq and Cyprus) countries in addition to Morocco, Algeria and Libya of North Africa.

Apart from these six bioclimatic zones, the Mediterranean Basin also has a mountainous bioclimatic zone which prevails in the territories of countries with high mountains (all European, Asian (except Syria), and three North African countries (Morocco, Algeria and Tunisia), Le Houerou, 1992).

Such very diverse bioclimatic zones bear a large number of botanical, and consequently zoological types, including: desert and sub-desert steppes, evergreen and deciduous mountainous shrublands and forests. Land-use also varies widely depending on climatic, edaphic, historical and socio-economic conditions (Le Houerou, 1981, 1992).

**Table 1.2** Geographical location of the hyper-arid (HA), arid (A), semi-arid (SA), sub-humid (SH), humid (H), and hyper-humid (HH) bioclimatic zones of the Mediterranean basin in territories of European, North African and Middle East (Asian) countries (Le Houerou, 1992)

| | Bioclimatic zones | | | | | |
|---|---|---|---|---|---|---|
| Countries | HA | A | SA | SH | H | HH |
| *A- Europe* | | | | | | |
| 1- Portugal | – | – | + | + | + | + |
| 2- Spain | – | + | + | + | + | + |
| 3- France | – | – | + | + | + | + |
| 4- Italy | – | + | + | + | + | + |
| 5- Yugoslavia (former) | – | – | + | + | + | + |
| 6- Greece | – | – | + | + | + | + |
| *B- Asia* | | | | | | |
| 1- Turkey | – | + | + | + | + | + |
| 2- Syria | + | + | + | + | + | – |
| 3- Lebanon | – | – | + | + | + | + |
| 4- Israel–Palestine | + | + | + | + | + | – |
| 5- Jordan | + | + | + | + | + | + |
| 6- Iran | + | + | + | + | + | + |
| 7- Iraq | + | + | + | + | + | + |
| 8- Cyprus | – | – | + | + | + | + |
| *C- North Africa* | | | | | | |
| 1- Egypt | + | + | – | – | – | – |
| 2- Libya | + | + | + | + | – | + |
| 3- Tunisia | + | + | + | + | + | – |
| 4- Algeria | + | + | + | + | + | + |
| 5- Morocco | + | + | + | + | + | + |

## 1.6 Biodiversity of the Mediterranean Basin: General Features

Located at the cross roads between Europe, Asia and Africa, the Mediterranean Basin has served as a melting pot and meeting ground for species of varying origins. Many elements in the course of history have colonized the basin (Blondel and Aronson, 1999).

The physical environment and climate of the Mediterranean Basin have changed radically since the Mesozoic with the result that biological composition of its different regions and migration routes of invading species have changed repeatedly. Opportunities for invasion and secondary speciation have been continually renewed. As a result, one can find species originating from such different biogeographical realms as Siberia, South Africa and even some relics of the Antarctic continent for several components of the soil fauna.

The Mediterranean Basin, thus, might be considered as a huge “tension zone” (Raven, 1964) lying amid the temperate, arid and tropical biogeographical regions which surround it, a zone where intricate interpenetration and speciation has been particularly favoured and fostered as compared to the more homogeneous regions to the north and south. As a consequence of its kaleidoscopic topographical, climatic and edaphic complexity. The Mediterranean Basin is exceptionally rich in regional or local plant and animal endemics at the levels of genera, species and subspecies (Medial and Verlaque, 1997).

The Mediterranean Basin was recognized by Myers (1990) as one of the 18 world "hot-spots" where exceptional concentrations of biodiversity occur, over and above general trend that species richness increases with decreasing latitudes, such that one finds more species of plants and animals in the Mediterranean than further north. Blondel and Aronson (1999) after Hammond (1995), Quezel (1985), Cheylan and Poitenvin (1998), Cheylan (1991), Covas and Blondel (1998) and Higgins and Riley (1988) tabulated the status of biodiversity of the Mediterranean basin as follow (Table 1.3).

**Table 1.3** Total number of species, number of endemic species and proportion of world number of the species of the vascular plants, reptiles, amphibians, mammals, birds, insects and butterflies of the Mediterranean Basin

| Group | Number of species | Endemism | Proportion of world number of species (%) |
|---|---|---|---|
| Vascular plants | 25,000 | 50 | 7.8 |
| Reptiles | 165 | 68.5 | 2.5 |
| Amphibians | 63 | 58.7 | 1.5 |
| Mammals | 197 | 25 | 4.2 |
| Birds | 366 | 17 | 3.8 |
| Insects | 150,000 | – | 1.9 |
| Butterflies | 321 | 46 | – |

The flora of the Mediterranean basin is one of the richest in the Old World. It includes more than 25,000 species of flowering plants (Quezel, 1985), reaching about 30,000 species and subspecies (Greuter, 1991), as well as some 160 or more species of ferns. This is about 10% of all known plant species on Earth, a figure estimated at between 238,000 and 260,000, although the land area of the Mediterranean Basin represents only 1.5% of the land surface of the Earth.

Gomez-Campo (1985) stated that the main reason for the Mediterranean's richness is plant species in not so much the variety of species in any given area as the remarkable number of endemics, many of which are restricted to a single or a few localities in sandy areas, islands, geological "islands" of unusual soil or rock type or isolated mountain ranges. The Mediterranean Basin, as pointed out by Medial and Verlaque (1997) is nearly as rich in endemics as the whole of tropical Africa, even though the latter is about 4 times larger and harbours about the same number of vascular plant species. Thus, the Mediterranean Basin is an important reservoir of plant diversity.

One of the features of the Mediterranean Basin flora is that annual species, and particularly the ruderals, are especially well represented, because this life-history strategy is adapted to perturbation, including the stress of 6 or more months of absolute summer drought particularly in hyper-arid and arid areas. A further interesting feature of the flora is that the distribution of endemic species varies greatly according to regions.

The types of natural vegetation found in the Mediterranean lands are not only controlled by the different bioclimatic types but also by local variations in temperature and rainfall. Branigan and Jarrett (1975) stated that there are nine vegetation types: broad-leaved evergreen forest, stunted and degenerate woodland, deciduous forest, aquatic grasses and reeds, coniferous forest, grasslands, including steppe, dry-steppe and semi-desert, desert and alpine vegetation.

Broad-leaved evergreen forest (high maquis, Dallman, 1998) can be regarded as the type most characteristic of Mediterranean vegetation. The dominant trees are evergreen oaks of various kinds, the most numerous being holm oak (*Quercus ilex*), rock oak (*Q. suber*) and kermes oak (*Q. coccifera*). Other species include: *Arbutus unedo, Erica arborea, Myrtus communis* and *Genista hispanica*.

Stunted and degenerate woodland (low maquis) is represented by three species of the genus *Cistus* (*C. salviifolius, C. crispus, C. ladanifer*). Commonly associated with *Cistus* spp. is the genus *Cytinus* whose parasitic plants grow on the roots of *Cistus* and are found under the shrub. Deciduious forest is represented by two species of *Quercus* (*Q. aegilops* and *Q. cerris*).

Aquatic grasses and reeds are mainly represented by *Phragmites australis*, *Typha domingensis* and species of *Cyperus, Juncus, Panicum, Echinochloa* etc.

Coniferous forests are dominated by *Abies pectinata*, *A. marocana*, *Pinus pinaster*, *P. sylvesiris*, *P. pyrenaica*, *P. brutia*, *P. halepensis*, *P. pinea*, *Cupressus semprevirens*, *Cedrus libani*, *Juniperus communis*, *J. thurifera* and *J. drupacea*.

Steppe vegetation evolves in climatic conditions characterized by wide seasonal variation in temperature and low rainfall. Trees are absent and the landscape is one of various species of grasses and bulbous plants. Following the rains of winter the steppe is a scene of luxuriant growth; but this lasts only for a couple of months and begins to wither with the approach of hot, rainless summer. In autumn only hardy bushes and thorny plants show sign of life, and in general the steppes looks barren ad scorched. The main vegetative cover of the dry steppes and semi-desert is formed of *Stipa tenacissima* and *Lygeum spartum*.

Desert vegetation occur in the most arid areas in Morocco, Algeria, Tunisia, Libya and Egypt (North Africa) as well as in the Middle East (Israel, Palestine, Jordan, Syria, Iraq. Iran. Saudi Arabia, Turkey and Cyprus). Plants of these deserts show a further degree of adaptation to heat and drought, few parts of the desert are entirely devoid of vegetative life, and there are more species of plants than one might expect (details of the plant life of these arid deserts are described in the following pages).

The alpine flora, on the other hand, has adapted itself to extreme cold and may be found on islands of rock even above the snow-line of the high mountains. The dominant and most common species of the alpine vegetation are: *Erigeron alpinus, Azolla procumbens, Salix retusa, Juniperus nana, Corchus vernus, Colchicum alpinum, Erica carnea* and *Leontopodium alpinum*.

In invertebrates, as in plants, the Mediterranean Basin is very rich in terms of species diversity; 75% of the total European insect fauna are found in the Basin (Baletto and Casale, 1991). However, this is a limited fraction (2%) of the world's estimated 8 million insect species (Hammond, 1995), a figure that could be overestimated. Levels of endemism are high for most groups of insects. In some isolated mountains and larger islands endemics can account for 15–20% of the insect fauna. A figure that rises to 90% in some caves.

Baletto and Casale (1991) tentatively estimated the number of insect species in the basin or 150,000 only 70% of which have been described and named. Dafni and O'Toole (1994) estimated that there are 3,000–4,000 species of bee, which makes the Basin a prominent center of diversity for this group. As many as 1,500–2,000 species of bees occur in Israel (arid land) alone (O'Toole and Raw, 1991).

Reptiles are at home in the dry, warm Mediterranean Basin and are much more abundant and diverse than amphibians. Reflecting the contrasted ecology and physiology of the two groups, the diversity of reptile species increases from north to south (and from west to east) in parallel with aridity gradients. For example, there are 20 species of reptiles in Italy (with no endemics) whereas in North Africa and the Near East the number of the reptile species are 59 and 84, respectively, out of these there are 26 endemic species in each. In contrast, amphibian species richness increases from south to north (and from east to west). Since the north-south and east-west aridity gradients favour reptiles in the southern and eastern parts of the Basin, whereas the more humid climate in the northern and western parts favour amphibians, regions that are rich in reptiles and tend to be poor in amphibians and vice-versa. The total number of amphibian species in Italy is 17 (6 are endemics) whereas in North Africa and Near East there are 12 and 15 species of amphibians with two endemics in each (Cheylan and Poitenvin, 1998).

Covas and Blondel (1998) reported that judging from the great number of birdwatchers who visit the Mediterranean each spring and summer, the diversity of their quarry must be very high. There may be as many as 366 bird species, there compared to only 500 for the whole Europe.

Three groups of avi-fauna are clearly dominant in the Mediterranean Basin. The largest includes 144 northern species characteristic of the forests, freshwater marshes, and rivers all over western Eurasia. The second consists of 94 steppe species most of which presumably evolved in the margins of the current Mediterranean area, notably in the "eremic" Saharo-Arabian region extending from Mauritania eastwards across North Africa, the Red Sea and the Arabian Peninsula and on to the semi-deserts of southern Asia. The third group encompasses species more or less linked to shrubland habitats, the so-called matorrals. Good examples are the partridge (*Alectoris* spp.) and the many species of warbliers (*Sylvia* spp. and *Hippolais* spp.). Given the extent and diversity of shrubland formations in the Mediterranean Basin, the number of bird species of this group is surprisingly small (42 species, or 11.5% of the regional bird fauna).

A low level of endemism is a characteristic feature of the avi-fauna of the Mediterranean Basin. Only 64 species (17% of the total) are endemics. This appear to have evolved, within the geographic limits of the Basin (Covas and Blondel, 1998).

The terrestrial mammal faunas of the Mediterranean Basin (approximately 197 species, 25% of which are endemics) are distributed within the four quadrants of the Basin as follows (Cheylan, 1991):

The richest are 106 species that occur in the eastern part, including 23 of Asian species that do not occur elsewhere in the Basin, 84 species in the North Africa part, 80 species in Aegean part, and 72–77 species in the western (European) part.

The mammal faunas are sensitive to physical barriers, which means that the mammals of southern Europe, the Middle East and especially the North African are quite distinct. Although only 14 km wide, the Strait of Gibraltar has effectively isolated Europe from Africa for non-volant mammals since its opening after the Messinian Crisis[2] (Blondel and Aronson, 1999). As a consequence, mammal faunas on the northern side of the sea are basically Euro-Siberian in origin with Wild Boar (*Sus scrofa*), deer (*Cervus, Capreolus*) and the Brown Bear (*Ursus arcios*) as typical elements.

A large number of the North African mammal fauna are Afro-tropical or Saharo-Sahelian. Small shrews and rodents (*Crociduru, Mellivora, Mastomys*, and *Acomys*) colonized western North Africa, mainly Morocco from the south. Example of species of tropical origin that colonized the Near East through the Nile Valley are the genet (*Genetta genetta*), a mongoose (*Herpestes ichneumon*), the bubal antelope (*Alcephalus busephalus*), spiny mice (Acomys), large fruit-eating bat (*Rousettus aegyptiacus*) and the ghost bat (*Taphopzous nu-diventris*). Occasional vagrants to North Africa of species from the Saharo-Arabian region include the Hunting Dog (*Lycaon pictus*), the cheetah (*Aciynonyx jubdatus*) and several species of gazelles (*Addax nasomaculatus, Gazella dama* and *Oryx damah*).

The mammals, of the Mediterranean Basin include few local or regional endemics. Notable amongst these is the amazing Etruscan shrew (*Suncus etruscus*) which grows to be no more than 3.6–5.2 cm long and 1.6–2.4 g in weight (Jurgens et al., 1996). It is typically Mediterranean species lives only within the thermo- and meso-Mediterranean life zones and is widespread wherever it finds suitable habitats warm enough to allow for survival during winter. Endemic rodents include several gerbils and no fewer than eight species of voles (*Pitymus* spp.).

## 1.7 Habitats and Vegetation of the Mediterranean Basin Arid Lands

### *1.7.1 Habitats*

Landscapes and landforms of the arid land (hot deserts) are, on the whole, almost identical despite differing in other aspects. Kassas (1955), Hills (1966), Adams et al.

[2]The Messinian Salinity Crises referred to as the Messinian Event as a period when the Med. Sea evaporated partly or completely during the Messinian period of the Miocene epoch 5.96 million years ago.

(1978) and Evenari (1985) classify the habitats of hot deserts, including those of the Mediterranean Basin into fourteen types, namely: (1) Desert mountains, (2) Desert plains, (3) Alluvial fans, (4) Alluvial blankets, (5) Pediment, (6) Inselberg, (7) Drywash, (8) Dry lakes (playas), (9) Sabkha, (10) Salina, (11) Desert flats, (12) Desert pavement, (13) Badlands and (14) Sand dunes.

The following is an outline of the general characteristics of these habitats (Adams et al., 1978).

1. *Desert Mountains*. Desert mountains are usually barren and angular and exhibit primary stages in erosion and weathering. Their surfaces are washed away by rain running off the slopes in gullies or rivulets, or stripped bars by sand particles carried in the wind. The nature of the bed rocks influences their ruggedness; and bare outcrops are common. Smooth rounded surfaces, are exceptional produced by the heating-cooling process of insolation splitting shallow leaves of rock away, leaving a smooth surface. Soil is rare on desert mountains, and where it exists, it is very shallow, accumulating in pockets between rocks in fissures, and providing the only suitable medium for plant life in these mountains. Wind action is also responsible for the poor soil cover, carrying away loose particles not protected by hollows.
2. *Desert plains*. The desert mountains usually rise abruptly from the plains, of which there are three principal types (see also Section 1.8.1.5):

   (a) Hammada: The hammada/is a stone – covered plain. Wind abrasion exceeds insolation, and soil is virtually absent, because any which does accumulate is immediately blown away. The surface is thus virtually devoid of vegetation, and when any is seen, it is usually in fissures protected from the wind. The bedrock type will influence the plant colonization potential.
   (b) Reg: The surface of a reg is also flat, but is covered with rock detritus or gravel. While the wind can remove the finer particles lying between rock fragments, it cannot reach the soil protected by the larger pieces, and these areas plants can grow. The surface soils are thicker than those of the hammada plain, but unless there is adequate rainfall, plants are usually small, or only temporary inhabitants.
   (c) Erg: The erg is not as flat as either the hammada or the reg for it is the classical "sand sea", identified by its undulating surface of crests and troughs. The sand forming the surface of an erg originates in mountains, regs or hammadas, often a considerable distance away, or from neighbouring desert zones. The undulations are wind-formed and mobile as a rule. Vegetation is sparse. especially on the more mobile surfaces, but where the surface is relatively stable, plants can be found, particularly in the troughs and hollows between the crests.

3. *Alluvial fans*. Alluvial fans are formed when the streams running off the mountains in deep gorges meet the plains below. The streams carry rock detritus and alluvium, which is quickly deposited into fan-shaped structures spreading out from the foot of the mountains. A gradation of rock particle sizes is found in the fan, with the largest boulders deposited near the mountain face, and the

alluvium spread further out. The water running down the gorges flows under the boulders, so that the finer alluvium is carried along underground. Alluvial fans store large volumes of water underground, and dense vegetation can be supported. In the individual water channels, the vegetation can be even thicker. The size of the plants is regulated by the depth of the sediments and their water storage potential.

4. *Alluvial blanket.* As the number of gorges down the mountain sides increases, the number of alluvial fans multiplies. Eventually the fans unite to form a continuous layer or blanket. Being on a larger scale than alluvial fans, the alluvial blankets, which are alternatively known as "bajada", have a greater vegetation potential.
5. *Pediment.* The mountain faces, as they are weathered and eroded, gradually retreat and the surface left behind at their feet (the pediment) is a gentlyinclined bare rock, hammada-like surface. Pockets of soil are the only suitable spots for plant colonization. The pediment surface extends from the mountain face until it disappears under the alluvial blanket or fan.
6. *Inselberg.* The bases of rock rising up from a pediment or protruding above the alluvial fan, are known as inselbergs. They are also encountered on desert plains.
7. *Drywash.* The drywash goes under various other names: in the United States it is known as "arroyo" in Africa and Arabia as "wadi", in Chile as "quebrada", in China as "chap" and in South Africa as "laagte". The drywash is a water drainage channel on the surface of an alluvial fan. It can vary in width from 3 to 30 m, and according to its age it can develop a braided or detailed branching drainage system. The soils associated with a drywash are of good quality and can support large plants. Erosion by water flow is common, and the vegetation tends to be denser on the banks than in the actual beds as a result. Dry river bed is another name of drywash (Evenari, 1985) which carries water only during floods, which can be torrential but are short-lived and occur at irregular intervals.
8. *Dry lake or Playa.* In the desert plains there are occasional depressions, where rainwater can accumulate when it runs off pediment slopes. A lake bed can form in the lowest part of a desert basin, and water can persist there for several weeks (though rarely for longer) after rain. The surface of the playa is usually flat, and can support little or no vegetation. There are two distinct types of desert dry lakes:

   (a) Clay Pan or Clay Flat: This is formed in the desert valleys. When such a pan dries out, the clay surface usually cracks and eventually pulverizes. The water table tends to be well below the surface, but very deep-rooting plant species can become established, provided that they are not inhabited by poor soil or water quality.
   (b) Salt Pan, Salt Flat or Salt Playa: Although the water table is usually within 3–4 m of the surfaces in such areas, plants cannot grow because of the high

salt content of the deposits. Salt crystals are found on the surface and these are among the most hostile surfaces of the desert.

9. *Sabkha.* Sabkhas are found in desert areas which lie close to the sea, and are depressions which are not fed by either streams or run-off water. The surface of the sabkha is within –2 m of a very salty water-table, and in the dry season it is covered by a salt crust which overlies a saline bog. A sabkha usually has sand dunes on its edges. It can become flooded by the sea during high tides, or flooded by rainwater after a storm. The salt crust can be broken up during periods of desiccating winds, and the crystalized salt can be blown away to contaminate the soil in neighbouring areas. While no plants can be found growing in a sabkha, a limited range of plants can grow on the periphery, if they are able to tolerate the high level of salinity.
10. *Salina.* When a dry salt lake or playa remains moist or contains water throughout the year, it is known as a "salina". Vegetation will grow on the perimeter, but the range of species depends on the quality of the water and the soil.
11. *Desert flat.* The desert flat is found in broad valleys where the surface slopes gently between a dry lake and the alluvial deposits. The slope of a flat is 1° instead of the 7° usual for alluvial fans. The surface deposits are finer than those of alluvial fans, and are therefore very suitable for plant establishment.
12. *Desert Pavement.* A desert pavement is a type of surface which develops on desert flats, alluvial fans and alluvial blankets. It is formed when the wind blows away the sand, silt and clay deposits. The surface is almost level, and consists of rounded pebbles between 15 and 75 mm long. When the pebbles are rolled together by wind and rain, and become tightly interlocked, the surface is known as "pebble armour" or "serir". Vegetation can become established on this surface, but is sparser than that found on alluvial fan soils. Deep-rooting species are particularly well-adapted to the desert pavement surface.
13. *Badlands.* Badland scenery is usually associated with hilly or mountainous land, which has been deeply scored by gullies created in the aftermath of the occasional heavy storms. The normal rainfall is too scant to support plants, whose roots would otherwise bind the surface and stabilize the slopes. Badlands are most common in highland areas composed of soft bedrock types. Where there are harder rocks which are better able to resist erosion, these remain as tall pillars or platforms, rising above the surrounding landscape.
14. *Sand dunes.* There are four major sand dune types:

    (a) Barchan: The barchan dune is crescent-shaped, with the tails of the dune pointing downwind. It is constantly on the move, and progresses across the desert either on its own or as part of a swarm, where numbers of dunes become linked and move together.
    (b) Siefs or Longitudinal Dunes: These dunes usually run in parallel lines up to ten kilometres in length. They appear to have originated as barchans which have become temporarily anchored by a plant or group of plants. These hold down the tail of the dune enabling the free end to move on its own. The dune then coalesces with other barchans and thus forms the longitudinal

dune. Another theory is that, conversely, the barchan is formed out of a sief when the speed of the winds flowing over the sief drops and friction on the sand causes turbulence which breaks up the continuous surface into smaller crescent dunes.

(c) Transverse Dunes: Whereas the longitudinal dune is formed parallel to the wind, the transverse dune develops across its path. These dunes are either an amalgamation of barchans, or they are formed when there is more sand than can be used by a swarm of barchans.

(d) Whaleback Dunes: Very large siefs. on which both barchans and smaller siefs can develop are called whaleback dunes. They are either an agglomeration of a number of siefs, or can be considered as erosional dunes, because of the rocky surface exposed between them. Water is stored within all the large dunes as they tend to move quite slowly, and plants can be seen growing on them, with the roots reaching down to the sub-surface storage areas. Nevertheless, vegetation is sparse, and consists either of deep-rooting species, or those plants whose roots can survive in the moving surface, taking their moisture from dew.

Dunes can be classified according to size, environment, growth stage, origins, shape and wind direction. There are simple compound or complex dunes, depending on the prevailing conditions. When complex or compound dunes coalesce together into larger units still, they form a "dune field".

Not all these habitat types can be found in every dry desert. Each desert habitat has a distinct type of plant communities associated with it. The nature of the habitat determine the precise manner in which plants, and animals are able to exploit their potential.

### *1.7.2 Vegetation Forms*

Based on the life-forms and habits of the plants forming them, Adams et al. (1978) classified the vegetation types of the Afro-Asian Mediterranean coastal deserts under the following two groups:

**Group A**: Ephemeral, succulent perennial and woody perennial vegetation forms or,

**Group B**: Accidental, ephemeral, suffrutescent perennial (undershrubs) and frutescent perennial (shrubs) vegetation forms.

These are described below:

**Group A**

**a. Ephemeral vegetation form**

Fifty to sixty percent of all desert plants are ephemerals. They have extremely short life cycles and are herbaceous, i.e. non-woody in character. Their growth is rapid:

the time from germination to death is telescoped into 6–8 weeks. Their activity has to coincide with the moist season: they deal with the dry season and drought quite simply by by-passing them altogether. Therefore, they have no need to develop xerophytic or drought-resistance properties. Their roots are shallow, they are small plants, they grow very fast and flower quickly. The seeds can remain dormant in the soil until the next important rains, and even if there is no rain for several successive years, the seeds are not damaged. 10 mm of rain may not sufficient to promote germination: some seedlings might appear after 1–5 mm of rain; germination is only completed after 25 mm of rain. However, even a cloud-burst producing 75 mm of rain may not automatically ensure germination, and from this evidence we may conclude that the seed contains some agent which inhibits germination unless the circurrtances are all favourable. Germination may also be related to the number of sunny days following rainfall, the two classes of ephemerals (winter and summer) have different temperature requirements and optima. For winter ephemerals, the optimum ranges between 15.5 and 18°C, while the optimum for summer ephemerals ranges between 26 and 32°C (Adams et al., 1978).

Ephemerals are the first colonizers of most kinds of desert terrain as their seeds may be spread by animals, insects and wind. When the plants die and decay, they are attacked by soil organisms and soil fertility is thereby gradually improved, enabling the establishment of other plant types. In good years the plants produce a larger number of seeds than are produced in poorer, drier years.

**b. Succulent perennial vegetation form**

The phenomenon of succulence occurs when the parenchyma or outer cells of the plant stem or leaf enlarge or proliferate. The evolution of this characteristic means that the volume of the plant's stem or leaves is increased, which enables the plant to store an increased volume of water within its structure. The deposition of a water-proofing layer of wax or similar substance on the external surface of the plant reduces the risk of loss of water from leaves or stern, and this, in combination with the cell proliferation, enables the plant to reduce its overall water requirement. Succulent perennials are either spiny, like the cacti, or non-spiny but with swollen leaves. The cacti of the American continent, the succulent euphorbias of the African dry zones and certain other succulents have a further quality which fits them to life in their dry environments: they are able to close their stomata during the heat of the day to avoid excessive loss of water through transpiration and open them during the coolness of the night instead.

**c. Woody perennial vegetation form**

Although ephemeral plants make up 50–60% of the total number of plants in the desert, "woody perennial" is the dominant plant type. The group is composed of a number of morphologically different forms of plants ranging from grasses and woody herbs, through shrubs to trees. They are all very hardy, for they have to cope with the major characteristics of the desert environment: drought, heat and wind.

They can be evergreen or deciduous; deciduous species can be either drought-deciduous in hot deserts and cold-deciduous in the case of the cooler deserts. They grow most actively after rain and become dormant either during periods of drought, or during the cold season. Many of the woody perennials are spiny or harsh textured (Adams et al., 1978).

The seeds of woody desert perennials have a particular germination characteristic in many cases. Many perennials produce seed which will only germinate if the seed coat is damaged in some way. This damage can occur in a number of natural ways; for instance by the action of stones and boulders pushed along wadi beds after sufficiently large volumes of water have accumulated, these stones having a grinding effect on the coats of such seeds as may be amongst them and cracking the coats open. Other seeds rely on the effects of the digestive juices in animal intestines to soften their coats, prior to germination. Once the seeds of woody perennials in the desert have germinated, they produce only a few leaves before they appear to stop growing. This is because activity shifts to the root structure, which grows deeper and deeper down into the soil, to penetrate the moister layers below. Once the roots are well-established, the plant starts to produce more leaves and expand above its surface growth (Adams et al., 1978).

### Group B

It is possible also to separate the vegetation of dry lands into four major orders: accidental form, ephemeral form, suffrutescent (sub-shrub) perennial form and frutescent (shrub) perennial form. Each of these orders has a number of classes, and the predominant orders are the ephemeral form and the suffrutescent perennial form. Each contains a sequence of classes, developing from the more primitive to the more sophisticated depending on soil quality (Adams et al., 1978).

#### a. Accidental form

Only ephemeral plants are included in this order (though not all ephemerals are of the accidental type). Growing only when rain occurs, the formation cannot be considered permanent in any way. It is determined by extremely erratic rainfall, as is found, for example, in the Western Desert of Egypt where only 10 mm of rain can he expected every 10 years.

#### b. Ephemeral form

There are three categories in this group: succulent-ephemeral, ephemeral grassland, and herbaceous ephemerals. When any of these are found, rainfall is known to occur annually. The plants are ephemeral in the main, but perennials can be found occasionally. There is therefore some moisture, but because the soils are not water-retentive, it does not last throughout the year. As aridity decreases and rainfall regularity improves, the ephemerals become displaced by perennials.

##### (i) *Succulent-ephemerals*

The growing season for succulent ephemerals is longer than the 6–8 weeks typical of the non-succulent ephemerals, for they have the ability to store some moisture in their tissues. They can tolerate the severe conditions prevalent in soils that develop on erosion pavements, such as pediments, regs and hammadas.

The typical plants can be further sub-divided into three types: winter ephemerals summer ephemerals, and salt-marsh ephemerals.

1. *Winter ephemerals* growing after winter rains such as in those deserts influenced by the Mediterranean climatic types. The species involved include: *Aizoon canariensis, Aizoon hispanicum, Mesembryanthemum crystallinum, M. forsskaolii, M. nodifiorum, and Zygophyllum simplex.*
2. *Summer ephemerals* growing after summer rains such as in areas influenced by the Indian Monsoon or tropical climate systems. Typical species include: *Salsola inermis*, *S. kali* and *S. volkensii.*
3. *Salt-marsh ephemerals* include: *Halopeplis amplexicaulis* and *Salicornia herbacea.*

**(ii) *Ephemeral grasslands***

This form can develop into grassland over large stretches of ground, and in particular on shallow sand drifts. The predominants include species of: *Aristida, Bromus, Cenchrus, Eragrostis, Poa, Schismus, Stipa* and *Tragus*.

**(iii) *Herbaceous ephemerals***

These herbaceous ephemeral plants are only found on soft deposits in good locations, where a water supply is preserved even if for only a short while, and include: *Bassia muricate, Caylusea hexagyna, Schouwia thebaica, Tribulus terrestris* etc.

**c. Suffrutescent perennial (sub-shrub) forms**

This is the most wid-spread form occurring where there is a permanent flora of perennial species, and it includes a perennial grassland form. There are three layers: a suffrutescent layer 300–1,200 mm high, a grassland layer of the same height and a ground layer. The suffrutescent flora usually predominates, and the ground layer of dwarf and prostrate perennials is sometimes augmented by some ephemerals.

**(i) *Succulent sub-shrubs***

These plants have evolved sufficiently to have an internal moisture reserve system. Where salt-marsh communities are found, the dominants include species of: *Arthrocnemum, Salicornia, Suaeda, Anabasis, Hammada,* and *Zygophyllum.*

**(ii) *Perennial grassland***

Soils which are capable of storing some water are the habitats of these plants. They are particularly valuable for sand-sheet and sand-dune stabilization, and the different soil types encourage different plant associations. This type is dominated by, for example, *Hyparrhenia hirta, Lasiurus hirsutus, Panicum turgidum, Pennisetum dichotomum* and *Poa sinaica.*

**(iii) *Woody perennials***

This order is a transitional order falling between the succulent and the grassland forms. Plant cover is always thin, in particular in the more arid locations and many of the species which grow on rocks-the chasmophytes and rhizophagolithophytes – are included here.

**d. Frutescent perennial (shrub) forms**

This form is typical of all the vast desert scrublands. There are three layers: frutescent (1,200–3,000 mm) suffrutescent (below 1,200 mm) and ground.

**(i) *Succulent perennials***

The cacti of the American deserts, and the succulent Euphorbias of tropical Africa and Arabia are part of this category. The saxaul, *Hammada persicum*, is also included.

**(ii) *Scrublands***

Scrubland can only be found in good locations where there is adequate soil and rainfall, and where there are mountains around to supply runoff water. Scrubland indicates the highest level of water reserves available in the desert, and is particularly relevant to semi-arid areas. The dominant plants include species of: *Acacia, Pistacia, Prosopis, Retama, Tamarix* and *Ziziphus.*

## 1.8 Environment and Plant Life of the Afro-Asian Mediterranean Coastal Lands

Climatic aridity, as stated before, affects wide areas of the Mediterranean Basin particularly in the Mediterranean territories of the North African and SW Asian countries. The European Mediterranean section, however, is mainly located in the humid and hyper-humid Mediterranean bioclimatic provinces with only very small parts of the Spanish and Italian territories only under the influence of arid climate. The following pages will describe the environment and plant life of the arid zone lands of the North African and South West Asian regions of the Mediterranean Basin.

### *1.8.1 North African Mediterranean Coastal Lands*

#### 1.8.1.1 Landforms

The region of North Africa extends along the southern Mediterranean. According to the World Atlas of Desertification (UNEP, 1992), Africa north of the Sahara includes Morocco, Algeria, Tunisia, Libya and Egypt, plus Western Sahara and Cape Verde. This is an area of 545.3 million hectares; 98.3% of which are drylands vulnerable to desertification and the hazards of drought. Hyper-arid territories represent 70.6% of the total area (Kassas, 1996).

Aridity and geographical location of the North African region have a prominent effect on its biological element that rival these of the tropics in importance.

The region is the home of the wild relatives of many food crops, medicinal plants and feed for animals. However, the biological diversity is continuously deteriorating in view of the human population explosion, modernization and innumerable human activities using improper technologies. Batanouny (1996) stated "One should consider the loss of say 10 species from the flora or fauna of the desert ecosystem is relatively, drastic and considerable as compared to the biomes with greater species richness". From an ecological point of view, the North African countries represent a transition from the Mediterranean climate to the Saharan one. This has a remarkable impact on the bio-diversity and its distribution along the climatic gradient from north to south.

Morocco, the most westerly of the Arab states of North Africa, extending from the Peninsula of Ceuta in the north to the borders of Spanish Sahara in the south. To the east and southeast it is bounded by Algeria; while to the west and north lie the Atlantic Ocean and the Mediterranean Sea. The total area of Morocco is 714,000 $km^2$ out of which 77% is under the influence of climatic aridity, 15.4% semi-arid and 7.6% humid and mountainous (Batanouny and Ghabbour, 1996).

The physical geography of the Moroccan territory is complex ranging, for example, from low, flat plains such as the Rharb to the wild and rugged grandeur of the Rif and the high Atlas.

Two major series of mountain ranges – a northern and southern – partially enclose a triangular plateau which has its base along the coast between Rabat and Essaouria and which is sometimes known as the Moroccan Meseta. To the south and south-east of the Meseta the most prominent feature is the high Atlas range, which extends north-southwards from the coast near Agadir. The ranges are cut off sharply at the coast, giving rise to very dramatic cliffs and headlands. The highest point in Morocco rises to 4,165 m (Branigan and Jarrett, 1975).

Algeria extends for a distance of about 1,000 km along the Mediterranean coast between Morocco and Tunisia. It has total estimated area of about 2,224,000 $km^2$, the largest of the territories of north-west Africa. The greatest part of Algeria (1,897,000 $km^2$) lies in the Sahara Desert and a substantial part actually lies to the south of the Tropic of Cancer.

Tunisia is by far the smallest of the North African countries with a total area of 164,000 $km^2$. About half of this area is desert, or at least semi-desert which occurs largely in the rain shadow of the Atlas. While the annual rainfall at, for example Casablanca is about 406 mm, at Gabes, which lies in approximately the same latitude, the corresponding figure is just over 167 mm.

Libya is almost entirely a Saharan country, and only along the coastal strip does a variation of the Mediterranean type of climate occur. Libya (total area = 2,105,000 $km^2$) is characterized by a remarkably smooth coastline with two most prominent features being the Gulf of Sirte which thrusts southwards almost to latitude 30°N and the E1-Maij Peninsula which sweeps northwards to the east of the gulf around Al-Gabal Al-Akhdar. The greater part of the main plateau of Libya carries deserts of the varying types (hammadas, reg, erg etc.).

Egypt occupies the northeastern corner of the North Africa region. It is roughly quadrangular, extending about 1,073 km from north to south and about 1,229 km from east to west. Thus the total area of Egypt is little more than 1,000,000 $km^2$ (1,019,600 $km^2$) occupying nearly 3% of the total area of Africa (Ball, 1939; Said, 1962; Zahran and Willis, 1992). Out of this area, the Sinai Peninsula, the Asian part of Egypt, occupies about 61,000 $km^2$ in the northeast.

Egypt is bordered on the north by the Mediterranean Sea, on the south by the republic of Sudan, on the west by Libya and on the east by the Red Sea and the Gulf of Aqaba. It extends over about 10° latitudes being bounded by latitude 22 and 32°N, i.e. lies mostly within the temperate zone, less than a quarter being south of the Tropic of Cancer. The whole country forms part of the great desert belt that stretches from the Atlantic across the whole of North Africa through Arabia. The Mediterranean coast of Egypt extends for 970 km from Sallum eastward to Rafah in three sections: the western coast (550 km), the middle (deltaic) coast (180 km) and the eastern (Sinai) coast (240 km) (Zahran and Willis, 1992).

#### 1.8.1.2 Climate

The North African Arid Zone by and large has a Mediterranean climate characterized by winter rains and long dry summers. Monod (1937) divided it into two zones: arid and desert. The Arid Zone is defined here as having steppe vegetation distributed in a continuous and diffused pattern whereas the Desert Zone has large tracts of barren land with shallow soils, its perennial vegetation being confined to the run-off networks and laid out in clustered "contracted" pattern. The limit between the two zones is approximately the 100 mm isohyet. The upper limit of the steppe vegetation is approximately the 400 mm isohyet. Thus, as a first approximation one has the very simple definitions: Desert Zone < 100 mm <Arid Zone < 400 mm. The areas of these zones in the five countries of the North Africa are shown in Table 1.4.

The following is a description of the different climatic particulars of the North African region.

**(i) Rainfall**

The average annual precipition varies from almost zero in the most aird part of the Sahara to 400 mm at the northern edge of the Arid Zone. Examples of the monthly and annual averages in Desert and Arid Zones are given in Table 1.5. It is clear

**Table 1.4** Areas of arid and desert zones in North African countries (after Le Houerou, 1985)

| | Non-arid | | Arid | | Desert | | Arid+desert | | |
|---|---|---|---|---|---|---|---|---|---|
| Country | Area | % | Area | % | Area | % | Area | % | Total area |
| Morocco | 197 | 31.7 | 120 | 19.3 | 303 | 49 | 423 | 68.3 | 620 |
| Algeria | 181 | 7.6 | 200 | 8.4 | 2,000 | 84 | 2,200 | 92.4 | 2,381 |
| Tunisia | 37 | 23.9 | 55 | 35.4 | 63 | 40.7 | 118 | 76.1 | 155 |
| Libya | 5 | 0.2 | 90 | 5.1 | 1,665 | 94.7 | 1,755 | 99.8 | 1,760 |
| Egypt | 0 | 0.0 | 30 | 3.0 | 970 | 97.0 | 1,000 | 100 | 1,000 |

**Table 1.5** Average monthly and annual rainfall (mm) in the desert and arid zones of the North African countries (after Dubief, 1963)

| Country | Zone | J | F | M | A | M | J | J | A | S | O | N | D | Year |
|---|---|---|---|---|---|---|---|---|---|---|---|---|---|---|
| *Morocco* | | | | | | | | | | | | | | |
| – Boudnib | DZ | 5.4 | 4.2 | 11.1 | 8.2 | 4.2 | 4.7 | 0.9 | 6.9 | 10.6 | 17.2 | 12.7 | 13.8 | 99.9 |
| – Tarfaya | | 6.8 | 4.4 | 3.0 | 0.5 | 0.1 | 0.0 | 0.0 | 0.2 | 5.3 | 1.1 | 12.9 | 8.9 | 43.0 |
| – Marrakech | AZ | 24 | 30 | 17 | 33 | 15 | 7 | 2 | 3 | 10 | 20 | 34 | 27 | 242 |
| – Agadir | | 37 | 28 | 29 | 20 | 4 | 1 | 0 | 0 | 6 | 21 | 40 | 43 | 226 |
| *Algeria* | | | | | | | | | | | | | | |
| – Bechar | DZ | 7.0 | 7.2 | 12.7 | 7.5 | 2.3 | 2.9 | 0.4 | 3.6 | 6.6 | 13.8 | 11.6 | 9.6 | 85.25 |
| – In Salaf | | 2.9 | 0.8 | 2.1 | 1.1 | 0.3 | 0.1 | 0.0 | 0.7 | 0.1 | 1.3 | 2.3 | 33.9 | 15.6 |
| – Aflo | AZ | 31 | 33 | 38 | 32 | 28 | 20 | 9 | 11 | 24 | 45 | 30 | 33 | 342 |
| – Ain Sefra | | 10 | 10 | 14 | 9 | 15 | 28 | 8 | 7 | 15 | 29 | 29 | 18 | 192 |
| *Tunisia* | | | | | | | | | | | | | | |
| – Tozeur | DZ | 10.0 | 8.3 | 13 | 10 | 5.0 | 3.0 | 0.2 | 2.0 | 7.0 | 9.0 | 12 | 10 | 89 |
| – Fort Sait | | 6.7 | 3 | 4.1 | 1.3 | 2.7 | 1.0 | 0.0 | 0.0 | 1.5 | 2.4 | 4.1 | 4.0 | 30.8 |
| – Sousse | AZ | 43 | 34 | 30 | 22 | 18 | 6 | 1 | 5 | 50 | 43 | 37 | 38 | 327 |
| – Gabes | | 22 | 17 | 21 | 10 | 9 | 1 | 0 | 2 | 14 | 30 | 34 | 15 | 175 |
| *Libya* | | | | | | | | | | | | | | |
| – Bani Walid | DZ | 6.7 | 3.7 | 8.3 | 6.0 | 5.6 | 0.3 | 0.0 | 0.0 | 7.0 | 8.5 | 7.0 | 14.4 | 71.0 |
| – Sabha | | 2.0 | 2.0 | 0.8 | 0.6 | 0.2 | 0.0 | 0.0 | 0.0 | 0.0 | 0.7 | 0.2 | 0.0 | 8.0 |
| – Triopoli | AZ | 61 | 34 | 26 | 15 | 5 | 1 | 0 | 0 | 15 | 37 | 43 | 76 | 313 |
| – Sirt | | 42 | 28 | 13 | 3 | 2 | 1 | 0 | 0 | 7 | 16 | 26 | 34 | 172 |
| *Egypt* | | | | | | | | | | | | | | |
| – Port Said | DZ | 18 | 12 | 9 | 5 | 3 | 1 | 0 | 0 | 0 | 3 | 11 | 16 | 78 |
| – Cairo | | 5 | 4 | 4 | 2 | 2 | 0 | 0 | 0 | 0 | 2 | 2 | 4 | 25 |
| – Alexandria | AZ | 48 | 24 | 11 | 3 | 2 | 0 | 0 | 0 | 1 | 6 | 33 | 56 | 184 |
| – Sallum | | 38 | 4 | 5 | 0 | 7 | 0 | 0 | 0 | 1 | 4 | 29 | 27 | 132 |

DZ = desert zone, AZ = arid zone.

that in the Arid Zone there are two gradients of variation in the amount of average rainfall: a latitudinal gradient and altitudinal gradient. The latitudinal gradient is positive northwards whereas the altitudinal gradient is usually of the order of 10–12% per 100 m difference of altitude. The latitudinal gradient varies from 0.25 to 1.0 mm/km since rains increase from 100 mm in the south at the edge of the desert zone to 400 mm at the northern limit of the Arid Zone. Altitudinally, rainfall doubles for an increase of 800–1,000 m in elevation. In the Desert Zone there are three gradients: latitudinally, longitudinally and altitudinally. The latitudinal gradient is similar to that of the Arid Zone rains decrease from north to south with an absolute minimum on the Tropic of Cancer (23° 27′N), close to zero. The gradient is of the order of 0.15–0.33 mm per kilometre, or 16–36 mm per degree of latitude between the 10- and 50-mm isohytes; it is steepest in the Gulf of Sirte and most gradual in southeastern Algeria. The longitudinal gradient is less obvious, but there is, by and large, an increase in aridity from west to east, from the Atlantic Ocean to the Red Sea; the eastern Sahara (Libyan Desert) between 15 and 32°E is virtually rainless.

The altitudinal gradient in the Saharan mountains is slightly less than in the Arid Zone.

In the *Arid Zone*, rains occur from September or October until April or May. However, there are four different regimes according to: (1) whether or not there are significant summer rains, and (2) if summer rains are negligible, whether the maximum occurs in autumn, winter or spring.

The first regime with summer rains occurs at higher elevations above 1,000–1,500 m in the highlands of Algeria and Tunisia. Summer rain may amount to 10–20% of the total amount (Aflou and Laghouat in Algeria: Kasserine and Sbeitla in Tunisia).

The three other regimes have a sharp summer minimum, close to zero, but a maximum in the autumn, usually in the lowlands; in spring in the highlands; and in winter in the eastern part of the region (Libya-Egypt) where 50% or more of the annual total falls in December and January.

There are thus: (a) an eastern regime with winter maximum; (b) a highland regime with spring maximum and (c) a mountain regime with summer rains.

In the *Desert Zone* it is something of an exaggeration to speak of seasonality of rainfall; however, one can distinguish several trends. A typical Mediterranean regime exists in the north with no summer rains. To the south of the 25th parallel there is a clear tropical trend with summer rains amounting to 25–50% of the annual total.

As in all arid zones in the world, interannual variability is inversely related to the annual average. At the upper limit of the Arid Zone (the isohyet of 400 mm) the maximum rainfall observed in any particular year is 4–5 times the minimum, and the coefficient of variability is of the order of 40%. At the border of the desert the maximum reaches 10–12 times the minimum and the coefficient of variability is of the order of 60–80%. Monthly variations are still higher and any particular month in any particular year may be abnormally rainy or absolutely dry, and there is no correlation between consecutive months. Thus, rainfall is totally unpredictable, unlike the dry tropics, where rains usually occur in a single well-determined short season. In the desert, variability is generally over 60%. However, in the southern Sahara, under a tropical regime, variability is usually much lower. Besides temporal variability of rainfall, there is also in arid lands a very great spatial variability, because rain-storms are highly local in their incidence. This has been observed by many authors (e.g. Kassas and Imam, 1954, 1959 in Egypt, and Evenari et al., 1971, 1982 in Israel).

### (ii) Temperature

The mean annual temperature varies between 17 and 20°C along the coast, from 18 to 22°C in the Arid Zone and from 20 to 25°C in the Sahara. From an ecological point of view the most important data are the mean minimum of the coldest month, $\overline{m}$, and the mean maximum of the hottest month, $\overline{M}$. Since these data best characterize the normal thermal environment for plant and animal life. These temperatures are quite variable according to latitude, altitude and continentality. The mean minimum temperature of the coldest month (January) varies from 10°C on the seashore of southern Morocco and northern Libya to –2°C in the arid highlands of Morocco and Algeria. Vegetation is particularly sensitive to $\overline{m}$ since this parameter is closely linked to frost and to the length of the winter rest period of plants. From the systematic study of plant communities and crop performance the following thermal thresholds have been selected (Le Houerou, 1970, 1975) that are now often used in the climatic classification of the region:

$\overline{m} > 9$°C: no frost at ground level, very warm winters; no rest period for vegetation.
$9 > \overline{m} > 7$: no frost under shelter, warm winters, no rest period for vegetation.
$7 > \overline{m} > 5$: 5–10 days of light frost under shelter, mild winters, virtually no rest period for vegetation.
$5 > \overline{m} > 3$: 10–20 days of light frost, temperature winters, rest period of 1–2 weeks for vegetation.
$3 > \overline{m} > 1$: 20–30 days of forst, cool winters, rest period of 2–4 weeks for vegetation.
$1 > \overline{m} > -2$: 30–60 days of serious frost, cold winters, rest period of 1–3 months for vegetation.
$-2 > \overline{m} > -5$: 60–120 days of hard frost, very cold winters, rest period of 3–5 months for vegetation.

The value of $\overline{m}$ decreases with elevation at a rate of about 0.5°C per 100 m of difference of level.

Very warm winters occur in the southern and western Sahara: this explains why many tropical species (both plants and animals) are found in southern Morocco. Cool and cold winters occur only in the highlands of Tunisia. Algeria and Morocco-cool above 600 m and cold above 1,200 m. The larger part of the region lies within the range of mild and temperate winters.

The average maximum of the hottest month, $\overline{M}$, is much less variable: 30°C in a narrow strip along the coast 35–38°C in the hinterland and 40–45°C in the Sahara: nowhere does it reach 46°C. The hottest parts are in southern Algeria (Adrar 45.6°C, Aoulef 44.7°C and Salah 44.7°C) and not in the Libyan–Egyptian desert as one might expect (Dakhla 38.5°C, El Kharga 39.4°C, Al Kufrah 39.1°C). The absolute maximum ranges from 45 to 55°C in the Sahara. The highest shade temperature ever measured anywhere on earth seems to be at AlAziziah (Libya) where on September 3, 1922 it attained 58°C.

#### (iii) Relative humidity

The diurnal average relative humidity exceeds 70% throughout the year on the coasts. In the interior Arid Zone it is of the order of 60–65% in winter and 35–40% in summer. In the Sahara it reaches 40–55% in winter and drops to 20 or 25% in summer.

#### (iv) Hot winds

Hot winds, to which special local names have been given – *sirocco. ghihli*, *chergui*, *khamsin* etc. – blow with a frequency of from 20–90 days/year, especially in spring and autumn. The temperature of the air ranges from 35 to 45°C and the humidity drops to between 5 and 15% during these dry spells. When they occur in spring these hot winds are very detrimental to crops and native vegetation, which may dry out in a very few days annihilating the hopes for a good crop or lush grazing resulting from a favourable rainy season.

#### (v) Evaporation and evapotranspiration

Evaporation is of the order of 2,200–2,500 mm/year in the Arid Zone and 2,500–3,000 mm/year in the Desert Zone. Potential evapotranspiration is of the order of 1,400–1,600 mm/year in the Arid Zone and 1,600–2,200 mm/year in the desert.

### 1.8.1.3 Geomorphology

The Arid Zone of North Africa provides classic examples of arid-land geomorphology. The ranges of hills and mountains of the Atlas Chain, show erosion forms in accordance with their structure and lithology. These ranges are separated by wide synclines in which are developed four huge Quaternary pediments often covered by a thick calcareous crust of Early and Middle Pleistocene age. As endomorphism is the general rule, the lower parts are occupied by Quaternary alluvia which are often saline.

The Sahara presents four main types of land forms:

(1) Rocky or stony hills with steep slopes.
(2) The *regs* (also called *serirs* in the eastern Sahara). which are gravelly-pebbly desert pavements covering subhorizontal plains denuded by aeolian erosion. These regs may have various origins: flat structural surfaces of rather soft rocks; old pediments reshaped by aeolian erosion; coarse alluvial deposits of detritus.

They are therefore classified as "autochthonous" and "allochthonous" (Monod, 1937).

(3) The hammadas are structural surfaces, more or less flat, covered by large flagstones of limestone, sandstone or basalt.

(4) The ergs or sand seas are bodies of dunes of various sizes and shapes from the small crescentic very mobile *barkhans* to the huge fixed *ghourds* 50–200 m high, with the intermediate *elb*, and *silk*. Soft sandstones are often curved in *yardangs* especially in the southeastern Sahara (Mainguet, 1972).

The rocky mountains and hammadas represent 10% of the desert area (0.5 million $km^2$), the regs cover some 68% (3.3 million $km^2$), and the ergs occupy 22% (1.1 million $km^2$).

#### 1.8.1.4 Soils

In the Arid Zone mature soils are exceptional and are found only under little-disturbed vegetation in protected areas (Le Houerou, 1985).

One can first distinguish the soils under open forest or woodland of *Pinus halepensis* or *Juniperus phoenicea.* On soft or hard limestones, these are usually shallow, of the rendzina type, with 3–5% (exceptionally 5–10%) of organic matter in the top layers. On marls and shales, soils are of the brown calcareous type, with a deep layer of accumulation of calcium carbonate: exceptionally they may be vertisols. Red or brown mediterranean soils are to be found on karstic limestone plateaux, for instance in eastern Libya.

Under steppe vegetation the mature soil is of the isohumic type with 2–3% of organic matter; the organic matter content decreases progressively with depth. There is often a layer of accumulation of calcium carbonate below 50 cm from the surface. There are huge areas of Pleistocene calcareous crusts on the pediment. These give way to shallow soils.

Depressions are filled with alluvia of various nature (sandy, silty, loamy and clayey). Many depressions have saline or alkaline soils, with a shallow saline water table and are covered with halophytic vegetation, mainly of Chenopodiaceae.

From a chemical viewpoint, the soils are usually alkaline with pH from 7.5 to 9.0. Calcium carbonate content is high. Sulphates are common, especially gypsum, and sodium chloride is almost always present. Phosphorus and potassium are, in general, adequate for plant growth but nitrogen is usually deficient. Trace-element deficiencies have been reported very rarely. From an ecological and agronomic point of view the main soil factor to be considered in the Arid Zone is the moisture regime which is closely related to ecosystem structure and productivity. In this respect the most significant soil characteristics are: (a) topographic or geomorphic position (run-off or run-on): (b) texture [permeability of the upper layer (water intake), infiltration rate, field capacity etc.]: (c) depth (storage capacity); and (d) toxicity.

The good productive soils are those which are able to store water during the short rainy periods, and to release it afterwards to the vegetation or crops during the long dry spells – that is soils which owing to their hydrodynamic characteristics are able to reduce the effects of drought.

### 1.8.1.5 Plant Life

The flora and vegetation of the Arid and Desert Zones of North Africa has been studied in depth since seventeenth century, e.g.: Zanoni (1675), Spottswood (1696), Shaw (1738), Poiret (1789), Vahl (1790–1794), Desfontaines (1798–1800), Broussonet (1795–1801), and Schousboe (1801a, b). The four volumes of Flora Atlantic by Desfontaines were published between 1798 and 1800 and included not less than 1,500 species. The bibliography now amounts to several thousand publications which include Oliver (1938), Tadros (1949), Hassib (1951), Kassas (1952), Täckholm (1956, 1974), Quezel (1965), Sauvage (1954), Monod (1958), Keay (1959), Le Houerou (1959, 1975, 1985, 1992), Meigs (1966), Ayyad (1973), Batanouny (1973), Ayyad and El-Ghareeb (1982, 1984), Zahran (1982a), Zahran et al. (1985), Zahran and Willis (1992), Dallman (1998) and Blondel and Aronson (1999). The location of the region at the contact between three continents and three floras: Eurosiberian, Mediterranean and Sudano-Decania (Palaetropical) makes its flora much richer and varied than one might expect. From the point of view both of flora and vegetation we may distinguish four different zones: a Mediterranean Arid Zone, a Mediterranean Desert Zone in northern Sahara and the higher mountains of the central Sahara, a Sahara-Sindian Desert Zone in the central Sahara and a Sudano-Deccanian Tropical Desert Zone in the southern and south-western Sahara (Le Houerou, 1992).

**I. Phanerogams**

Approximately 3,000 species of the flowering plants have been recorded in the different habitats of the North African Mediterranean region with about 50% in the Arid Zone and 50% in the Desert Zone. However, as the Desert Zone is 10 times larger than Arid Zone, it means that the former is substantially poorer in species than the latter. For example, in Egypt (Boulos, 1995), there are 1,095 species in the Arid Zone with surface area of only 10,000 km$^2$ whereas the Desert Zone with surface area of 990,000 km$^2$ comprises 990 species. The origin of the flora of the region is mostly Mediterranean to the north, Tropical to the south and Sahara-Sindian in the central Sahara.

Ozenda (1958) estimated that in the central Sahara the percentages of the various elements to be: 28% (Mediterranean), 47% (Sahara-Sindian), 17% (Tropical) and 8% (Pluriregional). In the southwestern Sahara these proportions are (Monod, 1958): 4, 14, 62 and 20%, respectively.

From the systematic point of view, the flora of the Arid Zone habitats of the North African Mediterranean region include some 600 genera. The number of species per genus average about 2.5% in the Arid Zone and 1.7% in the Desert Zone (Le Houerou, 1959). Some 100 families are represented, the main ones being the Asteraceae, Brassicaceae, Fabaceae, and Poaceae which together account for 40% of the flora. Other important families are: Apiaceae, Boraginaceae, Caryophyllaceac, Chenopodiaceae, Cistaceae, Euphorbiaceae, Geraniaceae, Lamiaceae, Polygonaceae. Rubiaceae, and Scrophylariaceae which together cover a further 40%.

The main genera, listed in decreasing order of number of species, are *Astragalus, Helianthemum, Silene, Linaria, Euphorhia, Centaurea, Ononis, Erodium, Galium, Medicago, Plantago, Bromus, Aristida, Stipagrostis, Rumex, Trifolium, Convolvulus, Atriplex, Cyperus, Salsola, Fagonia, Tamarix, Launaea* and *Reseda.*

Endemism is high: about 300 species (that is, 10% of the flora) of which half are found in the western part of the region (Mauritania, Morocco and Algeria), and half in the eastern part (Tunisia, Libya and Egypt). There are some forty endemic genera, mostly found in the Sahara north of the twentieth parallel. They include: *Ammodaucus, Ammosperma, Argania, Battandiera, Chlamidophora, Cladanthus, Eremophyton, Foleyola, Fredolia, Gaillonia, Koniga, Lifago, Lonchophora, Mecomishus, Megastoma, Monodiella, Muricaria, Nucularia, Oudneya, Pegolettia, Perralderia, Pscuderucaria, Randonia, Rhetinolepis, Rupicapons, Saccocalyx, Spitzelia, Stephanochilus, Tourneuxia, Traganopsis,* and *Warionia* (Le Houerou, 1985).

The southern Sahara is characterized by tropical families such as: Apocynaceae, Asclepiadaceae, Caesalpiniaceae, Capparidaceae, Celastraceae, Menispermaceae, Mimosaceae, Moraceae, Nyctaginaceae, and Zygophyllaceae whereas typical Holarctic families almost disappear: Apiaceae, Brassicaceae, Caryophyllaceae, Chenopodiaceae and Rosaceae.

Tropical genera well represented include: *Abutilon, Acacia, Althaea, Balanites, Boerhavia, Boscia, Cadaba, Calotropis, Cassia, Chloris, Commicarpus, Corchorus, Enneapogon, Fagonia, Ficus, Glossonema, Grewia, Hibiscus, Hyparrhenia, Hyphaene, Indigofera, Lasiurus, Leptadenia, Maerua, Melhania, Oropetium, Pentzia, Rhynchosia, Salvadora, Schoenefeldia,* and *Tephrosia.*

### II. Cryptogams

The cryptogamic flora is not very well known, approximate number is shown in Table 1.6.

The total number of cryptogamic species known is thus 1,311 – that is, 22% of the number of phanerogams, which is much less than in temperate climates where cryptogam species are more numerous than phanerogams.

### III. Main vegetation zones

In relation to the main bioclimatic sub-zones prevailing in the North African Mediterranean region, Le Houerou (1992) described eight vegetation zones, namely: the upper arid subzone, middle arid subzone and lower arid subzone (Arid Zone), upper desert subzone, middle desert subzone, lower desert subzone (Desert Zone) in addition to the two zones in the True Desert Zone and the Saharan mountains. The following is a short description of each of these vegetation zones:

**Table 1.6** Number of species of the six cryptogamic groups (from Le Houerou, 1992)

| Group | Species number |
|---|---|
| 1. Algae | |
| a. Cyanophyceae | 45 |
| b. Xanthophyceae | 300 |
| c. Chlorophyceae | 73 |
| 2. Fungi | 60 (of which 30 are free-living) |
| 3. Lichens | 90 |
| 4. Mosses | 702 |
| 5. Liverworts | 20 |
| 6. Vascular cryptogams | |
| a. Ferns | 20 (oasis and mountains) |
| b. Horsetail | 1 |

## A. The arid zone

This zone corresponds with the arid Mediterranean climate, having average annual rainfall ranging from 100 to 400 mm: it is divided into upper, middle and lower arid sub-zones.

### a. *Upper arid subzone*

The primeval vegetation is forest along the coast in the warm and mild winter (varieties of the climate, $m > 5°C$) dominated by *Tetraclinis articulata* or *Juniperus Phoenicea*, with tree and shrub species: *Ceratonia siliqua. Chamaerops humilis, Olea europaea, Periploca laevigata, Pistacia lentiscus, Rhus pentaphyllum*, and *Withania frutescens*. In western Morocco *Argania sideroxylon* is codominant with *Tetraclinis* (Safi and Essaouira areas). In eastern Libya *Tetraclinis* is absent and the dominant species is *Juniperus phoenicea* with some remnants of *Arbutus pavarii, Ceratonia siliqua, Cistus* sp. *Cupressus sempervirens, Pistacia lentiscus, Olea europaea* etc. In the hinterland of the cool and cold winter varieties of the climate ($m < 5°C$) the primeval forest is dominated by *Pinus halepensis* with a number of companion species, such as: *Artemisia atlantica, Calycotome villosa, Capparis spinosa, Coridothymus capitatus, Cistus libanotis, Juniperus oxycedrus, Phillyrea media, Pistacia atlantica, Retama sphaerocarpa, Rhus tripartitum, Stipa tenacissima,* and *Thymus hirtus*.

Only about 70,000 ha of this zone can be classified as forest; 2 million hectares are covered by evergreen open bushland or shrubland of the garigue type. This garigue vegetation is mostly located on the shallow soils of the hills of the Arid Zone. On the pediments and plains it has almost everywhere, been cleared for grazing and/or cropping. This generalized deforestation has resulted in an extension of steppe vegetation over huge areas (Quezel, 1985; Le Houerou, 1959, 1969a). This change is supported by evidence from prehistorical, historical and present-day data drived from palynology, geographic descriptions from antiquity through the nineteenth century, and modern surveys using remote-sensing techniques, via, sets of aerial photographs taken several years or decades apart. The resulting steppe vegetation is kept more or less stable by heavy and permanent pressure from livestock. The main steppic vegetation types resulting from deforestation are the followings:

(i) *Alfa grass (*Stipa tenacissima*) steppe* is usually located on shallow soils (hills and pediments with a superficial thick calcareous crust) and results from deforestation by woodcutting and burning, without cultivation. *Stipa tenacissima* is a forest species which can only regenerate in the shade (Le Houerou, 1969a). In the steppe it survives only by rhizomatous expansion of its tussocks. The species cannot withstand cultivation for more than 2–3 years. Alfa grass steppes occupied some 12 million ha at the end of the nineteenth century, but has now reduced by about 50% owing to over-exploitation (alfa grass is exported for high-quality paper), repeated burning for grazing, and clearing for cultivation. In Tunisia, for example, the area of alfa grass steppe has reduced by an average of 10,000 ha/year since the beginning of this century. The mechanism of this degradation of vegetation and soil was described in detail by Le Houerou (1969a, b) as part of the general process of steppization. The main companion species of alfa grass in the upper arid bioclimatic zone are: *Artemisia campestris (*sandy soils*), A. herba alba (*silty soils*), Astragalus incanus* subsp. *nummularoides, Atractylis humilis subp. caespitosa, A. serratuloides (*shallow soils*), Avena bromoides, Centaurea tenuifolia, Dactylis hispanica, Fumaria ericoides, F. thymifolia, Helianthemum cinereum* subsp. *rubellum, H hirtum, H virgatum* subsp. *ciliatum, Hippocrepis scabra, Phagnalon rupestre, Pituranthos scoparius, Stipa parviflora,* and *Thymus hirtus*.

   Degradation of the alfa grass steppe in turn leads to various chamaephytic steppes. The most widespread of these is the type dominated by *Artemisia herba-alba*.

(ii) Artemisia herba-alba *steppes* cover huge areas of silty soils. and are among the best Arid Zone pastures in the region. This type of steppe is found from Spain throughout the whole of North Africa to the Near and Middle East (Syria, Israel–Palestine, Jordan, Iraq, Iran. Afghanistan). Soils are always silty, with often a hard pan of calcium carbonate between the surface and a depth of 100 cm. The usual companion species are: *Ajuga pseudoiva, Anabasis oropediorum, Asteriscus pygmaeus, Astragalus armatus (*gypseous soils*), Atractylis humilis* (highly calcareous soils), *A. serratuloides* (shallow soils), *Carrichtera annua, Eruca vesicaria, Eryngium dichotomum, E. ilicifolium, Lygeum spartum* (gypseous soils)*, Noaea mucronata* (shallow soils)*, Matthiola fruticosa, Plantago ovata, Poa bulbosa, Salsola vermiculata* var. *villosa* (gypseous and slightly saline soils)*, Stipa capensis, and S. parviflora* (shallow soils).

   When the *Artemisia herba-alba* steppe is cleared for cultivation and then abandoned it gives way to a post-cultivation steppe characterized by the dominance of *Artemisia campestris* and *Cynodon dactylon* which, after a period of 5–10 years, returns to the *Artemisia herba-alba* steppe.

(iii) *Sandy Steppes* of *Artemisia capestris* are derived also from evergreen bushland or shrubland, but on sandy soils. The co-dominant species are: *Alkanna tinctoria, Anacyclus cyrtolepidioides, Argyrolobium uniflourm, Astragalus caprinus, Carduncellus pinnatus, Centaurea dimporpha, Cutandia divericata, Dianthus crinitus, Echinops spinosus, Echiochilon fruticosum, Eragrostis papposa, Eragrostis trichophora, Hedysarum spinosissimum, Helianthemum sessiliflorum, Launaea resedifolia, Lotus creticus, L. pusillus, Minuartia geniculata, Plantago albicans, P. colorata,* and *Stipa lagascae.*

**b. *Middle arid subzone***

The primaeval vegetation is still an evergreen open forest or bushland of *Juniperus phoenicea* with some of the dominant species of the Upper Arid Subzone, such as *Cistus libanotis, Olea europaea, Rhamnus lycioides*, and *Rosmarinus officinalis*. However, the floristic composition is different, with an increased number of Arid Zone species such as: *Coridothymus capitatus, Celsia ballii, Centaurea tenuifolia, Fagonia cretica, Genista microcephala, Launaea acanthoclada, L. arborescens, Pituranthos scoparius, Reseda duriaeana,* and *Scrophularia arguta.*

This evergreen shrubland still occupies over 1.5 million ha on the hills of the Middle Arid Subzone, but has been eliminated over most of the zone and replaced by steppe communities physiognomically similar to those of Upper Arid Subzone owing to the dominance of the same species: *Artemisia campestris A. herba-alba, Lygeum spartum,* and *Stipa tenacissima*. However, the floristic composition is again different with the addition of pre-Saharan species such as: *Acacia tortilis* subsp. *raddiana, A. gummifera (Morocco), Cleome arabica, Gymnocarpos decander, Hammada schmittiana, H. scoparia, Lygos raetam, Rhanterium suaveolens, Stipagrostis obtusa, S. plumosa*, and *Zygophyllum album* any of which can be co-dominant.

**c. *Lower arid subzone***

Some remnants of the evergreen shrublands are still found in remote places, including: *Argania sideroxylon, Cistus libanotis, Juniperus phoenicea, Pistacia atlantica, Rhus tripartitum*, and *Rosmarinus officinalis*. This tends to show that the Lower Arid Subzone may well have been wooded land several centuries ago. Some of those remnants (*Juniperus phoenicea*) have actually been found below the 100-mm isohyet in particular niches (cliffs), especially in southern Jordan and in the Sinai (Long, 1957; Danin, 1972).

The Lower Arid Subzone is almost entirely covered with a chamaephytic steppe vegetation where the main species are: *Acacia gummifera, A. raddiana, Anabasis aphylla, A. articulata, Anarrhinum brevifolium, Astragalus armatus* subsp. *tragacanthoides, Atractylis serratuloides, Calycotome intermedia, Chenolea arabica, Cleome arabica, Diplotaxis harra, Echiochilon fruticosum, Erodium arborescens, Fagonia glutinosa, F. kahirina, Farsetia aegyptiaca, Forsskaolea tenacissima, Fredolia aretioides, Gymnocarpos decander, Hammada scoparia, H. schmittiana, Helianthemum ellipticum, Herniaria fontanestii, Linaria aegyptiaca, Lycium arabicum, Lygos raetam, Marrubium deserti, Nitraria retusa, Noaea mucronanta, Ononis natrix* subsp. *falcata, Pergularia tomentosa, Reaumuria vermiculata, Rhanterium suaveolens, Salsola sieberi, S. tetrandra, S. vermiculata* var, *brevifolia, Suaeda mollis, Thymelaea hirsuta, T. microphylla, Traganum nudatum,* and *Zygophyllum album.*

One important ecological and vegetational fact is that this bioclimatic zone contains the southernmost extensions of the alfa steppe.

**B. The desert zone**

The Desert Zone corresponds geographically to the northern Sahara between the isohyets of 25–100 mm.

**a.** ***Upper desert subzone (50–100 mm isohyte)***

In contrast to the Arid Zone, here, vegetation on shallow soils is disposed in a contracted pattern along the drainage system with large barren interfluves on the pediments, whereas on deep soils and especially on sandy material the vegetation pattern remains diffuse and of the steppe type.

The dominant perennial species are chamaephytes (small shrubs) except on sand dunes where perennial grasses play an important role.

The main chamaephytic species are: *Agathopora alopecuroides, Anabasis articulata, A. setifera, Anthyllis henoniana, Anvillae radiata, Astragalus pseudotrigonus, Atriplex mollis, Bubonium graveolens, Calligonum comosum, Cornulaca monacantha, Euphorbia guyoniana, Fagonia microphylla, Fredolia aretioides, Globularia arabica, Gymnocarpos decander, Hammada schmittiana, H. scoparia, Helianthemum confertum, H. kahiricum, Limoniastrum feei, L. guyonianum, Lygos raetam, Moltkia ciliata, Pituranthos battandieri, Pulicaria crispa, Salsola tetragona, S. vermiculata* var. *microphylla, S. cyclophylla, Tamarix boveana, T. pauciovulata, Traganum nudatum, Warionia saharae, Withania adpressa,* and *Zilla spinosa.*

On sand dunes characteristic and dominant species are: *Calligonum comosum, Cyperus conglomeratus, Ephedra alata subsp. alenda, Euphorbia guyoniana, Genista saharae, Panicum turgidum, Pennisetum dichotomum, P. elatum, Stipagrostis acutiflora, S. plumosa, S. pungens,* and *S. scoparia.*

**b.** ***The middle desert subzone***

The middle desert subzone vegetation is distributed in a contracted pattern – that is along runnels and wadis and in depressions. There is, however, diffuse vegetation in the sand dunes and on the cliffs. The regs and hammadas are barren, without perennial vegetation. A small number of tropical species (10%) mix here with the dominant Mediterranean flora. Characteristic species are: *Acacia raddiana, Achillea fragrantissima, Antirrhinum ramosissimum, Artemisia judaica, Asteriscus graveolens. Atriplex leucoclada, Balanites aegyptiaca, Caylusea hexagyna, Fagonia arabica, F. bruguieri, Hyoscyamus muticus, Iphiona mucronata, Lasiurus hirsutus, Launaea spinosa, Leptadenia pyratechnica, Maerua crassifolia, Matthiola livida, Morettia canescens, Moringa peregrina, Ochradenus baccatus, Pituranthos chloranthus subsp. intermdius. Pulicaria crispa, Randonia africana, Reaumuria hirtella, Reseda villosa, Trichodesma africanum, Zilla spinosa. Ziziphus saharae,* and *Zygophyllum coccineum.*

**c. *The lower desert subzone***

The lower desert subzone vegetation is restricted to main wadis and depressions benefitting from run-off or having shallow water tables. Vegetation is a mixture of Mediterranean and tropical species in approximately equal proportions.

Among the main Mediterranean species are: *Alhagi graecorum, Anabasis articulata, Astragalus vogelii, Brocchia cinerea, Caylusea hexagyna, Cornulaca monacantha, Foleyola billottii, Hammada schmittiana, Lavandula coronopifolia, L. stricta, Morettia canescens, Moricandia foleyi, Nucularia perrini, Plantago ciliata, Polycarpaea fragilis, Randonia africana, Reseda kahirina, R. villosa, Salsola haryosma, S. vermiculata, Salvia pseudojaminiana, Shouwia thebaica, Tamarix aphylla, T. brachystylis, T. leucocaris, T. nilotica,* and *Trigonella anguina.*

The main tropical species are: *Acacia asak, A. ehrenbergiana, A. raddiana. A. seyal, Aerva persica, Aristida mutabilis, Aristida sahelica, Balanites aegptiaca. Calotropis procera. Capparis decidua, Cassia aschrek. C. italica, Cenchrus biflorus, Chrozophora bracchiana, Cleome droserifolia, Crotalaria aegjptiaca, C. saharae, Desmostachya bipinnata. Hyphaene thebaica, Indigofera argentea, Lasiurus hirsutus, Leptadenia pyrotechnica, Maerua crassifolia, Panicum turgidum, Pergularia tomentosa, Psoralea plicata, Salvadora persica, Solenostemma arghel, Tephrosia leptoscachya,* and *Ziziphus mauritiana.*

**C. The true desert (less than 10 mm isohyte)**

The true desert is virtually rainless. Perennial vegetation is linked to water tables often due to fossil aquifers (as in the Libyan-Egyptian desert), or to exogenous water-courses.

Occasional showers on limited areas may produce an ephemeral burst of short-lived annuals such as: *Brocchia cinerea, Monsonia nivea, Plantago ciliata, Stipagrostis plumosa*, and *Trigonella anguina.*

Perennial vegetation on water tables is dominated by: *Acacia albida, A. nilotica, Alhagi graecorum, Cornulaca monacantha, Desmostachya bipinnata, Hyphaene thebaica, Imperata cylindrica, Phragmites australis, Phoenix dactylifera, Saccharum ravennae, Salvadora persica, Traganum nudatum* and *Zygophyllum album.*

**D. The Saharan mountains**

The higher elevations of the Saharan mountains above 1,800–2,000 m have a vegetation similar to the Middle Arid Subzone, with strong Mediterranean affinities. Characteristic species of this arid montane vegetation are: *Agrostis tibestica, Argyrolobium uniflorum, Artemisia campestris, A. herba-alba, A. tilhoana, Asplenium quezeli, Atriplex halmius, Avena tibestica, Ballota hirsuta, Campanula bordesiana, Crambe kralickii, Cupressus dureziana, Dichrocephala tibestica, Ephedra tilhoana, Erigeron trilobus, Festuca tibestica, Ficus gnaphalocarpa, F. ingens, F. salicifolia, Globularia alypum* var, *vescetirensis, Helianthemum lippii, Helichrysum monodianum, Hyparrhenia hirta, Lavandula antinea, Luzula tibestica, Marrubium deserti, Micromeria biflora, Myrtus nivellei, Olea laperrini,*

*Oryzopsis coerulescens, Pentzia monodiana, Periploca laevigata, Phagnalon tibesticum, Pistacia atlantica, Pituranthos scoparius* var, *fallax, Rhus tripartitum, Salvia chudaei, Senecio hoggariensis, Silene mirei, Stipa parviflora, Teucrium polium,* and *Varthemia sericea.*

**IV. Main vegetation types**

The arid zone habitats of the North African Mediterranean region are distinguished by 7 vegetation types (Le Houerou, 1992), namely: Forest, Matorral (garrigue), erme, steppe, pseudo-steppes, desert savanna and sand-dune open woodland.

**a. *Forest***

Even within the Arid Zone, forests (defined as vegetation having trees over 5 m high and over 100 single stems per hectare) occur. They are, however, very rare and located in the upper bioclimatic subzone (300–400 mm of average annual rainfall). They are almost always open forest of *Pinus halepensis* (Aleppo pine), *Tetraclinis articulata, Argania sideroxylon* (in southern Morocco only) or *Juniperus phoenicea.* There is always an understorey of shrubs, except in certain types of Argania woodland or parkland. Such forests probably occupy not more than 100,000 ha in the Arid Zone in Morocco, Algeria, Tunisia and Libya (Cyrenaica).

**b. *Matorral or Garrigue***

The matorral[3] (American equivalent: chaparral) consists of evergreen tall shrubs or small trees is found on the hills between isohyets of 200 and 400 mm, and covers some 30,000 km$^2$ in the North African Arid Zone. Most of the plant species of the garigue (garrigue = matorral) are spiny cushion shrubs, e.g. *Euphorbia acanthothamnos* and *Sacropoterium spinosum,* drought-deciduous and dimorphic plants, e.g. *Euphorbia dendriodes, Calycotoma villosa,* aromatic plants of the mint family, e.g. *Rosmarinus officinalis* and *Lavandula stoechas* and bulbs, e.g. *Narcissus tazetta, N. bulbocodium, Pancratium arabicum, P. maritimum, Crocus biflorus* etc. (Dallman, 1998).

**c. *Erme***

Erme is a low-growing community of forbs and grasses consisting of unpalatable plants, including Liliaceae (*Asphodelus, Urginea*), thistles (*Carduncellus, Carthamus, Carlina, Centaurea, Onopordon* etc.), and other unpalatable species such as *Cleome arabica, Euphorbia* spp., *Ferula* spp., *Peganum harmala, Thymelaea hirsuta, Thapsia garganica* etc. ... and an annual grass. *Stipa capensis*, is often dominant. Erme is a post-pastoral vegetation type resulting from overgrazing around villages, wells ... etc. It is estimated to cover some 20,000 km$^2$ in the

---

[3]The spelling of matorral and garrigue could be also mattoral and garigue, respectively (Branigan and Jarrett, 1975; Dallman, 1998; Blondel and Aronson, 1999).

region, mostly in the Arid Zone. When similar heavy grazing pressure is exerted in desert areas, the ground usually remains barren.

**d.** ***Steppe***

Steppe is the typical vegetation type of the Arid Zone of North Africa, and extends into the Desert Zone; it covers some 400,000 km$^2$ between the isohyets of 100 and 400 mm. It also occurs between the isohyets of 50 and 100 mm in depressions with permeable soils which receive run-on water. Steppe is a short, open treeless vegetation, with various proportions of bare ground, physiognomically dominated by perennial species. Plant cover may be anywhere between 5 and 80%. One generally differentiates three main types of steppes:

**(i)** *Gramineous steppe* is dominated physiognomically by perennial grasses such as *Stipa tenacissima* (alfa grass), *Lygeum spartum* (esparto grass). *Stipa* spp., *Panicum turgidum, Stipagrostis pungens, Stipagrostis* spp. Alfa grass steppe covers some 60,000 km$^2$ in Morocco, Algeria, Tunisia and Libya (Tripolitania only). Esparto grass steppes cover about 30,000 km$^2$ from Morocco to Egypt.

**(ii)** *Chamaephytic steppe* is characterized by the dominance of low shrubs or half-shrubs 0.2–0.5 m high. This type of steppe covers 250,000–300,000 km$^2$ in our area.

The main dominant species are: *Anthyllis henoniana* on shallow soils; *Artemisia campestris* and *A. monoica* on sandy soils, *Artemisia herba-alba* on silty or shallow soils, sealed on the surface; *Gymnocarpos decander* on shallow soils; *Hammada scoparia* on silty soils, *Noaea mucronata* on shallow soils; *Rhanterium suaveolens* on loose sandy soils; and *Thymelaea hirsuta* on a variety of soils.

**(iii)** *Crassulescent steppe* is dominated by fleshy halophilous shrubs mostly belonging to the Chenopodiaceae. It occurs over some 30,000 km$^2$ in the Arid and Desert Zones.

The main dominant species in approximate order of decreasing halophily are: *Halocnemum strobilaceum, Halopeplis amplexicaulis, Arthrocnemum glaucum (A. macrostachyum), Salicornia fruticosa, Salsola baryosma, S. sieberi, S. tetragona, S. tetrandra, S. vermiculata, Suaeda fruticosa, S. monoica, Atriplex halimus, A. glauca. Traganum nudatum, Zygophyllum album* and *Z. coccineum.*

**(iv)** *Non-halophilous succulent steppe* is only found in the western part of the region, along he Atlantic Ocean in southern Morocco. Dominant and characteristic species are non-halophilous succulents such as: *Aizoon theurkauffii, Euphorbia balsamifera, E. echinus, E. resinifera, Kalanchoe faustii, Senecio anteuphorbium* and *Zygophyllum waterlottii.*

This type of steppe shows a strong tropical affinity in spite of the fact that the region receives winter rains; winter temperatures are mild (never dropping below 10°C) and atmospheric humidity is high. The flora itself is peculiar to this part of the region and is usually referred to as “Canarian” or “Macaronesian”.

**e. *Pseudo-steppe***

This has taller-growing types of vegetation than steppes (0.5–3 m high); and could also be called open bushland or open shrubland. Dominant shrubs cover 5–20% of the ground with interspersed smaller shrubs and annuals. The main tall shrubs are: *Acacia gummifera, A. raddiana, Argania sidoroxylon. Balanites aegyptiaca, Calligonum comosum, Capparis decidua. Ephedra alata, Limoniastrum guyonianum, Lygos raetam, Maerua crassifolia, Nitaria retusa, Periploca laevigata, Rhus tripartitum, Tamarix* spp and *Ziziphus lotus*.

This type of vegetation occurs in the best-watered locations: wadi terraces, areas subject to floods, various depressions, sand dunes and cliffs. It is totally different from matorral in botanical composition, structure physiognomy and habitat.

**f. *Desert savanna***

This is found in the Sahara along the main wadis, in wadi beds. terraces, spreading areas and in the karstic (limestone) depressions. It consists of a layer of scattered trees interspersed with tall perennial grasses. The most common trees are: *Acacia raddiana, Balanites aegyptiaca, Maerua crassifolia, Pistacia atlantica,* and *Ziziphus spina-christi*. The main grasses are *Cymbopogon schoenanthus, Lasirus hirsutus, Panicum turgidum,* and *Pennisetum dichotomum*. Small shrubs occur such as *Aerva persica, Solenostemma arghel*, and *Zilla spinosa*.

**g. *Sand-dune open woodlands***

These are extremely rare and composed of small trees and shrubs 5–8 m high such as *Calligonum azel, Ephedra alata* subsp. *alenda, Genista saharae, Leptadenia pyrotechnica and Lygos raetam*. These are found in remote places in the Grand Erg Oriental and in the Idehan Ubari (Fezzan).

#### 1.8.1.6 Representative Sectors

The preceding pages have presented a general account of the major vegetation forms growing naturally under the prevailing climate of the North African coastal lands. Details on the environment and vegetation types of these Mediterranean coastal areas will be described from three representative sectors: the Jefara and Al-Gabal AlAkhdar sectors in Libya and the Mariut sector of Egypt.

**I. Libyan coastal sectors**

The Mediterranean coastal land of Libya extends for about 2,400 km from the Tunisian – Libyan border in the west to the Libyan – Egyptian border in the east. It is a remarkably uniform coastal land along which there are natural harbours, some very rocky and steep, others are low and swampy. Coastal sand dunes and salt depressions represent a characteristic features of the Libyian Mediterranean coastal land (Branigan and Jarrett, 1975).

Ecologically, the Libyan Mediterranean coastal belt is categorized into two natural regions: the north-western and the north-eastern. The northwestern region lies in the NW part of Tripolitania and on the whole it is the best favoured and the most prosperous part of the whole country. The core of this region is the Jefara sandy plain bounded on the north by the low marshy coast and to the south by the purple cliffs of Gebel Nefusa. Cultivation is possible along this coastal strip wherever the rainfall is comparatively high. The remainder of Jefara is semi-desert, with Esparto-grass (*Stipa tenacissima*) maintained by semi-nomadic grazing. Cultivation depends partly upon winter rain and partly upon water drawn from wells especially in summer. Olives, citrus fruits, almond, vegetables and barley are produced. The chief crop, however, is the date. In some other areas nearby parts of the Jefara coast where irrigation is possible, wheat and tobacco area also grown.

The north eastern region lies within the north protecting bulge of Cyrenaica, around Al-Gabal Al-Akhdar (the green mountain). On this mountain, forests of cypress (*Cupressus* spp.), ilex (*Ilex* spp.) and Juniper (*Juniperus* spp.) predominate.

Further south, these forests give way to another vegetation type called garigue (widespread in the Mediterranean basin) and then to a wide treeless grassland steppe.

Though the natural flora of the coastal and inland deserts of Libya has been studied by several authors, e.g. Maire (1952–1977), Keith (1965), Boulos (1971, 1972, 1977, 1979a, b), Greuter et al. (1984–1986, 1989), Ali and Jafri (1976), Jafri and El-Gadi (1978), Pratov and El-Gadi (1980), Qiser and El-Gadi (1984), and El-Gadi (1989) etc.; yet few published reports have dealt with its vegetation ecology: Edrawi and El-Naggar (1995), Ebrahim (1999), El-Kady (2000), Al-Sodany et al. (2003), and El-Morsy (2008).

### A. Jefara coastal sector

#### a. *Climate*

The Jefara coastal belt extends for about 400 km along the north western Mediterranean borders of five Libyan coastal governorates, namely Zuwarah (in the west) followed eastwared with Subratha, Azzawarah, Tripoli and Al-Khoms. The average area of the Jefara coastal plain is about 4,032 km$^2$ (Hassan, 1975). It has a typical Mediterranean semi-arid climate (El-Morsy, 2008). The mean annual rainfall ranges between 238 mm (Zuawarah) to 296 mm (Al-Khoms). The main bulk of the precipitation (> 97%) usually occurs in the September–April period. The mean annual temperature ranges between 19.8 and 19.1°C with monthly means up to 27.6°C (August) and down to (12.1°C) (January). Mean annual relative humidity ranges between 68 and 72.6% with monthly means ranging from 76% (June–August) to 65.2% (December–January).

#### b. *Habitats and flora*

The Jefara coastal plain is organized, ecologically; into five habitats: sand dunes, rocky ridges, salt marshes, wastelands and field crops. Each of these habitats has its own vegetation type with their characteristic and associate species. The flora comprises about 293 species: 62 (21.1%) monocots and 231

(78.84%) dicots belonging to 154 genera and 46 families (El-Morsy, 2008). The Compositae has the, relatively, highest number of species (56, 19.1%) followed by Gramineae (38, 12.47%), Leguminosae (22, 7.51%), Chenopodiaceae (18, 6.48%), Cruciferae (14, 4.78%), Liliaceae and Caryophyllaceae (13, 4.44% each), Umbelliferae and Plantaginaceae (8, 2.73% each), Cyperaceae, Labiatae and Solanaceae (7, 2.39% each), Boraginaceae (6, 2.05%), Polygonaceae, Papaveraceae, Geraniaceae, Zygophyllaceae and Euphorbiaceae (5, 1.71% each). Each of the remaining 28 families is represented by 1–4 species.

Most taxa (155 species) are annuals; there are 133 perrnnials and only 5 biennials. The life-form spectrum of these floristic elements shows that: 53.6% are therophytes, 18.8% are chamaephytes, 13.3% are cryptophytes, 12.6% are hemicryptophytes and 1.7% are nanophanerophytes. More than 72% (42 species) of these elements belonging to the Mediterranean region, the remaining species (81, 28%) are distributed as follow: 25 species (8.5%) are cosmopolitan, 13 (4.44%) are Saharo-Sindian, 11 (3.74%) are palaeotropical, 8 (2.73%) are pantropical, 6 (2%) are naturalized and cultivated and 5 species (1.7%) are neotropical.

#### c. *Vegetation types*

The vegetation types of the five habitats are formed of 42 communities: 9 in the sand dunes, 5 in the rocky ridges, 15 in the salt marshes, 5 in the wastelands and 8 in the field crops (El-Morsy, 2008).

The following is a short account of these communities with respect to their characteristic and associate species.

#### 1. *Sand Dune habitat*

Chains of coastal sand dunes extend along the Jefara coastal belt forming more or less a continuous bars. This habitat is characterized by the storing rain of water forming a freshwater layer above one of seawater. Psammophytes (sand-loving plants) constitute the main bulk of the sand dune vegetation. The dominant species, are all perennials: *Anabasis articulata, Artemisia lampestris, Asparagus stipularis, Cyperus capitatus, Euphorbia paralias, Lotus cytisoides, Lygos raetam, Silene succulenta* and *Stipagrostis ciliata*. There are 26 associate species: 15 perennials and 11 annuals: *Aeluropus lagopoides, A. littoralis, Alhagi graecorum, Allium roseum* var. *tournexii, Calycotome villosa* var. *tournexii, Ecballium elaterium, Elytrigia juncea, Iris sisyrinchium, Muscari racemosum, Pancratium maritimum, Polygonum equisetiform, Salsola longifolia, Sporobolus spicatus, Stipagrostis lanata* and *Urginea maritima* (perennials), and *Apium garveolens, Cakile maritima, Cutandia dichotoma, Daucus guttatus, Lagurus ovatus, Parapholis incurva, Plantago ovata, Pseuderucaria teretifolia, Reseda pruinosa, Rumex crispus, Schoenefeldia gracilis* (annuals).

#### 2. *Rocky Ridge habitat*

Six perennial species predominate the rockey ridge habitat: *Deverra tortuosa Globularia alypum, Lycium europaeum, Lygeum spartum, Medicago marina* and

*Thymus capitatus*. The associate species are 58: 47 perennials, 3 biennials and 8 annuals. The common perennial associates include: *Asparagus aphyllus, Cornulaca monacantha, Cynara caringera, Echinops hussonii, E. spinosissimus, Echiochilon fruticosum, Echium sericeum, Glaucium flavum, Helianthemum stipulatum, H. virgatum, Herniaria hemistemon, Kickxia aegyptiaca, Limonium axillare, Ononis natrix, Pancium repens, Phagnalon nitridum, Reaumuria hirtella, Salsola longifolia, Sarcocornia fruticosa, Scilla peruviana, Scorzonera undulate and Ziziphus lotus* etc. the biennial associates are: *Centaurea calcitrapa, Onopordum ambiguum, and O. alexandrinum.* The associate annuals are: *Beta vulgaris, Schenopheldia gracilis, Astragalus hamosus, Centaurea glomerata, Lagurus ovatus, Limonium amplexicaulis, Neurada procumbens, and Scabiosa arenaria.*

### 3. *Salt Marsh habitat*

The salt marsh habitat occupies the shoreline as well as depressed salt-affected areas within the sand dunes and rocky ridges. Its halophytic vegetation is formed of 15 communities dominanted by: *Arthrocnemum macrostachyum, Atriplex glauca, A. halimus, A. portulacoides, Cressa cretica, Cyperus laevigatus, Halocnemum strobilaceum, Inula crithmoides, Juncus acutus, J. rigidus, Limoniastrum monopetalum, Sporobolus spicatus, Suaeda pruinosa, Tamarix nilotica* and *Zygophyllum album.* The few associate species are mainly halophytes including: *Aeluropus lagopoides, A. littoralis, Imperata cylindrica, Limonium narbonense, Reaumuria hirtella, Salsola longifolia, Silene succulenta,* and *Spergularia marina* (perennials) and *Mesembryanthemum crystallinum,* and *M. nodiflorum* (annuals).

### 4. *Wastelands*

The wastelands are actually the non-cultivated areas occupying the zone landward to the salt marsh habitat. The number of the floristic elements here is, the highest (> 150 species) among the other habitats of the Jefara coastal area: about 70 species are perennials, 75 annuals, and a few biennials (5 species).

Five perennial xerophytes (drought tolerant and drought resistant species) predominate: *Artemisia monosperma, Cornulaca monacantha, Nicotiana glauca, Peganum harmala and Thymelaea hirsuta.* The perennial associates are mostly xerophytes with few halophytes. The xerophytes include: *Alkanna lehmani, Anabasis articulata, A. setifera, Androcymbium gramineum, Artemisia compestris, A. monosperma, Asparagus aphyllus, A. stipularis, Asphodelus ramosus, Aster squamatus, Bellevalia sessiliflora, Calycotome villosa, Calotropis procera, Centaurea dimorpha, Citrullus colocynthis, Convolvulus prostratus, Conyza stricta, Cynodon dactylon, Cyperus rotundus, Deverra tortusa, Echinops hussonii, E. spinosus, Echiochilon fruticosum, Echium sericeum, Erodium bryoniaefolium, E. glaucophyllum, Glaucium flavum, Haplophyllum tuberculatum, Helichrysum lacteum, Launaea angustifolia, L. nudicaulis, Lycium europaeum, Lygeum spartum. Marrubium vulgare, Matthiola livida, Medicago sativa, Muscari comosum, Paronychia argentea, Phagnalon nitidum, Plantago albicans, Polygonum equisetiforme, Salvia lanigera, Solanum incanum, Sonchus maritimus, Thymus capitatus, Verbascum letourneuxii, Verbena tenuisecta* etc. The perennial associate halophytes are: *Atriplex glauca, A. portulacoides, Cressa cretica Imperata cylindrical, Reaumuria hirtella, Salsola longifolia, Suaeda pruinosa Tamarix nilotica and Zygophyllum album.* Five associated biennials have been recorded in this habitat: *Carthamus lanatus, Centaurea calcitrapa, Echium plantagineum, Onopordon ambiguum* and *O. alexandrinum.* The annual associates include: *Amarauthus lividus, A. tricolor, Asphodelus tenuifolius, Bassia indica, Beta vulgaris, Chenopodium ficifolium, C. murale, Conyza aegyptiaca, C. bonariensis, Crepis libyca, Diplotaxis acris, D. harra, Echium sabulicola, Euphorbia peplis, Filago desertorum, Geranium rotundifolium, Hyoscyamus desertorum, Lactuca serriola, Malva parviflora, Matthiola longipetala, Papaver rhoeas, Phalaris minor, Reichardia tingitana, Setaria verticillata, Silene villosa, S. viviani, Sisymbrium irio, Solunum nigrum, Sonchus oleraeceus, Spergularia diandra, Tribulus terrestris, Urospermum picroides, Urtica urens, Vaccaria hispanica, Volutaria lippii, Xanthium spinosum* etc.

### 5. *Field crops*

The cultivation of fruit trees (olive, citrus, grapes, date palm etc.), and vegetables is a minor component of the inland zone of the Jefara coastal plain. The natural vegetation growing in this habitat has about 150 species, mostly annuals (106 species) with a few (4) biennials: *Carthamus lanatus, Echium plantagineum, Onopordum ambiguum and O. alexandrinum* and 40 perennials. Among the perennials, 8 species are widespread and abundant, five of these are halophytes growing in the wetter areas: *Arundo donax, Cyperus conglomeratus, C. longus, C. rotundus* and *Phragmites australis*. Other abundant perennials are xerophytes: *Convolvulus arvensis, Echiochilon fruticosum* and *Thymelaea hirsuta*. Except *Cressa Cretica* (halophyte), all of the associated perennials are xerophytes, e.g.: *Alkanna lehmanii, Allium roseum* var. *tourneuxii, Androcymbium gramineum, Asphodelus ramosus, Aster squamatus, Astragalus kahiricus, Bellevalia sessiliflora, Centaurea dimorpha, Conyza stricta, Cynodon dactylon, Echinops hussonii, E. spinosissimus, Erodium bryonifolium, E. crassifolium, Euphorbia retusa, Gundelia tournefortii, Launaea fragilis, L. nudicaulis, Lotus glaber, Ononis natrix, Oxalis corniculatus, O. pes-caprae, Paronychea argentea, Plantago albicans, Polygonum equisetiforme, Salvia deserti, S. lanigera, S. officinalis, Spergularia marina, Verbascum letourneuxii, Ziziphus spina-christi* etc. The asscociated annuals classified under 3 groups.

(a) Annuals active all year (11 species): *Amaranthus lividus, Bassia indica, Chenopodium galucum, C. murale, Lactuca serriola, Malva parviflora, Pseuderucaria teretifolia, Reichardia tingitana, Setaria verticillata, Solanum nigrum and Sonchus aleraceus,*
(b) Summer active annuals (15 species): *Amaranthus graecizans, A. hybridus, A. tricolor, Ammi majus, A. visnaga, Conium maculatum, Dactyloctenium aegyptium, Datura innoxia, Echiochilon colona, Eleusine indica, Emex spinosa, Polypogon monospelienis, Portulaca oleracea, Salsola kali* and *Sesbania sericea* and,

(c) Winter active annuals (about 80 species): e.g. *Adonis dentata, Aegilops kostchyi, Anagallis arvensis Anthemis melampodina, Anthemis retusa, Asphodelus tenuifolius, A. viscidulus, Avena fatua, A. sativa, Brassica tournefortti, Bromus catharticus, B. diandrus, Calendula arvensis, Capsella bursa-pastoris, Carduus argentatus, C. getulus, Centaurium pulchellum, Coronopus didymus, Cotula anthemoides, C. cinerea, Crepis aculeata, Cutandia dichotoma, Eruca longirostris, E. sativa, Euphorbia helioscopia, E. peplis, E. prostrata, Filago contracta, F. desertorum, Fumaria densiflorae, Geranium rotundifolium, Hordeum marianum, Ifloga spicata, Kickxia spura, Koelpinia linearis, Lamium amplexicaulis, Lepidium sativum, Linaria tenuis, Lobularia libyca, Lolium perenne, Matthiola longipetala, Medicago intertexta, M. polymorpha, Melilotus indicus, Ononis serrata, Papaver dubium, P. hybridum, P. rhoeas, Phalaris canariensis, P. minor, Picris altissima, Plantago lagopus, P. ovata, Poa annua, Polycarpon tetraphyllum, Reseda pruinosa Rumex crispus, R. dentatus, R. vesicarius, Scabiosa arenaria, Sencio glaucus subsp. coronopifolius, Setaria virdus, Silene villosa, S. viviani, Stellaria pollida, Trifolium rupsinatum, Vicia articulata, V. monantha, V. sativa, V. villosa* etc.

### B. Al-Gabal Al-Akhdar coastal sector

#### a. *Geomorphology and climate*

Al-Gabal Al-Akhdar (the green mountain) upland is a plateau formed as a result of tectonic elevation of a primary plain of marine accumulation with height up to 878 m. The area of Al-Gabal Al-Akhdar has been studied ecologically by Al-Sodany et al. (2003) in a 30-km N–S transect between El-Hamamah; near sea level (33° 53′N, 21° 39′E) to El-Bydda (600 m 33° 17′N, 21° 40′E). This coastal area is characterized by four geomorphological units: coastal plain, coastal hills, inland plateau and wadis.

1. The coastal plain extends for about one km landwards from the sea. It is composed of marine accumulation and three main habitats: saline sand flats, sand dunes (up to 50 m high) and sand flats.
2. The coastal hills (up to 100 m high) generally confined to marginal parts of the coastal plain.
3. The inland plateau which appears in the form of three terraces: the first up to 400 m high at Ras El-Hilal, the second up to 600 m high at El-Bydda and the third up to 880 m high at Slentah.
4. Wadis, dissecting the plateau, running in N-S and covered with a shallow sandy-loam soil.

   The Mediterrannean climate of Al-Gabal Al-Akhdar area is classified under 3 bioclimatic zones according to the elevation of land above sea level. In El-Hamamah just at sea level, the aridity index ($Q$) is 1.3 i.e. arid climate prevail. This changes to a semi-arid ($Q = 2.0$) climate in El-Wesaetah at 300 m and then to subhumid ($Q = 2.8$) in El-Bydda at 600 latitude.

The annual means of temperature, rainfall, relative humidity and wind velocity in these three bioclimatic zones are: 19.9, 17.4 and 16.4°C, 323.6, 417.5 and 567.1 per year, 73.0, 65 and 68% and 42.7, 65 and 68 km/h (Al-Sodany et al., 2003).

**b. *Plant life***

The flora of Al-Gabal Al-Akhdar coastal area is formed of about 119 species: 43 annuals (36%) and 76 perennials (63.9%), belonging to 105 genera and 44 families. The compositae is represented by the, highest number of species (15) followed by Gramineae (13), Leguminosae (12), Liliaceae and Labiatae (7 each), Euphorbiaceae, and Solanaceae (4 each), Cruciferae, Cistaceae, Chenopodicceae, and Iridaceae (3 each), Cyperaceae, Ranunculacea, Cupressaceae and Anacardiaceae (2 each). The remaining families represented by a single species. Regarding the life-form spectera, therophytes have the highest contribution (36.1%) followed by cryptophytes (19.3%), phanerophytes (17.6%), chamaephytes (13.5%) and hemi-cryptophytes (13.5%). Most of these taxa (105) belong, floristically to the Mediterranean region, and include nine endemic species: *Arbutus pavarii* (Ericaceae), *Cupressus sempervirens* var. *horizontalis* (Cupressaceae), *Arum cyrenaicum* (Araceae), *Cyclamen rholfsianum* (Primulaceae), *Romulea cyrenaica* (Iridaceae) and *Bellis sylvestris* var. *cyrenaica, Crepis libyca, Onopordum cyrenaicum* and *Cynara cyrenaica* (Compositae). The monoregional Mediterranean taxa are 55, the biregionals are 19 whereas the pluriregional are 31 species. The monoregional species belonging to other floristic regions: 4 belong to the Saharo-Arabian region (*Brassica deserti, Conyza aegyptiaca, Solanum nigrum and Trisetaria acrochaeta*), and one is a tropical plant (*Oxalis pes-carpae*). There are seven cosmopolitan species: *Conyza bonariensis, Convolvulus arvensis, Cynodon dactylon, Scripus maritimus, Chenopodium ambrosoides, Polypogon mospeliensis* and *Sonchus oleraceus*. The remaining species belong to the Sudanian, Euro-Sibarian and/or Irano-Turanian floras either biregaonally or pluri-regionally (Al-Sodany et al., 2003).

The four geomorphological units of Al-Gabal Al-Akhdar coastal sector comprise 10 habitats as follows (Al-Sodany et al., 2003):

1. Three habitats in the coastal plain: saline sand flat, coastal sand dunes and sand flats,
2. Two habitats in the coastal hills: seaward slope and leeword slope,
3. Two habitats in the plateau: first terrace and second terrace, and
4. Three habitats in the wadi: south-east slope, north-east slope and wadi bed.

Six vegetation groups, named after the dominant species, are recognized in these habitats, these are:

- Group I has two co-dominant species: *Juniperus phoenicea – Sarcopoterium spinosum*, occupying a wide elevation gradient and recognized in all habitats except sand dunes,

- Group II also has two co-dominant species: *Suaeda vermiculata, Crucinella maritima* along the seaward direction of the coastal hills,
- Group III is dominated by *Lygos raetam* in the coastal sand flats,
- Group IV is dominated by *Pancratium maritimum* and *Ammophila arenaria* in the coastal sand dunes.
- Groups V and VI are dominated by *Crucinella maritima* and *Limoniastrum monopetalum*, respectively, in the saline sand flats of the coastal plain.

The highest number of the associates (113 species) occur in group I with 70 perennials, and 43 annuals. Among the perennials are: *Arbutus pavarii, Callycotome villosa, Ceratonia siliqua, Cupressus sempervirens* var *horizontalis, Euphorbia characias, E. dendroides, Lycium europaeum, Myrtus communis, Olea europaea, Phlomis floccosa. Pinus halepensis, Pistacia lentiscus, Quercus coccifera, Lygos reatam, Rhus tripartita, Rosmarinus officinalis, Tamarix africana,* and *Thymelaea hirsuta (*phanerophytes*), Centaurea ragusina Cistus incanus, Conyza bonariensis, Fumana thymifolia, Globularia alypus, Helianthemum salsifolium, H. stipulatum, Helichrysum stoechas, Malva sylvestris, Thymus capitatus* and *Withania somnifera (*chamaephytes*), Ajuga iva, Anchusa undulata, Asteriscus spinosus, Convolvulus arvensis, Marrubium vulgare, Medicago marina, Reichardea picroides, Satureja thymbra, Stachys tournefortii, Thapsia garganica and Tolpis virgata* (hemicryptophytes), *Arum cyrenaicum, Asparagus aphyllus, Asphodelus ramosus, Barleria robertiana, Bellevalia mauritanica, Cynodon dactylon, Dactylis glomerata, Eleocharis palustris, Gagea fibrosa Gladiolus trinervia, Ranunculus asiaticus, Romula cyrenaica, Scirpus maritimus, Smilax aspera* and *Urginea maritima* (cryptophytes). All of the annual associates (40 species) belong to group I and only four of them have also been recorded with other groups: *Euphorbia helioscopa* with groups II and IV, *Ononis pedula* with group II and III, *Medicago ridigula* with group II, *Poa annua*, and *P. bulbosa* with groups II and VI. The other annuals include: *Aegilops kotschi, Ammi majus, Avena barbata, Beta vulgaris, Biscutella didyma, Brassica deserti, Briza maxima, Bromus alopeuros, B. redbens, Carduus argentus, Carthamus lanatus, Chrozophora tinctoria, Conyza aegyptiaca, Crepis libyca, Cynara cyrenaica, Cynosurus coloratus, Euphorbia peplis, Hordeum muriun, Lathyrus aphaca, Lotus tetragonolobus, Onopordum cyrenaicum, Plantago arenaria, P. lagopus, Polypogon monspeliensis, Rumex simpliciflorus, Scorpiurus muricatus, Solanum nigrum, Sonchus oleraceus, Trifolium tomentosum, Trisetaria linearis, Urtica urens* and *Vicia laxiflora.*

## II. Egypt's coastal sector (Mariut)

### a. *General remarks*

The Mediterranean coastal land of Egypt extends for about 970 km from Sallum in the Libyan-Egyptian border (eastwards) to Rafah in the Palestine border. It is a narrow coastal belt assigned to the dry arid climatic zone of Koppen's (1931) classification system (as quoted by Trewartha, 1954) and the

Mediterranean bioclimatic zone of Emberger (1955). However, the bioclimatic zone of UNESCO/FAO (1963) indicates that it is a subdesertic warm climate.

Geographically, the Mediterranean coastal land of Egypt can be divided into three sectors (Zahran et al., 1985, 1990): western (Mariut), middle (Deltaic) and eastern (Sinai) sectors. The first two sectors belong to the North African Mediterranean coast while Sinai sector belongs to the South West Asian Mediterranean coast.

#### b. *Location and climate*

The Mariut coast of Egypt extends along the Mediterranean Sea for about 550 km from Sallum eastwards to Abu Qir (about 30 km east of Alexandira). It is the northern coast of the Western Desert of Egypt that narrows or widens according to the position of its southern boundary – the Western Desert Plateau – with an average width, of about 20 km. Its remarkable feature is the prevalence of ridges formed of oolitic limestone, often 20 m or more high extending parallel to the sea-shore for long distances (Ball, 1939). Commonly one line of ridges skirts the coast closely, while another runs parallel to it a few km inland, and there is sometimes a third ridge between the second and the edge of the Western Desert Plateau.

Rainfall occurs mainly during the October–March period (60% or more); the summer is virtually dry. The highest amount usually falls during either January or December and varies appreciably along the coast from 119.7 mm/year in Sallum and 144.0 mm/year in Mersa Matruh[4] and 192.1 mm/year in Alexandria. The annual mean temperature maxima and annual mean minima are: 25.3, 24.3, 24.9, 13.3, 14.3 and 14.9°C, respectively. Relative humidity is lower in Salum (mean annual = 69%) than in Mersa Matruh (67%) and higher in Alexandria (72%). The reverse in true for evaporation being highest in Sallum (mean annual = 7.2 mm/pich/day) followed by that of Mersa Matruh (6.5 mm pich/day) and Alexandria (3.8 mm pich/day).

Winds along the Mariut coast are generally strong and violent, and dust storms and pillars are not unusual. Dry hot dust-laden winds from the south known as khamasin blow occassionaly for about 50 days during spring and early summer. There can also be strong, winds blow strongly with an average velocity of about 20–23 km/h during winter and early spring. Wind speed decreases in May and June, but July is windy. Shaltout (1983) stated that, the ends of summer records many calm days and the average wind speed drops to 15 km/h. In Burg El-Arab area (about 50 km west of Alexnadria), Ayyad (1973) estimated the mean annual evapotranspiration to be 995 mm.

It is worth to state that the climate of Mariut coast of Egypt and, according to Murray (1951), has not changed since Roman time (2,000 years ago). Zahran and Willis (2008) stated: Sutton (1947) quotes records of annual rainfall made by Thurnburn (1847–1849) and brought up to 1970 as follows: 1847–1849 = 191 mm, 1881–1886 = 209 mm, 1901–1906 = 47 mm, 1921–1926 = 178 mm, 1939–1941 = 161 mm, 1951–1956 = 187 mm and 1960–1970 = 207 mm.

---

[4]Mersa Matruh is located at about 350 km west of Alexandria.

### c. *Land use*

The Western Mediterranean coastal land of Egypt is called the Mareotis District, being related to Mariut Lake. In the past this lake was a fresh-water one. Kassas (1972a) states "Strabo (66-24BC) records that Lake Marea is filled by many canals from the Nile through which a greater quantity of merchandise is imported". De Cosson (1935) notes that the lake was rather deep fresh water and adds: "There seems to be little doubt that 2,000 years ago it was of greater extent than in modern times. The Canopic Nile Branch and the other canals that fed the lake gradually silted and its water receded. Thus, Lake Mariut was in Graeco-Roman times a fresh-water lake, the water of which was used for irrigating the fields. This source of freshwater gradually diminished and by the end of the twelfth century the lake became saline".

Kassas (1970) infers that, in this coastal region, agriculture and horticulture became established by a resident population of cultivators. The farms depended partly on irrigation from an ancient branch of the Nile (the Canopic) that extended for some distance west of the present site of Mariut, but the location of farms far beyond the reach of this branch indicates that effective methods of dryland farming were used. According to Kassas (1972a), the Mareotis district was an area of prosperous cultivation particularly vineyards, and was densly inhabited. Good wine was produced in such quantities that Mareotis wine was laid down for keeping over long periods. By the tenth century, the district gradually declined and the vineyards were replaced by desert. It is unlikely that there have been major climatic changes during the last 2,000 years that could have caused the deterioration of the area. There is also evidence that the fresh water of Lake Mariut and its arm that extended westward for 79 km was used for irrigating farms and orchards fringing the shores of the lakes and banks of its western arm. These strips of irrigated agriculture must have been limited in extent because of the topography.

Earlier this century some attention was given to the Mareotis region. The extension of a railway westward from Alexandria to Mersa Matruh, and the plantation of vine, olive and date palm at Ikingi (20–25 km west of Alexandria) were "early steps towards regeneration" (De Cosson, 1935). Several attempts have been made to reintroduce a variety of orchard crops in Mareotis: vine (*Vitis vinifera*), fig (*Ficus carica*), date palm (*Phoenix dactylifera*), olives (*Olea europaea*), carbo (*Ceratonia siliqua*), almond (*Prunus amygdalus*) and pistachio (*Pistacia vera*) (Kassas, 1972b).

At present the main land uses of Mareotis are grazing and rain-fed farming (or irrigated by underground and run-off water). The main annual crop is barley (*Hordeum vulgare*). Figs are successful on calcareous coastal dunes and olives, almonds and pistachio in inland alluvial depressions. Irrigated agriculture of pasture, grain crops and fruit trees (mainly vine) is spreading after the extension of irrigation canals from the Nile up to 60 km west of Alexandria (Ayyad, 1983).

## d. *Plant cover*

### (i) Floristic analysis

The Western Mediterranean coastal belt is by far the richest part of Egypt in it floristic composition owing to its relatively high rainfall. The number of species in this belt makes up about 50% of the total of the Egyptian flora which is estimated to be about 2,000 (Oliver, 1938) about 2,080 species (Täckholm, 1974), 2,094 species by Boulos (1995). However, Boulos (1999, 2000, 2002, 2005, 2009) recorded a total of 2,145 species of which 44 species are cultivated. Most of these species are therophytes that flourish during the rainy season, giving the coastal belt a temporary flush as a grassland desert. During the longer dry period, only the characteristic woody shrubs and perennial herbs are evident; these constitute the scrub vegetation of the area, scattered sparsely in parts and grouped in denser more distinct patches in others (Tadros, 1956).

Hassib (1951) describes the percentage distribution of both annual and perennial species among the life-forms in this coastal belt as follows: neither mega- and meso-phanerophytes nor epiphytes are represented. But there are the micro- and nanophanerophytes (3.2%), stem succulents (0.1%), chamaephytes by (9.2%), hemicryptophytes (11.7%), geophytes (11.9%), hydrophytes and helophytes (4.0%), therophytes (58.7%) and parasites (1.1%). Maquis vegetation that characterizes the other Mediterranean countries is not represented in Egypt. The prevailing life-form of perennials is chamaephytes; nanophanerophytes are less.

Xerophytes make up about 90% of the total number of species in this coastal belt; most are therophytes (67%), followed by geophytes (11%), halophytes and helophytes (11%), chamaephytes (6.6%), micro-and nanophanerophytes (3%), parasites (1.2%) and stem succulents (0.1%). The common xerophytes include: *Achillea santolina, Ammophila arenaria, Anabasis articulata, Euphorbia paralias, Gymnocarpos decander, Hammada scoparia, Helianthemum lippii, Lygos raetam, Ononis vaginalis, Pancratium maritimum, Plantago albicans, Thymelaea hirsuta* and *Thymus capitatus.*

The halophytes include about 45 species. Algae are well developed in the rock coastal areas but apparently absent from the loose soil. The submerged phanerophytes include: *Cymodocea major, Posidonia oceanica* and *Zostera notei.* Terrestrial halophytes include *Arthrocnemum macrostachyum, Atriplex* spp., *Juncus acutus, J. rigidus, Limoniastrum monopetalum, Nitraria retusa, Salicornia fruitcosa, Suaeda fruticosa, S. pruinosa, Tamarix nilotica* and *Zygophyllum album.*

The helophytes and fresh-water hydrophytes represent about 4% of the total number of flora of this coastal belt. They include: submerged species (e.g. *Ceratophyllum demersum, Potamogeton crispus),* floating species (e.g. *Eichhornia crassipes, Lemna* spp.), reeds (e.g. *Phragmites australis* and *Typha domingensis),* and sedges (e.g. *Cyperus* spp., *Scripus* spp.).

### (ii) Habitats and vegetation types

In spite of the relative simplicity of the relief and the apparent uniformity of the climate, the plant habitats in the region present some diversity. For the causal observer, however, the physiognomy of the vegetation seems monotonous over large tracts of land, owing to the prevailing life-form of the perennial plants, being mostly chamaephytes and to a less extent nanophanerophytes with scattered distribution. The only variation in the physiognomy is the change from the short vernal (spring) aspect of the vegetation to the longer aestival (summer) aspect (Tadros, 1956).

The distribution of plant communities here is controlled by topography, the origin and nature of the parent material and the degree of degradation influenced by human manipulation (Ayyad and El-Ghareeb, 1984). Generally, the vegetation of this coastal belt belongs to the *Thymelaeion hirsutae* alliance with two associations:

1. *Thymelaea hirsuta, Noaea mucronata* association with two variants dominated by *Achillea santolina,*
2. *Anabasis articulata, Suaeda pruinosa* association (El-Ghonemy and Tadros, 1970).

The local distribution of communities in different habitats is linked primarily to physiographic variations. According to these variations two main sets of habitats may be distinguished – one on ridges and plateaux and the other in depressions. Ridge and plateau habitats may be further differentiated into two main types. The coastal ridge is composed mainly of snow-white oolitic (calcareous) sand grains overlained by dunes in most places whilst inland are less calcareous ridges and the southern tabeland. The southern tabeland is characterized by the dissection of the landscape into an extensive system of wadis which drain into the Mediterranean Sea and form a distinct type of habitat. Inland siliceous dunes are sporadically distributed on the southern tableland and support a community different from that of calcareous dunes on coastal ridge. Habitats of depressions differ according to the relative proximity of the water-table to the surface and consequently to the level of salinity and extent of waterlogging. Five main types of ecosystems may be recognized (Ayyad and El-Ghareeb, 1984; Zahran and Willis, 1992, 2008; Salama et al., 2005).

1. Sand dunes (coastal calcareous and inland siliceous);
2. Rocky ridges and plateaux with skeletal shallow soils;
3. Saline depressions;
4. Non-saline depressions;
5. Wadis.

#### 1. *Sand dunes*

Along the Western Mediterranean coast lies a chain of intensely white calcareous granular sand dunes. They are formed of loose oval pseudo-oolitic grains, each composed of a series of successive coats of calcium carbonate. These dunes form a fairly continuous ridges with an undulating surface and present a type of habitat notable for its monotony. However, such monotony does not invariably mean that either the soil or the vegetation lacks variety. Owing to proximity to the sea, the dunes are more humid and exposed to the immediate effect of the northerly winds. They are also reached by sea spray (Ahmed and Mounir, 1982). Certain sections of the coast are devoid of dunes.

A short distance from the beach, fresh water is frequently obtained by digging carefully in the sand to a depth 3–4 m. This fresh water is undoubtedly rain water, which, having a lower specific gravity than saline water below, can form a layer above it; there may be a hard pan of limestone rock underlying the sand which prevents percolation of rain water, the sand acting as a reservoir of fresh water.

Plants growing in sand dunes are highly specialized and many have the ability to elongate vertically on burial with sand (Girgis, 1973). They are also subject to partial exposure of their underground organs, often without being seriously affected. The coarse grain and loose texture of the sand result in poor water-retention because of rapid percolation. Many psammophytes develop extensive superficial roots that make use of dew.

The vegetation of these sand dunes has been studied by Oliver (1945), Tadros (1953, 1956), El-Sharkawi (1961), El-Ghonemy (1973), Girgis (1973), Ayyad (1973), Ayyad and El-Bayyoumi (1979), Ayyad and El-Ghareeb (1984) etc. Bordering the sea, a community of *Ammophila arenaria* and *Euphorbia paralias* can be usually distinguished on the mobile young calcareous sand dunes. Associates include *Lotus polyphyllos* and *Sporobolus virginicus*.

The vigorous growth made by *Ammophila* when sand covers it dominates the mobile dunes. It is a pioneer species in invading mobile coastal dunes and is consequently extensively used for stabilizing sand-dunes. On the older, advanced and higher dunes, where the sand may be consolidated in parts. *Crucianella maritima* and *Ononis vaginalis* dominate. Associated species include *Ammophila arenaria, Cakile maritima, Centaurea pumila, Echinops spinosus, Echium sericeum, Elymus farctus, Euphorbia paralias, Hyoseris lucida, Launaea tenuiloba, Lotus polyphyllos, Lygos raetam, Pancratium maritimum, Plantago albicans, Reseda alba, Salvia lanigera,* and *Silene succulenta*. In the more advanced stages of dune stabilization, communities of *Crucianella maritima, Echinops spinosus, Elymus farctus, Euphorbia paralias, Pancratium maritimum* and *Thymelaea hirsuta* become successively more common. When the coastal ridge is fairly exposed a community of *Globularia arabica, Gymnocarpos decander, Helichrysum conglobatum* and *Thymus capitatus* predominate.

The inland siliceous dunes are dominated by communities of *Plantago albicans, P. squarrosa* and *Urginea maritima*.

### 2. *Rocky ridges*

Two (or sometimes three) ridges run south of the sand dune zone extending parallel to the Western Mediterranean coast of Egypt and are separated from the sea by the sand dunes. These ridges are composed of oolitic sand and shell debris, often 20 m or more high with smooth rounded summits. The outer ridge closely skirts the coast while the second one runs parallel with it at a distance of a few km inland. The third ridge, when present, is between the second one and the edge of the Western Desert.

The vegetation of these rocky ridges is an association of *Thymelaea hirsuta* and *Gymnocarpos decander* (Tadros, 1956). However, local variation in the nature of the position and degree of slope lead to parallel variations in the distribution of the vegetation. The characteristic species of this community include *Aegilops kotschyi, Arisarum vulgare, Bupleurum nodiflorum, Carduus getulus, Chenolea arabica, Erodium cicutarium, Limonium tubiflorum, Lotus corniculatus, L. ceticus, Lygeum spartum, Malva aegyptia, Medicago minima, Moricandia suffruticosa, Orlaya maritima, Plantago notata, Reaumuria hirtella, Reichardia orientalis, Scorzonera alexandrina, Stipa capensis, S. parviflora* and *Teucrium polium.*

Rocky sites with low moisture availability are dominated by communities of *Globularia arabica* and *Thymus capitatus* while sites with fairly deep soils and high moisture availability are dominated by communities of *Asphodelus microcarpus, Herniaria hemistemon, Plantago albicans* and *Thymelaea hirsuta.* In sites of intermediate rockiness and moisture availability, *Echinops spinosus, Helianthemum stipulatum, Noaea mucronata, Pituranthos tortuosus* and *Scorzonera alexandrina* are abundant (Ayyad and Ammar, 1974).

These communities extend to the plateau of the south tableland. Two other communities dominated by *Hammada scoparia* and *Anabasis articulata* are found on degraded shallow skeletal soils subjected to active erosion. Associate species of this community include *Asphodelus microcarpus, Atriplex halimus, Carthamus mareoticus, Noaea mucronata, Pituranthos tortuosus, Verbascum letourneuxii* and *Zilla spinosa. Salsola tetrandra, Suaeda fruticosa* and *Suaeda pruinosa* are poorly represented. Bushes of *Capparis spinosa* and *Ephedra alata* often grow in vertical rock.

### 3. *Saline depressions*

The saline depressions (littoral salt marshes) are a common habitat of the Western Mediterranean coastal belt. Tadros (1953) recognized two series of salt marshes. One is formed from depressions directly adjacent to the dune strips. The salinity of this series results from the evaporation of seepage water, where the water-table is exposed or near the surface and where there is poor drainage. The soil is mostly calcareous-sandy due to the encroachment of sand from the neighbouring dunes. In certain places in these salt marshes, low bushes of *Arthrocnemum macrostachyum* and *Halocnemum strobilaceum*, others eventually become buried under moist conditions, forming dense black rotten material from which frequently the smell of hydrogen sulphide can be detected. The second series of salt marshes is formed from the dried bed of Lake Mariut lying between the two ridges. The causes of salinity are essentially as in the first series, but the soil texture is different, having a considerable proportion of silt, regarded as having been derived from the Nile during its previous connection with the lake.

The littoral salt marsh vegetation of the Western Mediterranean coast of Egypt has been described by several authors: e.g. Oliver (1938), Hassib (1951), Tadros (1953, 1956), Migahid et al. (1955), Ayyad and El-Ghareeb (1982, 1984), Ahmed and Mounir (1982) and Zahran and Willis (1992, 2008).

Apart from the communities of the swamp vegetation dominated by *Phragmites australis, Scripus tuberosus* and *Typha domingensis*, the halophytic vegetation is characterized by some 11 communities:

1. *Salicornia fruticosa-Suaeda salsa* community. This usually occupies the zone on the more elevated banks with less submerged saline soil. Associate species are *Phragmites australis* and *Salicornia herbacea.*
2. *Juncus rigidus* community. This occupies lower parts of the marsh with high moisture content where the calcareous and fraction dominates the soil texture. Associated plants include *Halimione portulacoides, Inula crithmoides, Juncus acutus, Limonium pruinosum* and *Sporobolus pungens.* In certain patches of this community, there are societies dominated by *Schoenus nigricans.*
3. *Sporobolus pungens* community. This occupies higher parts of the marsh, especially where calcareous sand is plentiful. The associate species are *Juncus rigidus* and *Limonium pruinosum.*
4. *Halocnemum strobilaceum* community. This community occurs over a wide range of fluctuations of salt concentration between the wet and dry seasons where there is a high proportion of fine fractions affecting soil texture. Associate species are *Arthrocnemum macrostachyum, Juncus rigidus* and *Salicornia fruticosa.*
5. *Salicornia fruticosa – Limonium pruniosum* community. This is present in somewhat more elevated and less saline parts than that of the *H. strobilaceum* community. Common associated species include *Inula erithmoides, Juncus rigidus, Parapholis marginata, Plantago crassifolia* and *Sphenopus divaricatus. Halimione portulacoides* and *Phragmites australis* dominate in some patches, the latter species being associated with depressed areas with high water content.
6. *Arthrocnemum macrostachyum-Limoniastrum monopetalum* community. This occurs on even more elevated substrates than the *S. fruticosa-L. pruinosum* community. Characteristic species are *Cressa cretica, Frankenia revoluta, Mesembryanthemum nodiflorum* and *Parapholis marginata.*
7. *Zygophyllum album* community. *Z. album* frequently forms an almost pure community on saline patches recently covered by drifted sand in shallow layers. It is also found in communities with other species in similar habitats.
8. *Lygeum spartum* community. This occurs in less saline parts with high organic matter content. Associate species are *Frankenia revoluta, Halimione portulacoides, Limoniastrum monopetalum* and *Limonium pruinosum.*
9. *Salsola tetrandra* community. This community is usually present on the elevated border of the dry saline beds of the marshy valleys. *S. tetrandra* is a very efficient at conserving soil against the wind blowing as well as being a soil builder. The associate species include *Anthemis cotula, Coris monspeliensis,*

*Frankenia revoluta, Haplophyllum tuberculatum, Limoniastrum monopetalum, Salicornia fruticosa, Sphenopus divaricatus, Suaeda fruticosa, S. pruinosa* and *Traganum nudatum.*

10. *Limoniastrum monopetalum-Lycium europaeum* community. This is another community rich in floristic composition. It may follow in succession the community dominated by *Salsola tetrandra.* Associate species include *Asphodelus microcarpus, Bassia muricata, Carthamus glaucus, Cutandia dichotoma, Echinops spinosus, Ifloga spicata, Lotus villosus, Noaea mucronata, Orlaya maritima, Plantago albicans, Reaumuria hirtella* and *Suaeda pruinosa.*
11. *Atriplex halimus-Picris radicata* community. This is the richest of all communities of the salt-affected land. It occurs on deep sandy loam at the edges and upper parts of valleys where the vegetation covers the soil almost completely. Associate species include *Anthemis microsperma, Chenolea arabica, Chrysanthemum coronarium (Glebionis cornaria), Koeleria phleoides, Lolium rigidum, Lycium europaeum, Medicago minima, Picris radicata, Salvia lanigera, Schismus barbatus* and *Stipa capensis.*

**4. *Non-saline depressions***

The non-saline depressions (barley fields) are the most fertile areas of the Western Mediterranean coastal belt of Egypt. These depressions are mainly limited to the plains south of the second ridge in the eastern section of the coast, but are widespread in the valley and plains of the western section.

The soils of these depressions (e.g. the Abu Sir depression), are variable (Ayyad, 1976). In some parts, highly calcareous soils are derived from drifted oolitic grains of the coastal ridge; in other parts alluvial, less calcareous, loamy soils are derived from the Abu Sir ridge.

There depressions provide favourable conditions for cultivation; extensive areas are occupied by barley, figs and olives. Farming operations promote the growth of a considerable number of species, mostly therophytes. Weeds of barley fields are recognized as the *Achilleetum santolinae mareoticum* association, with subassociation of *Chrysanthemetosum coronariae* and *Arisaretosum vulgare*, composed of the following characteristic species: *Achillea santolina, Anagallis arvensis, Calendula aegyptiaca, Carthamus glaucus, Convolvulus althaeoides, Echinops spinosus, Echium sericeum, Eryngium creticum, Hordeum murinum, Koeleria phleoides, Lathyrus cicera, Muscari comosum* and *Vicia cinerea.*

According to Ahmed and Mounir (1982), there are still other species of different communities occasionally present in the barley fields, e.g. *Atriplex halimus, Trifolium tomentosum* and *Suaeda fruticosa.* These species may indicate possible affinities with other associations. The "accidental" species recorded include: *Anchusa hispida, Anthemis cotula, Asteriscus graveolens, Avena sterilis, Beta vulgaris, Bupleurum subovatum, Crucianella maritima, Echiochilon fruticosum, Emex spinosus, Filago spathulata, Francoeuria crispa, Gagea fibrosa, Helianthemum stipulatum, Hippocrepis bicontorta, Hymenocarpus nummularius, Hyoseris lucida, Ifloga spicata, Koniga arabica, Limonium tubiflorum, Lotus creticus, Malva parviflora, Moricandia nitens, Ononis vaginalis, Orlaya maritima, Ornithogalum trichophyllum, Papaver hybridum, Reseda alba. Salvia aegyptiaca, Scorzonera alexandrina, Silene villosa, Thesium humile* and *Verbascum letourneuxii.*

The vegetation belongs to the *Plantagineto-Asphodeletum microcarpae* associations. The *Anabasis articulata* community is found on more or less sandy soils with low contents of calcium carbonate, a *Zygophyllum album* community where the soil content of calcium carbonate and salinity are higher. A *Plantago albicans* community occurs where salinity is lower and an *Asphodelus microcarpus-Thymelaea hirsuta* community on fine-textured soils (Ayyad, 1976). The characteristic species include *Alkanna tinctoria, Brachypodium distachyum, Brassica tournefortii, Bupleurum subovatum, Carthamus glaucus, Centaurea glomerata, Linaria haelava, Lolium perenne, Malva parviflora, Medicago littoralis, Onopordum alexandrium, Orobanche ramosa, Papaver rhoeas, Polygonum equisetiforme, Raphanus raphanistrum, Reseda alba, R. decursiva* and *Zygophyllum album.*

### 5. The wadis

The landscape of the Western Mediterranean coastal land of Egypt is dissected by a drainage system (wadis) originating from a southern limestone plateau lying parallel to the Mediterranean Sea. The plateau reaches a maximum elevation of about 200 m above sea level at Sallum and slopes gently to the coastal plain west of Mersa Matruh (from 10 to 20 m above sea level). These wadis drain northwards into the Mediterranean Sea. An ecological account of one of these wadis (Wadi Habis) is given below.

#### 5.1. *Wadi Habis*

Wadi Habis (31° 24′N, 27° 03′E) is of ecological and historical interest. In this wadi there are archaeological remains of apparently Graeco-Roman age (about 300 B.C.–600 A.D.). The Graeco-Roman occupation of the wadi was restricted to its mouth and its immediate vicinity.

According to El-Hadidi and Ayyad (1975), Wadi Habis is characterized by nine habitats: fallow saline areas, fallow non-saline area, barley fields, olive orchards, wadi bed, lower position of slopes, middle slopes, upper slopes and plateau.

The saline fallow areas are dominated by *Reseda decursiva* and *Asphodlus tenuifolius*. The abundant associates are *Carthamus glaucus* and *Onopordum alexandrinum* while other associates include *Centaurea glomerata, Chrysanthemum coronarium, Echium sericeum, Glaucium corniculatum, Malva parviflora, Papaver rhoeas, Paronychia argentea, Plantago albicans. Salvia lanigera, Senecio desfontainei* and *Trigonella maritima.*

In the non-saline fallow areas, the most abundant species are: *Chrysanthemum coronarium, Picris sprengeriana* and *Trigonella maritima*. Other associates include *Asphodelus tenuifolius, Chenopodium murale* v. *microphyllum, Emex spinosus, Eragrostis pilosa, Erucaria pinnata, Lolium rigidum, Matthiola longipetala, Schismus barbatus, Silene apetala* and *Trifolium tomentosum.*

The barley fields support about 40 species dominated by *Chrysanthemum coronarium, Convolvulus althaeoides, Launaea nudicaulis* and *Plantago albicans.* Other associates include: *Achillea santolina, Adonis dentata, Anagallis arvensis, Arisarum vulgare, Avena sterilis* ssp. *ludoviciana, Beta vulgaris, Brassica tournefortii, Echinops spinosus, Echium setosum, Erodium laciniatum, Lamarckia aurea, Lathyrus aphaca, Linaria haelava, Lotus creticus, Medicago littoralis, Noaea mucronata, Papaver hybridum* and *Senecio desfontainei.*

The olive orchards of the frontal section are characterized by a dense cover of weeds which may be distinguished into two main synusiae. The upper is dominated by grasses such as *Hordeum murinum* subsp. *leporinum, Lolium rigidum* and *Lophochloa cristata* and the lower by *Achillea santolina, Astragalus boeticus* and *Matthiola longipetala* subsp. a*spera.* Other associates include: *Anchusa milleri, Emex spinosus, Euphorbia parvula, Filago desertorum, Fumaria bracteosa, Glaucium corniculatum, Hippocrepis cyclocarpa, Reichardia orientalis, Roemeria hybrida, Schismus barbatus, Scorpiurus muricatus* var. *subvillosus* and *Spergularia diandra.*

The vegetation of the wadi bed is sparse, but the number of speices is high. In this habitat, fine soil material has little chance to settle owing to the high velocity of the water stream during the rainy season. The wadi bed is filled mainly with large boulders, the sparse vegetation being largely restricted to shallow soil accumulation between rock fragments. Common perennials in the wadi bed are *Allium erdelii, Echium sericeum, Euphorbia terracina* and *Salvia lanigera.* Less common ones include: *Allium aschersonianum, A. barthianum, Arisarum vulgare* var. *veslingii, Cynara sibthorpiana, Lygos raetam, Scorzonera alexandrina, Silybum marianum* and *Suaeda pruinosa.* Common annuals include: *Astragalus boeticus, Erodium gruinum* and *E. hirtum.* Less common annuals include: *Aizoon hispanicum, Chenopodium murale* var. *microphyllum, Emex spinosus, Fumaria bracteosa, Mesembryanthemum nodiflorum, Minuartia geniculata* var. *communis, Polycarpon succulentum, Polygonum equisetiforme, Rumex vesicarius, Spergula fallax, Spergularia diandra* and *Trifolium formosum.* More than two-thirds of the taxa recorded in the wadi bed are Mediterranean. The lower gentle slopes support meadow-like vegetation of annual species; the most common are: *Astragalus hamosus, Hippocrepis bicontorta, Medicago littoralis, M. truncatula* and *Spergula fallax.* Perennial associates include: *Allium barthianum, Asphodelus microcarpus, Cynara sibthorpiana, Salsola lonifolia, Salvia lanigera, Scorzonera alexnadrina, Silybum marianum* and *Traganum nudatum.*

On the middle slopes the vegetation is dominated by shrubby species including: *Artemisia inculta, Gymnocarpos decander, Limonium sinuatum* and *L. tubiflorum* and grasses such as *Hyparrhenia hirta* and *Stipa capensis.* Other associates include: *Allium erdelii, Asparagus stipularis* var. *tenuispinus, Avena sterilis* subsp. *ludouiciana, Brassica tournefortii, Bromus rubens, Carduus getulus, Erucaria pinnata, Hammada scoparia* (= *Hyloxylon scoparium*), *Limonium thouini, Lycium europaeum, Mesembryanthemum nodiflorum, Noaea mucronata, Phalaris minor, Picris sprengeriana, Pituranthos tortuosus, Plantago albicans, P. squarrosa, Reichardia orientalis, Salvia verbenaca, Spergula fallax, Spergularia diandra, Suaeda pruinosa, Traganum nudatum, Trifolium scabrum, T. stellatum* and *Umbilicus horizontalis*.

The upper slopes are usually steep and almost completely devoid of soil cover. They support a typical cliff vegetation dominated by *Asparagus stipularis, Capparis orientalis, Ephedra aphylla, Lycium europaeum, Periploca angustifolia, Phlomis floccosa* and *Umbilicus horizontalis*. Common perennials include: *Allium barthianum, Asphodelus microcarpus, Echinops spinosus, Gymnocarpos decander, Hammada scoparia, Hyparrhenia hirta, Micromeria nervosa, Noaea mucronata, Scorzonera alexandrina* and *Thymus capitatus*. Common annuals include *Echium setosum, Mesembryanthemum forsskaolii (= Opophyllum forsskaolii), Picris sprengerana, Reichardia orientalis* and *Thesium humile* var. *maritima*. Less common are: *Anagallis arvensis, Arisarum vulgare* var. *veslingii, Astragalus asterias, Carthamus glaucus, Convolvulus althaeoides, Cutandia dichotoma, Echium sericeum, Fagonia cretica,Globularia arabica, Helianthemum ciliatum, Hippocrepis cyclocarpa, Leontodon hispidulus (= Crepis bulbosa), Limonium thouini, Lotus creticus, Malva parviflora, Medicago aschersoniana, Pallenis spinosa, Plantago crypsoides, Pteranthus dichotomus, Ranunculus asiaticus, Salvia lanigera, S. verbenaca* and *Valantia hispida*.

In the plateau of the wadi, the vegetation is dominated by *Gymnocarpos decander, Hammada scoparia* and *Phagnalon rupestre*. In this habitat the fewest associate a species have been recorded, including *Artemisia inculta. Asparagus stipularis* var. *tenuispinus, Atractylis prolifera, Echinops spinosus, Ephedra aphylla, Filago desertorum, Globularia arabica, Helianthemum ciliatum, Lycium europaeum, Micromeria nervosa, Noaea mucronta, Periploca angustifolia, Reichardia orientalis, Reseda decursiva, Rumex vesicarius, Salvia lanigera,* and *Thymus capitatus*.

#### 1.8.1.7 Wetlands

Wetlands are swampy ecotones that function as downstream recipient of water from natural sources, e.g. seas, rivers, wadis etc. as well as human activities, e.g. agricultural, industrial and sewage. The ecological importance of the wetlands stems from their hydrologic attributes being rich in their biological, genetic and chemical resources. Their productivity is high and they comprise valuable habitats for fisheries wildlife and birds (Shaltout and Al-Sodany, 2008).

The coastal wetlands of the North African countries are both Atlantic (in the west) and Mediterranean (in the north). According to Flower (2001), these wetlands consist of ten lakes, three in each of Morocco and Tunisia and four in Egypt. The three lakes of Morocco (Sidi Bou Rhaba, Zerga and Bokka), are located along the Atlantic coast, those of Tunisia (Chitane, Ichkeul and Korba) and Egypt (Mariut, Edku, Burullos and Manzala) are located along the Mediterranean coast.

Studies conducted on the various aspects of these lakes are many and include: Andreossy (1799, *Notes on Lake Manzala*, non-published), Muschler (1912), Montasir (1937), El-Masry (1961), Abu Al-Izz (1971), Al-Kholy (1972), Samaan (1974), Banoub (1979), Khedr (1989, 1999), Zahran and Willis (1992, 2008), Khedr and Zahran (1999), Shaltout and Al-Sodany (2008), Flower (2001), Flowers et al. (2001), Birks and Birks (2001), Birks et al. (2001), Peters et al. (2001), Fathi et al. (2001), Ramadani et al. (2001), Shaltout and Khalil (2005), Khalil and Shaltout (2006), Eid (2008) etc.

Unfortunately the ten lakes of the North African countries were and are still under increasing threats due to human impacts through the uncontrolled water withdrawal, eutrophication, pollution, destruction of biodiversity etc. the CASSARINA[5] project described ecosystem changes in nine of these lakes over the last century and related these to environmental and human stressors in the catchments (Flower, 2001).

As the present chapter of this book is concerned with the North African Mediterranean coastal land, our report will be restricted to six lakes distributed along that coast of Tunisia (Chitani, El-Ichkeul and Korba) and Egypt (Edku, Burullos and Manzala).

## A. Tunisian lakes

### 1. *Lake Chitane*

This is a very small (0.025 $km^2$), shallow (40–100 cm depth) lake with fresh and soft water (salinity = 0.04–0.12%). It is located on the north of Gebel Chitane at 150 m altitude, 37° 11′N and 9° 10′E overlooking the Mediterranean Sea to the north. The mean annual rainfall over the area of the lake is about 500 mm with maximum and minimum temperatures of 39.5 and 11°C, respectively.

Lake Chitane is protected and was declared nature reserve on 1993. Its emphemeral outflow to the NW has a perimeter fence some 50–100 m from the lake protecting its catchment. Up-slope from the lake, subsistence farming is practised on a small[6] and a small aquifer supplies two small freshwater springs feeding the upper mire bog and eventually the lake. Water flowing through the mire is currently being exploited for crop irrigation. The sandstone aquifer and the peat bog are acid and this confers acidity to the lake (Ramadani et al., 2001).

The plant life of Lake Chitani is quite distinct. It is an important site for the very rare floating hydrophytes; the water lily (*Nymphaea alba*) and the North African species: *Nitella opaca* (Birks et al., 2001). Other characteristic aquatic and terrestrial flora are: *Isoetes velata, Potamogeton natans, Myriophyllum alterniflorum, Juncus acutiflorus, J. heterophyllus, J. bulbosus, J. effuses, Lemna minor, Typha angustifolia, Lotus ornithopodioides and Cotula* spp.

### 2. *Lake El-Ichkeul*

Lake El-Ichkeul (37° 2′N, 9° 48′E) is a large (89 $km^2$) shallow (10–80 cm), brackish (salinity = 1.66–3.7%) basin that has experienced major disturbance in recent decades. It is located in a coastal area where the mean annual rainfall is about 578 mm and maximum and minimum air temperatures are about 44 and 11°C,

[5] CASSARINA = Change and Sustainability: Aquatic Ecosystem Resilience in North Africa.

[6] Mire = Deep mud.

respectively. It is connected to the Mediterranean Sea through the Oued Tinja and Lac de Bizerte (Ramadani et al., 2001).

Although Lake Ichkeul is surrounded by the productive agricultural land (mainly cereals), of the Mateur region, the main changes stem from twentieth century hydrological modifications of the five main inflowing rivers. Following canalization, all but one of these inflows has been barraged since 1984 so that freshwater inputs are now much reduced (Ramadani et al., 2001). In the past, water usually flowed out from the lake during dry periods. Since 1986, Sluice gates on the Tinja River outflow have control over seawater inflow but there are conflicts between fisheries and biodiversity interests. Recently, the salinity of Ichkeul Lake has increased so that hypersalinity persisted in the mid-1990s (Karaiem and Ben Hamza, 2000). This lake was designated a Biosphere Reserve on 1977 and World Heritage Site in 1996. The marshland and the lake used to receive 200,000 over-wintering and migrating birds annually, mainly ducks, coots, grebes cormorants, spoonbills, storks, herons, and waders. The geese have declined in the 1990s mainly because of the loss of marshland.

The plants of Ichkeul Lake are of ecological interest. The exposed lake shores are dominated by *Sarcocornia fruticosa (Arthrocnemum fruticosum), Juncus subulatus, Ammi visnaga and Frankenia laevis*. In the fore-shore depressions, there are communities dominated by *Tolypella glomerata, Callitriche* sp., *Potamogeton pectinatus, Zannichellia palustris, Enteromorpha intestinalis* and *Chaetomorpha linum*. Other flora recorded in the lake are: *Ruppia cirrhosa, Trifoluim stellatum, Scirpus miritimus* and *S. littoralis*. However, according to Ramadani et al. (2001), the two species of *Scirpus* were absent during the 1998 monitoring and the *Ruppia cirrhosa* was well established in the centre of the lake. Small residual patches of *Phragmites australis* indicated the potential for *Phragmites* regeneration if salinity is reduced. Until the late 1980s, *Phragmites* formed an extensive marginal fringe around most of the lake shore but by the mid-1990s, this had almost disappeared, the zone being marked in 1997–1998 with dead, occasionally *Balanus* encrusted bases of stems.

3. *Korba lake*

Lake Korba (36° 46′N, 11°E) is an elongated small (0.32 km$^2$) hypersaline (salinity = 2–7.9%) lagoon with a depth ranging between 15 and 85 cm, in an area receiving about 450 mm annual rainfall and maximum and minimum temperatures of 40 and 11°C, respectively (Ramadani et al., 2001). It occupies a long narrow shallow channel on the east coast of the Cap Bon peninsular. A partially vegetated sand dunes separates it from the Mediterranean, approximately 100 m distant. Water quality is affected at the south part by pollution from the town of Korba and also by sea water penetration. At two points, the sand dune is breached forming temporary connection with the Mediterranean. Seasonally, fresh water inflows to the north end of the lake from the nearby Chiba and Boulidin stream occur but water abstraction for agriculture is intense. Ramadani et al. (2001) stated that in recent years, saltwater has encroached into the freshwater-table in the Korba region. On the lake's west side, soil is irrigated for crops in summer months and is separated from the lake by *Salicornia europaea* salt marsh community. At the center of the lake, the depth varies from 30 to 60 cm and the shallower northern part consists of hard-packed grey clays, probably ancient and of marine origin. A large part of the lake bed is covered by a dense mat of green filamentous algae.

In Korba Lake, the recorded aquatic macrophytes include: *Ruppia cirrhosa, Posidonia oceanica,* and *Chara, lamprothamnium*. All these taxa tolerate sea-water. The littoral salt-marsh is dominated by *Juncus maritimus, Blysmus compressus*, and *Cyperus* spp. together with *Scirpus maritimus* and *Typha angustifolia*. The presence of *Typha* may suggest some fresh-water influence. On the other hand, the presence of Gramineae fruits may originate from *Phragmites australis* or other grasses of salt-marsh or sand dune habitats likely to be growing around the lake. The growth of *Zygophyllum album*, *Mesembryanthemum crystallinum* and *Verbena* sp. may indicate the presence of saline sand soil.

## B. Egyptian lakes

In Egypt, the northern coast of the Nile Delta close to the Mediterranean sea is characterized by three shallow lakes: Edku (in the west), Burullos (in the middle) and Manzala (in the east). They are separated from the sea by narrow to very narrow strips of land and are connected with it through narrow outlets (straits). These straits are either remnants of the mouths of old deltaic branches of the Nile or merely gaps in weak sections of the bars known as tidal inlets (Abu Al-Izz, 1971). These deltaic Mediterranean lakes receive the main bulk of the drainage waters from the Nile Delta (Zahran and Willis, 2008). Mean annual rainfall ranges between 196.7, 196.7 and 112.2 for the three lakes respectively (Ramadani et al., 2001).

The Nile deltaic Lakes have been subjected to major disturbances since ancient times caused by several factors, such as continual degradation and deposition, the accumulation of the remains of vegetation, the blowing of sand and the construction of levees. All these factors and others have caused a decrease in their areas starting 1799 (Al-Kholy, 1972; Butzer, 1976). For example, the area of Lake Edku was about 80,000 feddans in 1799 decreased to 17,000 feddans on 1970 and those of Lake Burullos and Lake Manzala were 270,000 and 470,000 feddans on 1799 decreased to 130,000 feddans and 300,000 feddans on 1970 respectively (Feddan = 4,200 $m^2$). Today, the loss in areas is rather greater.

Ramadani et al. (2001) stated that the hydrological regime of these deltaic lakes results from a balance of freshwater runoff from the agricultural regions in the south and seawater from the north. The undisturbed margins are extensively vegetated, mainly by reedwamps (*Phragmites australis* and *Typha domingensis*). These two reeds are also frequent on the shorelines and islets. Water hyacinth (*Eichhornia crassipes*) is proliferating everywhere. Productive agriculture around the lakes, date palm and sugar cane plantations, cereals and leguminous crops, has been encouraged by the increased supply of fresh Nile water for irrigation in recent decades. The water quality of all lakes is locally affected by sewage and agro-chemicals drainage water mainly from southern agricultural drainage regions. According to Goodman and Meininger (1989), many Palearctic bird species usually migrate via the Deltaic Mediterranean Lakes in internationally significant numbers, and hence changes in their nature are important conservation issues.

1. *Edku lake*

Lake Edku (31° 15′N, 30° 15′E) lies parallel to the Mediterranean Sea at about 36 km east of Alexandria with an area of about 126 km$^2$ and depth between 50 and 200 cm. It receives drainage water through two main drains, as well as, saline water from the sea to which it is connected by the Boughaz channel (El-Masry, 1961; Ramadani et al., 2001). It is surrounded by a productive agriculture to the south, by ongoing land reclamation activities to the east, and by housing and industry to the west side where much reclamation has occurred since the nineteenth century. The northern border of the lake is a sand ridge that separates the lake from the sea and supports date-palm groves, subsistence agriculture and several villages.

The wetland vegetation comprises submerged, floating, reed swamp, sedge meadow as well as halophytic elements. *Phragmites australis* usually dominates the reed swamp habitat with abundant growth of *Typha domingensis*. The floating type is dominated by *Eichhornia crassipes* associated with *Jussiaea repens, Lemna minor*, *L. gibba, Azolla nilotica, Wolfia hyalina, Alternanthera sessilis* and *Spirodela polyrrhiza*. Submerged vegetation is dominated by *Ceratophyllum demersum* associated with *Najas armata, Potamogeton pectinatus, P. crispus, and Ruppia maritima*. The sedge-meadow vegetation is dominated by *Juncus acutus* associated with *Cyperus articulatus* and *Scirpus litoralis* and some halophytes including: *Inula crithomides, Arthrocnemum macrostachyuns, Atriplex portulacoides, Salicornia fruticosa*, and *Zygophyllum album* (El-Masry, 1961; Birks et al., 2001).

2. *Lake Burullos*

Burullos wetland is located along the Mediterranean coast bordered from the north by the sea and from the south by the agricultural lands of north Delta (Kafr El-Sheikh Governorate). It lies in a central position between the two branches of the River Nile: Rosetta in the west and Damietta in the east (31° 36′N and 31° 07′E to 31° 22′N, 30° 33′E to 31° 22′N, 31° 07′E). It has a total area of about 460 km$^2$ including the lake and approximately 75 islets with a shoreline of about 65 km. The area of the sand bar is about 165 km$^2$ (Ramadani et al., 2001; Shaltout and Khalil, 2005).

Lake Burullos is connected with the Mediterranean Sea at its NE side through the El-Burullos outlet (Bougaz El-Burullos) which is about 250 m wide and 5 m deep. The depth of the lake ranges between 40 cm in its middle sector and near the shores and 200 cm near the outlet to the sea.

Lake Burullos has probably suffered less disturbance than Lake Edku but it is subjected to land reclamation, particularly along its southern and eastern edges. The northern border is a low sand ridge that is currently under development with a town and a major road linking Rosetta and Damietta. This lake is a Ramsar Site and plays host to large populations of migratory and resident water-birds. Valuable cover is provided by extensive beds of reed vegetation formed mainly from *Phragmites australis* and *Typha domingensis*. Water hyacinth (*Eichhornia crassipes*) is abundant. On the northern side of the lake there are extensive patches of the submerged weed *Potamogeton pectinatus* which are considered important refuges for fish fry.

Six major types of habitats characterize Burullos wetland: salt marshes, sand formations, lake cuts, drains, the lake proper and it's islets. According to Shaltout and Khalil (2005), the flora of these habitats comprises 197 species: 100 annuals and 97 perennials (including 12 hydrophytes). These species belong to 139 genera and 44 families. The grasses have the highest contribution to the total flora (18.1%) followed by Compositae (13.6%), chenopods (10.1%), legumes (7.0%) and crucifers (6.0%). Twelve species are wide spread and recorded in about 75% of the habitats, including 7 perennials (*Phragmites australis, Arthrocnemum macrostachyum, Halocnemum strobilaceum, Sarcocornia fruticosa, Cynodon dectylon, Suaeda vera* and *Tamarix nilotica*) and five annuals (*Salsola kali, Senecio glaucus subsp. coronopifloius, Mesembryanthemum nodiflorum, Polypogon mospeliensis* and *Spergularia marina*). The following is a brief account of distribution of species between the six major habitats (Shaltout and Khalil, 2005).

### 1. Salt marshes

In the salt affected lands of Burullos wetlands, a total of 51 species have been recorded, including: 30 perennials and 21 annuals. The most abundant species are: *Arthrocnemum macrostachyum, Halocnemum strobilaceum, Juncus acutus, Phragmites austrarlis, Tamarix nilotica* and *Salsola kali*. Others are rare such as: *Cyperus rotundus, Atriplex canescens, Cyperus alopecuroides, Cynomorium coccineum, Polypogon monspeliensis, Chenopodium murale* and *Sonchus oleraceus*. Species not found elsewhere are: *Alternaria sessilis, Echinops spinosus, Frankenia revoluta, Imperata cylindrica, Ricimus communis, Scirpus holoschoenus, Tamarix aphylla, Carex divisa, Astragalus peregrinus, Amaranthus lividus, Frankenia pulverulenta* and *Rapistrum rugosum*.

### 2. Sand formations

A total of 45 species have been recorded in this habitat: 18 annuals and 27 perennails. Species not recorded from elsewhere are: *Cistanche phelypaea, Convolvulus lanatus, Orobanche cernua, Panicum turgidum, Silene succulenta, Bromus catharticus, Cakile maritime* and *Fagonia arabica*. The common species are: *Arthrocnemum macrostachyum, Halocnemum stroblilaceum, Zygophyllum album* and *Salsola kali*, whereas the following are rare: *Cynodon dactylon, Cressa cretica, Cynanchum acutum, Alhagi graecorum* and *Sphenopus divaricatus*.

### 3. Lake cuts

A total of 29 species have been recorded in this area: 12 annuals and 17 perennials. Only *Rumex pictus* is unique to this habitat of the lake ecosystem. The common species are: *Arthrocnemum macrostachyum, Halocnemum strobilaceum,*

*Sarcocornia fruticosa, Suaeda vera, Zygophyllum album, Juncus rigidus, Launaea nudicaulis, Salsola kali* and *Spergularia marina*. Some species are rare, such as: *Tamarix nilotica, Cynodon dactylon, Suaeda pruinosa, Juncus rigidus, Alhagi graecorum, Chenopodium album, Trigonella stellata, Reichardia tingitana* and *Amaranthus viridis*.

### 4. Drains

#### 4.1. *Terraces of the drains*

A total of 87 species have been recorded in this habitat: 50 annuals and 37 perennials. Species not found elsewhere are: *Polypogon viridis, Salix tetrasperma, Silybum marianum, Brassica rapa, Coronopus didymus, Coronopus squamatus, Juncus bufonius, Raphanus raphanistrum* and *Trifolium alexandrinum*. The common species are: *Arthrocnemum macrostachyum, Suaeda vera, Salsola kali* and *Senecio glaucus* subsp. *coronopifolius*. The rare species are: *Cyperus rotundus, Typha domingensis, Phyla nodiflora, Saccharum spontaneum, Atriplex portulacoides, Ifloga spicata, Cyperus difformis* and *Emex spinosa*.

#### 4.2. *Slopes of the drains*

A total of 69 species have been recorded in the habitat: 45 annuals and 24 perennials. Species not found elsewhere are: *Plantago major, Amaranthus hybridus, Coriandrum satium, Gnaphalium luteo-album, Lathyrus marmoratus, Phalaris paradoxa, Sisymbrium irio, Sonchus macrocarpus* and *Trifolium resupinatum*. The common species are: *Phramgites australis, Arthrocnemum macrostachyum, Sarcocornia fruticosa, Suaeda vera, Salsola kali, Senecio glaucus* subsp. *coronopifolius* and *Sonchus oleraceus*. The rare species are: *Paspalidium geminatum, Atriplex halimus, Ipomoea carnea, Ranunculus sceleratus, Cichorium endivia* subsp. *pumilum, Hordeum marinum, Medicago polymorpha* and *Anagallis arvensis*.

#### 4.3. *Water-edges of the drains*

A total of 59 species have been recorded: 19 annuals and 40 perennials including 6 hydrophytes. Species not found elsewhere are: *Clerodendrum acerbianum, Sida alba, Medicago itnertexa* var. *ciliaris, Rorippa palustris, Setaria verticillata* and *Setaria viridis*. The common species are: *Phragmites australis, Sarcocornia fruticosa* and *Azolla filiculoides*. Species recorded as rare are: *Halocnemum strobilaceum, Inula crithoides, Cynanchum actum, Suaeda maritima, Centaurea calcitrapa, Sphaeranthus suaveolens, Tamari tetragyna* and *Ammi visnaga*.

#### 4.4. *Open-water of the drains*

A total of 14 speies were recorded in this habitat. The common speices are: *Phragmites australis, Eichhornia crassipes, Ceratophyllum demersum, Azolla filiculoides and Echinochloa stagnina. The rare species are: Arthrocnemum macrostachyum, Sarcocornia fruticosa, Lemna perpusilla, Potamogeton crispus* and *Salsola kali.*

## 5. Lake proper

### 5.1. *Lake shores*

A total of 65 species have been recorded in this habitat: 22 annuals and 43 perennials including 5 hydrophytes. Species unique are: *Orobanche ramosa* var. *schweinfurthii, Persicaria senegalensis, Vigna luteola, Chyrsanthemum coronarium* and *Poa annua*. The common species are: *Phragmites australis, Arthrocnemum macrostachyum, Sarcocornia fruticosa, Tamarix nilotica, Juncus acutus, Spergularia marina* and *Polypogon monspeliensis*. The species recorded as rare are: *Zygophyllum album, Cyperus rotundus, Persicaria salicifolia, Phyla nodiflora, Cyperus articulatus, Conyza bonariensis, Melilotus indicus, Potamogeton pectinatus, Eichhornia crassipes, Lemna perpusilla, Ludwigia stolonifera* and *Wolffia hyalina*.

### 5.2. *Open-water of the lake*

Sixteen perennial species have been recorded in this habitat including 10 hydorphytes. Species only recorded in this habitat are: *Ceratophyllum submersum, Najas marina* var. *armata* and *Najas minor*. The common species are: *Phragmites australis, Typha domingensis, Potamogeton pectinatus, Eichhornia crassipes* and *Ceratophyllum demersum*. The rare species are: *Cyperus alopercuroides, Echinochloa stagnina, Lemna perpusila, Ludwigia stolonifera, Wolffia hyaline* and *Potamogeton crispus*.

## 6. Lake islets

A total of 89 species were recorded in this type of habitat: 45 annuals and 44 perennials including 5 hydrophytes (see Khedr and Lovett-Doust, 2000). Species only recorded from this habitat are: *Allium roseum, Asparagus stipularis, Cyperus laevigatus, Lycium schweinfurthii, Pancratium maritimum, Phoenix dactylifera, Urginea undulata, Adonis dentata, Astragalus boeticus, Brassica tournefortii, Calendula aegyptiaca, Cutandia dichtoma, Echinochloa colona, Erodium laciniatum, Euphorbia peplis, Filago desertorum, Launaea capitata, Lobularia arabica, Paronychia arabica, Parapholis incurva, Portulaca oleracea, Ranunculus marginatus, Silene villosa, Spergula fallax, Sinapis arvensis* subsp. *allionii* and *Sporpolus pungens*. The common species are: *Phragmites australis, Arthrocnemum macrostachyum, Sarcocornia fruticosa, Inula crithmoides* and *Juncus acutus*. The rare species are: *Suaeda pruinosa, Cressa cretica, Aster squamatus, Saccharum spontaneum, Cyperus articulatus, Limoniastrum monopetalum, Tamarix tetragyna, Mesembryanthemum crystallinum, Anagallis arvensis, Potamogeton pectinatus, Eichhornia crassipes, Ceratophyllum demersum* and *Lemna perpusilla*.

3. *Lake Manzala*

Lake Manzala (31° to 31° 30′N and 31° 50′ to 32° 15′E) is the largest of the northern deltaic lakes of Egypt. With an area of about 1,400 $km^2$ (Khedr, 1989). It lies in the northern quadrant of the delta between the Mediterranean Sea to the north, the Suez Canal, Port Said and Ismailia Provinces to the east, the Damietta Branch of the River Nile and the provinces of Sharkiya and Dakahliya to the west. Rough waves do not occur in Lake Manzala, because it is too deep (0.7–1.5 m). The lake is joined to the Mediterranean Sea by the Straits Ashtum El-Gamil and other openings; the northern section of the lake is, therefore saline. The southern border of the lake has many inlets through which water drains from the surrounding provinces (Port Said, Ismailia, Damietta, Sharkiya and Dakahliya). Sewage water from Cairo also reaches Lake Manzala through the Bahr El-Baqar drain. Fresh water reaches the lake from the downstream part of the Damietta branch of the River Nile. Thus, Manzala Lake receives three types of water: freshwater, sea water and drainage water. Abu Al-Izz (1971) stated "The flow of drainage water rate into the lake diminishes the salinity to between 0.8 and 1.0%" compared to the Mediterranean Sea at 3.3–3.9%. Zahran and Willis (1992) noted that the water of Manzala Lake was used for drinking during times of flood. Since the establishment of the Aswan High Dam in 1965 no more floods occur and, thus, the salinity of the lake's water is increasing. Now the huge amounts of drainage and searage waters reaching the lake every day have considerable effects on its chemical, physical and biological characteristics.

Due to the establishment of Port Said/Damietta coast road associated with landfill and drainage activity, much of the north west area of Lake Manzala is now wholly or partly reclaimed or converted into fish farming lagoons. The south-western region is probably the least disturbed where dense stands of reed-swamp vegetation *(Phragmites australis* and *Typha domingensis*) occur. These stands are cut and harvested for different uses. Water hyacinth colonizes large area of the lake. The reed-beds support a rich bird life community as well as the fisheries (Ramadani et al., 2001).

The vegetation of Lake Manzala has been reported long ago by Montasir (1937) who recognized three vegetation types: hydrophytic, halophytic and helophytic. The dominant helophytes were *Cryperus* spp., *Juncus* spp., *Phragmites australis* and *Typha domingensis* and the dominant hydrophytes were *Ceratophyllum demersum, Eichhornia crassipes, Lemna* spp. and *Potamogeton crispus*. Khedr (1989) reported that, apart from the dominant reeds, the waters are characterized by five dominant hydrophytes: *Eichhornia crassipes, Ludwigia stolonifera, Najas marina* subsp. *armata, Potamogeton pectinatus* and *Ruppia maritima*. Other species include: *Alteranthera sessilis, Ceratophyllum demersum, Cyperus articulatus, Echinochloa stagnina, Epilobium hirsutum, Juncus subulatus, Leersia hexandra, Lemna gibba, L. minor, Leptochloa fusca, Nymphaea caerulea, Panicum repens, Paspalidium geminatum, Paspalum distichum, Persicaria salicifolia, P. senegalensis, Pistia stratiotes, Scripus litoralis* and *S. maritimus*. Recently, Khedr (1999) recognized eleven aquatic communities in Lake Manzala dominated by: *Ceratophyllum demersum, Najas marina* subsp. *armata, Potamogeton crispus* and *Ruppia maritima* (submerged), *Eichhornia crassipes, Ludwigia stolonifera* and the fern *Azolla filiculoides* (floating) and *Echinochloa stagnina, Phragmites australis, Scirpus maritimus* and *Typha domingensis* (emergent). The associate species, in addition to those recorded by Khedr (1989), include *Nymphaea lotus* and *Spirodela polyrhiza* (floating) and *Marsilea aegyptiaca, Rorippa palustris* and *Veronica anagallis-aquatica* (emergent).

From a TWINSPAN *two-way classification* of the 100 stands, Khedr (1999) suggested splitting its communities into 8 groups which seem ecologically meaningful. In terms of their indicator species, these groups are:

*Group A.* Indicated by *Ludwigia stolonifera* and *Azolla filiculoides* characteristic of stagnant freshwater in the western and southern parts of the lake.

*Group B.* Indicated by *Eichhornia crassipes, Echinochloa stagnina* and *Azolla filiculoides*, and characteristic of the polluted parts of the lake at the mouth of the drains in the south-western shores.

*Group C.* With *Potamogeton pectinatus* as indicator species, more abundant than floating emergent species in all parts of the lake.

*Group D.* With *Najas marina* subsp. *armata* and *Ceratophyllum demersum* as indicator species, both of which are dominant in the western and middle parts of the lake. They form very dense monospecific stands that hinder navigation.

*Group E.* Indicated by *Typha domingensis* which is variably distributed in the lake with relatively low abundance in the northern section close to the sea because, unlike *Phragmites australis*, it is not very salt-tolerant.

*Group F.* With *Scirpus maritimus* as indicator species in the swamps and around the islands of the lake, particularly in the shallow parts.

*Group G.* Indicated by *Phragmites australis*, the most frequent species in the lake; occurring in all parts, even those which are saline and polluted.

*Group H.* Indicated by *Ruppia maritima* which is dominant in the shallow parts and lagoons in the northern section of the lake near the Mediterranean Sea where salinity is relatively high.

## *1.8.2 South-West Asian Mediterranean Coastal Lands*

### 1.8.2.1 General Remarks

The south-west Asian Mediterranean coastal lands extend along the northern Mediterranean coast of the Sinai Peninsula in Egypt and the coasts of the countries of the Levant Lands,[7] i.e. Palestine, Israel, Lebanon and Syria. Except Lebanon all of these countries have territories in the arid zone of the Mediterranean Basin where the annual amount of rainfall rarely exceeds 350 mm and vegetation is sparse

[7]Levant Lands are the lands of the East or Orient where sun is seen to rise (Branigan and Jarrett, 1975).

and devoid of arboreal dominants. However, according to the map of the world distribution of arid regions (UNESCO/FAO, 1963), only the northern parts of Israel and Syria are marked as semi-arid. It is area of striking contrasts in its topography and climate. Its highest mountain peak, in Edom (Transjordan) rises up to 1,700 m above sea level; its lowest depression, the Dead Sea region, falls to 396 m below sea level. Its northern districts enjoy a typical Mediterranean climate while its southern parts are desolate desert (Zohary, 1962). Since vegetation is the major component determining the nature of ecosystems, the area under consideration may be subdivided from the vegetation point of view into deserts and steppes, the demarcation line running between the 70- and 100-mm isohyets. Admittedly, the term steppe is not a good one, since the distinction between deserts and steppes is difficult, and terms like semi-desert, semi-steppe, steppe-desert and shrub-steppe are confusing and ill-defined (Zohary, 1973). However, the 70–100 mm isohyets seem to be the lower threshold for rain-fed vegetation, whereas below it vegetation is restricted to runnels and turns out to be "run-on" vegetation (*a mode contracté*).

The biodiversity (flora and fauna) of the SW Asian Mediterranean coastal lands has always been a most attractive subject to explorers and scientists since the sixteenth century. Among the earliest collectors and explorers the name of Rauwolf (1583) is noteworthy. He journed through Palestine, Syria and Mesopotamia in the years 1573–1575. The results of this botanical findings were published by Gronovius in the latter's Flora Orientalis (1755). Strand (1756) was the first to compile a *Flora Palestine* with the guidance of Carl von. Linné (Linnaeus). These were followed by the valuable book *Iter Palaestinuns* by Hasselquist published in the years 1749–1752, and the Flora *Aegyptiaco-Arabica* by Forsskääl published in 1775.

During the first half of the nineteenth century the flora and fauna of the Levant countries and the Sinai became the focus of attention of many eager explorers, botanists and collectors. Among these we may mention: Delile (1809–1812), Fresenius (1834), Aucher-Eloy (1843), Kotschy (1861), Boissier (1867–1888), Tristram (1884), and Hart (1891). During the last decades of the nineteenth century, a series of ecological investigations in Syria, Palestine and Sinai were carried by Post (1896), Bornmuller (1898, 1912), Post and Autran (1899), and Post and Dinsmore (1932–1933). These were followed by Eig (1938, 1948), Eig and Zohary (1939), Zohary (1932, 1941, 1962), Hassib (1951), Bodenheimer (1957), El-Hadidi (1969), Waisel (1971), Täckholm (1974), Danin (1983), Orshan (1985), Ayyad and Ghabour (1986), Zahran and Willis (1992, 2008) and Boulos (1960, 1995, 1997, 2009).

#### 1.8.2.2 Climate

The climate of the SW Mediterranean Coastal deserts is of the Mediterranean type. A major factor determining it is the western jet stream blowing over the Mediterranean from November to March. During the rest of the year, however, there is an easterly flow pattern in the upper atmosphere (Kendrew, 1961; Meigs, 1962; Fisher, 1978; Orshan, 1985). In winter a zone of small low-pressure depressions is formed which cause a succession of distributed cyclonic condition. In summer and very early autumn cyclonic depressions rarely affect the Middle East. A general account on the precipitation and temperature is given below.

### (i) *Precipitation*

The fact that rainfall in the SW Mediterranean coastal desert occurs under conditions of instability means that much of it may be heavy but of short duration and extremely capricious, both in periods of onset and distribution. In its more arid parts it may be patchy, with the size of patches sometimes not exceeding a few square kilometers. Average of rainfall are also misleading in many cases, since variation from the mean may exceed 100%. The coefficient of variation of annual rainfall in Israel increases with decreasing amounts of rainfall, reaching a figure as high as 73% in Eliat (Sharon, 1972).

Although the exact distribution of rainfall varies from year to year, the general trend is as follows. Rainfall occurs first in early autumn when the dry summer air masses are displaced by damper and more unstable currents from the west. Light showers may occur during September and October, with heavier and more prolonged falls at the end of October and November. The latter are generally followed by relatively fine periods. The real rainy season begins at the end of December and January or February are generally the months with maximum rainfall. By the middle of May rain has practically ceased, and no rain normally fall until September or October (Table 1.7).

Apart from rainfall, dew is also an important form of precipitation. The number of average annual dew nights measured in the Negev highlands for the years 1963–1966 amounted to 195 and the amount of dew to 33 mm, which is more than a third of the total amount of rainfall (Evenari et al., 1971, 1982). Although dew is generally ineffective in changing the soil moisture content, since it evaporates rather quickly, it has an important effect on lower plants and certain animals (Lange et al., 1970; Shachak et al., 1976).

### (ii) *Temperature*

Summer temperatures of the SW Asian Mediterranean deserts are generally high (Table 1.7). The warmest months are July and August with monthly average temperatures around 30°C. July is generally the hottest month inland, August in localities closer to the coast.

Winter temperatures are moderate; the coldest month is January with average winter temperatures reaching as high as 15°C, where they rarely reach the freezing point. Temperatures lower than freezing and occasional snow may occur, but they are generally of a short duration.

Of all criteria for climatic and biotic classifications of the SW Asian Mediterranean coastal desert, annual rainfall seems to be the best. The 350 mm isohyet corresponds roughly to the borderline between the arid and non-arid territories, whereas the 100 mm isohyet corresponds to the borderline between arid/desert and semi-arid/steppe territories.

**Table 1.7** Seasonal monthly range (means) and average annual rainfall and temperature of SW Asian countries having arid land territories in the Mediterranean basin (after Orshan, 1985)

| | Rainfall (mm) | | | | | Temperature (°C) | | | | |
|---|---|---|---|---|---|---|---|---|---|---|
| | Seasons | | | | | Seasons | | | | |
| Countries | Wn | Sp | Sm | Au | Average annual | Wn | Sp | Sm | Au | Average annual |
| 1. Aleppo (Syria) | 64–88 | 14–34 | 0.0 | Tr-44 | 373 | 3–9 | 10–21 | 26–32 | 11–25 | 18 |
| 2. Beersheba (Israel) | 39–44 | 3–32 | 0.0 | 0.3–27 | 204 | 12–14 | 14–23 | 25–26 | 18–25 | 20 |
| 3. El-Arish (Sinai, Egypt) | 16–18 | 1–16 | 0.0 | 0.0–14 | 95 | 11–13 | 19–28 | 23–26 | 17–24 | 19 |

Wn = Winter, Sp = Spring, Sm = Summer, Au = Autumn, Tr = Traces.

### 1.8.2.3 Floristic Regions and Vegetation Types

The south-west Asian Mediterranean coastal land is a meeting place of four-plant geographical (floristic) regions: Mediterranean, Irano-Turanian, Saharo-Arabian (all belonging to the Holarctic Region) and Sudanian (Palaeotropic Region). Various territories of these coastal lands are dominated by elements of one of the latter three regions with elements of the other more or less represented in them. It is therefore not easy to delimit the exact borders of each territory, and more or less gradual transition areas between them are quite common (Eig, 1932; Zohary, 1973).

The Mediterranean region, which lies outside the desert proper has, however, a marked influence on it, since its elements are well represented in the flora of its various territories. The Irano-Turanian region is, generally, characterized by its continentality and a relative low amount of rainfall. It is characterized by two stressful seasons limiting plant growth – a cold winter when low temperatures are the limiting factor for plant growth and development, and a hot dry summer when water is the limiting factor. However, in our area with its Mediterranean-type climate, the winter stress period is practically missing. The 100-, 150- and 350-mm isohyets run more or less parallel to the boundaries between the Irano-Turanian, Mediterranean and Saharo-Arabian territories, respectively. The vegetation characterizing the Irano-Turanian territory is a dwarf-shrub steppe.

The Saharo-Arabian region is characterized by its higher temperatures both in winter and summer, and its lower annual rainfall. Under the most extreme conditions the amount of rainfall approaches, and in certain years even reaches, zero. The vegetation dominating the Saharo-Arabian territories is generally restricted (except in sandy areas) to runnels and water-courses, where it is dominated by dwarf-shrubs with annuals as a background. Savanna-like vegetation type characterizes this climatic region (Orshan, 1985).

The Sudanian Region occupies part of Tropical Africa, north of the equatorial rainforest region, as well as the south-eastern corner of Arabia, southern Iran and Baluchistan. The Tropical flora of Palestine contains partly Omni-Sudanian and Eritreo-Arabian elements, and partly bi-and pluri-regional Tropical groups (Zohary, 1962).

Apart from the above mentioned floristic regions, Zohary (1962) considers that the aquatic elements (15 species) of this section of the Mediterranean coastal land come from the Eurosiberian Region.

Ecologically, the arid land territories of the SW Asian Lands of the Mediterranean coast have been classified by Orshan (1985) into five major vegetation types: Steppes (semi-deserts), deserts, Savanoid deserts, sand formations and salinas. The following is a brief account on these vegetation types.

#### (i) *Steppes (semi-deserts)*

Due to the destructive effects of man on vegetation in the Levant countires during millennia, it is not easy in many cases to determine the nature of the climax plant communities. This is especially true in semi-desert areas or in the transition zones between the typical Mediterranean areas and those of semi-deserts or steppes.

Remnants of trees which are found here and there in more protected areas suggest that a kind of a steppe-forest with widely scattered trees dominated these areas (Long, 1957). The main species seem to have been *Pistacia atlantica, Amygdalus* spp., *Rhamnus* spp., *Pyrus* spp., and others. Between these trees or shrubs, vegetation is suggested to have been more or less similar to that dominating the area at present, but richer in grasses which had been removed by selective grazing, Boyko (1949) suggested, however, that it had been dominated mainly by perennial grasses.

Remnants of such steppe-forests or groups of trees are found in the upper slopes of the Jordan Valley in Israel and Jordan (Zohary, 1973; Feinburn and Zohary, 1955), in central Sinai and on the higher mountains of southern Sinai where *Pistacia khinjuk* replaces *P. atlantica* on igneous rocks (Zahran and Willis, 1992). They are also found in the central Negev (Danin et al., 1975), in the Jordanian Desert (Zohary, 1940), in the Syrian Desert (Pabot, 1954) and in northwestern Iraq (Guest, 1966).

With increasing aridity the trees are replaced by shrubs – *Ziziphus lotus* in the upper Jordan Valley and the Syrian Desert (Zohary, 1962), and by *Lygos raetam* and *Rhus tripartita* in the upper Jordan Valley. Under higher amounts of rainfall the lower plant layer is dominated by herbaceous perennials or in more arid habitats by dwarf-shrubs or half-shrubs. These plant communities are generally rich in species and of relatively high cover. Among the perennial herbaceous plants *Echinops polyceras, Psoralea bituminosa* and *Carlina* spp., as well as *Hordeum bulbosum*, should be mentioned. The plant communities dominated by these plants occupy generally phosphorus-rich soils and cover the greater part of Golan Heights and the upper Jordan Valley. On such soils this herbaceous marginal vegetation invaded the more humid areas after the destruction of their climax vegetation producing the best grazing areas in the region. It is suggested that the continuous grazing over millennia also prevented the regeneration of *Pistacia atlantica* and other components such as species of *Amygdalus, Pyrus, Crataegus* and others of the supposed climax communities of the area. With increasing aridity the following are the species which dominate the main dwarf-shrub communities:

*Sarcopoterium spinosum*. Although this is a Mediterranean plant it dominates marginal plant communities together with various co-dominants such as *Thymelaea hirsuta, Ononis natrix, Phlomis brachyodon, Noaea mucronata, Salvia syriaca* and others in the transition zone between typical Mediterranean vegetation and the semi-desert (Feinburn and Zohary, 1955; Danin et al., 1975).

*Artemisia herba-alba*. This dominates extensive areas in the Negev, the Jordan Valley, the Tih Plateau in central Sinai, the upper elevations of the igneous rock mountains in southern Sinai where it is accompanied by *Pyrethrum santolinoides*, and the Jordanian and Syrian deserts (Pabot, 1954). With increasing aridity in the northern Negev *Artemisia* is accompanied successively by *Thymelaea hirsuta, Noaea mucronata, Salvia lanigera* and *Gymnocarpos decander* (Danin et al., 1975).

*Poa bulbosa* is a constant, and in many cases high-cover, companion of the *Artemisia herba-alba* community. It is reported as a dominant plant of extensive areas in the

Syrian Desert after eradication of the shrubs which, apart from being grazed, are used extensively for fuel.

*Hammada scoparia*. The plant community dominated by this species generally occupies loessial and loess-like plains as well as plateaux and sometimes hillsides under the most arid steppe conditions in central Sinai, the northern Negev, and the Syrian Desert.

*Salsola villosa* and *Reaumuria Palaestina*. The plant community dominated by *Salsola villosa* generally occupies soils derived from chalks and marls in the Jordan Valley; so do plant community dominated by *Reaumuria palastina*, which occupies similar soils in the Jordan Valley and the northern Negev in probably more saline habitats (Zohary, 1973; Danin et al., 1975).

*Achillea fragrantissima*. Small-to medium-sized wadis and runnels in the steppes of the Negev and the Judaean Desert are dominated by the *Achillea fragrantissima* community. The larger wadis in the Negev and northern Sinai are occupied by the *Retama raetam-Thymelaea hirsuta* community. Other communities, dominated by *Achillea fragrantissima* and accompanied by *Zilla spinosa*, occupy the wadis in the higher elevations of southern Sinai on igneous rocks (Orshan, 1985).

*Achillea santolina*. In the Negev Desert with rainfall amounts of 150–300 mm where dryfarming cultivation is practised, a community dominated by *Achillea santolina* is common in the fields (Zohary, 1973; Danin et al., 1975).

### (ii) *Deserts*

With increasing aridity below 80–100 mm of annual rainfall – the vegetation cover decreases until it becomes contracted (Monod, 1958) and restricted to more favourable habitats, e.g. runoff-fed depressions and runnels, rocky habitats and steep northern slopes. Under such conditions the effect of parent rock increases in importance. Accordingly, the differences between the plant communities dominating soils derived from dolomite and limestone, and soils derived from chalks and marls, are greater than under less arid conditions.

The main non-sandy and non-granitic desert areas in the SW Asian Mediterranean Coastal desert are found in Sinai, Israel and also, to a more limited extent, in Syria. The hard rocky hills in the Negev and Sinai, made up mainly of limestone and dolomite, are dominated by *Zygophyllum dumosum*, accompanied in various localities by *Reaumuria* spp., *Gymnocarpos decander* and other species. In the transition areas between deserts and semi-deserts it is also accompanied by *Artemisia herba-alba* (Zohary, 1962).

Plant communities dominated by *Anabasis articulata* occupy the regs of central Sinai, the central Negev, Jordan and Syria. It is accompanied by various species, such as *Gymnocarpos decander, Zilla spinosa, Fagonia* spp. *Anvillea garcini, Asteriscus graveolens* and others (Danin et al., 1975; Zahran and Willis, 1992). *Halogeton alopecuroides* and *Anabasis setifera* dominate in the Negev and Sinai deserts on chalks and sometimes marls which are rather saline, and yield soils more saline than those developed from harder rocks. *Salsola tetrandra* is the leading species of the vegetation dominating the marly chalks in the lower Jordan Valley, the Negev and Sinai (Orshan, 1985).

### (iii) *Savanoid deserts*

Sudanian elements penetrate rather deeply northwards in the SW Asian Mediterranean coastal desert through the Rift Valley and the territories adjacent to it. This is probably because of the combination of higher temperatures, especially higher winter minima, on the one hand, and available water throughout the year in the deeper soil layers of the larger wadis on the other. Such a combination of high temperatures and available soil moisture allow trees like *Ziziphus spina-christi, Hyphaene thebaica, Acacia* spp., *Balanites aegyptiaca* and others as well as shrubs like *Calotropis procera, Capparis decidua, Ochradenus baccatus, Lavandula coronopifolia* and others, to penetrate the Rift Valley.

### (iv) *Sand formations*

Sand formations are considered to be more suitable for plant growth under arid conditions than non-sandy ones. Their plant cover and productivity are generally markedly higher. By contrast, under amounts of annual rainfall exceeding 350 mm, where rain water leaches the whole profile, the total amount of soil moisture available for land growth is lower in sandy than in non-sandy areas due to the lower water-holding capacity of the sand. Therefore, their total productivity is lower and they are considered to be worse habitats under such climatic conditions (Orshan, 1985).

The sand of the Mediterranean coastal plain, washed out by the Nile River and its tributaries, deposited in the Mediterranean Sea and subsequently carried by the prevailing current, redeposited on the coasts of northern Sinai and Palestine, and blown inland by the prevailing winds, has been leached of some of its mineral components. The following sandy habitats differ from each other in the nature of their vegetation (Kassas, 1955; Zohary, 1973; Orshan, 1985).

(a) *Mobile sand dunes*. The more or less mobile sand dunes are generally devoid of vegetation. However, the more protected bases of these dunes are occupied by the *Stipagrostis scoparia* community, which is poor in species and of low plant cover.
(b) *Sand flats*. The wider depressions between the dunes, as well as more extensive sand fields, are dominated by few plant communities. The most widespread of them in northeastern Sinai and the northern Negev is *Artemisia monosperma, Convolvulus lanatus, Panicum turgidum* and *Pennisetum divisum* as important components. In northwestern Sinai plant communities dominated by *Zygophyllum album* as well as by *Cornulaca monocantha* and *Convolvulus lanatus* prevail.

### (v) *Salinas*

Saline habitats are a common feature in the SW Asian Mediterranean deserts. The causes for salt accumulation are: (a) a water table which is high enough to cause an upward movement of salts and their accumulation in the upper salt layers;

(b) poor drainage; and (c) inundation by sea water along the coasts. Generally, the desert salinas are formed by various interactions of these factors which, together with topography, form a diversity of habitats and microhabitats differing from each other with respect to moisture and salt content, and occupied by various communities. These communities are generally poor in species, and throughout the area under discussion are made of more or less the same floristic stock. On the Mediterranean coast of Sinai the plant communities dominated by the following species are arranged in a sequence of belts according to salinity and moisture content (Orshan, 1985; Zahran and Willis, 1992, 2008).

(1) A highly saline and moist belt (a high water table of saline water) with *Halocnemum strobilaceum; and Arthrocnemum macrostachyum*;
(2) A moderately saline and moist belt (a high water table of brackish water) with *Suaeda vermiculata; and Frankenia hirsuta*;
(3) A slightly saline and moist belt (a high water table of brackish to non-saline water) with *Juncus maritimus; and Phoenix dactylifera*;
(4) A deeper water table of brackish to saline water with *Nitraria retusa*;
(5) An outer belt of salinas where soil salinity is low with *Zygophyllum album*.
    On the coast of the Dead Sea the following belts were formed with decreasing salinity (Zohary and Orshan, 1949).

(1) *Tamarix tetragyna* and *Arthrocnemum macrostachyum*;
(2) *Suaeda palaestina* on the northern and *S. monoica* and *Tamarix nilotica* on the southern coast;
(3) *Suaeda fruticosa, Seidlitzia rosmarinus, Atriplex halimus* or combinations;
(4) *Nitraria retusa;*
(5) Moist habitats with poor drainage and a high water table of brackish water with *Juncus rigidus, Phragmites australis, Inula crithmoides, Aeluropus litoralis* and others.
    Such salinas are sometimes fringed by plant communities dominated by what are probably saline ecotypes of *Prosopis farcta, Alhagi graecorum,* and *Desmostachya bipinnata*.
    The aquatic vegetation comprises 15 species (hydrophytes) classified under: submerged species, e.g. *Ceratophyllum demersum, Myriophyllum spicatum,* and *Potamogeton schweinfurthii*, floating species, e.g. *Lemna gibba,* and *L. minor* and emergent species, e.g. *Cyperus longus, Phragmites australis,* and *Typha domingensis*.

#### 1.8.2.4 Representative Coastal Areas

The following pages include an account on the vegetation ecology of two coastal areas in the arid land territories of the SW Asian Mediterranean Coast. The first area occurs in the Asian Mediterranean coast of Egypt: the Sinai Mediterranean coast, and the second one represents the Palestine-Israel Mediterranean coastal land.

## Sinai Mediterranean Coastal Area

### a. *The physical environment*

The Sinai Peninsula, the Asian extension of Egypt, is a triangular plateau in the NE part of the country with its base, extending along the Mediterranean Sea for about 240 km from Port Said eastward to Rafah on the border with Palestine. The apex of Sinai Peninsula, in the south, is called Ras Muhammed where the eastern coast of the Gulf of Suez meets the western coast of the Gulf of Aqaba.

The Peninsula (61,000 $km^2$ in area) can be divided ecologically into three regions: northern, central and southern. Having coastal lands extending along the Mediterranean Sea and those of the Gulfs of Suez and Aqaba of the Red Sea, the Sinai Peninsula will be considered in two chapters of this book. Its northern Mediterranean region is described in this chapter whereas the other regions will be included in Chapter 2.

The northern Mediterranean Coastal region of the Sinai is the eastern section of the Mediterranean coastal land of Egypt. It has an area of about 8,000 $km^2$ and is bordred by the northern limits of the central region in the south, the Mediterranean Sea in the north, Palestine in the east and the Suez Canal in the west. This coastal belt of the Sinai Peninsula consists of a wide coastal plain sloping gradually northwards. It is characterized by three geomorphological units: coastal sand dunes, Lake Bardawil and Wadi El-Arish. The coastal sand dunes (80–100 m high) is extending for several km landward forming a continuous series parallel to the sea. Abu Al-Izz (1971) stated that these coastal sand dunes absorb and store rain water, and the low lands between them are considered a permanent source of fresh water that can be tapped by digging shallow wells. The best quality and highest volume of water is in the delta of Wadi El-Arish (Hume, 1925). The water supply of the wells varies and its quantity depends upon rainfall. The depth of the wells can be a little as 3 m or as much as 60 m. Lake Bardawil is a shallow lake that extends for about 98 km along the western section of the Sinai northern Mediterranean coastal land. It is elliptical is shape with total area of about 6,900 $km^2$, and is not continuously covered with water because during summer it becomes separate ponds and small lakes. There is no silt since this lake is far from the old branches of the Nile Delta (Zahran and Willis, 1992). Lake Bardawil is highly saline because of its close connection with the Mediterranean Sea. The low sand bar which divides it from the sea is often covered by sea water.

Wadi El-Arish is important geomorphological unit of the northern Sinai. Its basin is about 20,000 $km^2$ with a length of about 250 km, narrowing in its upper reaches as it cuts across the El-Tih Plateau (the middle region of the Sinai), where the old pilgrimage road to Mekka used to be. Wadi El-Arish can be divided into 3 sections: upper, central and lower (coastal). The upper section is about 100 km with a gradient of 6:1,000; in the central section the wadi descends from 400 to 150 m in about 100 km, i.e. with a gradient of 2.5:1,000. The coastal section covers the final northern 150 km where the wadi has a gradient of 3:1,000.

Generally, the climate of the Sinai belongs to the dry province, that can be divided into two main climatic zones: arid and hyperarid. The arid zone occurs in the northern Mediterranean coastal region (Ayyad and Ghabour, 1986) with relatively shorter dry period (attenuated) and annual rainfall ranging between 100 and 200 mm, some times up to 304 mm in Rafah. The air temperature is up to 30°C in summer and as low as 9°C in winter. The mean annual maximum and minimum relative humidity are 79 and 56%, respectively.

**b. *Vegetation types***

The terrestrial vegetation types of the Sinai Mediterranean Coastal desert is described below under these titles:

(i) Vegetation of the coastal belt, and
(ii) Vegetation of shoreline-landward transects.

(i) *Vegetation of the coastal belt*

The natural vegetation of the coastal belt is very sparse; three main habitats can be recognized: Sabkhas, littoral sand dunes, plain dunes and inland dunes (Zahran and Willis, 1992). Sabkhas are present in the northern section near the sea, and can be distinguished into four basic types according to the distribution of plants: (1) salt-encrusted sabkhas; (2) wet sabkhas; (3) dry sabkhas; and (4) drift-sand-covered sabkhas. Salt-encrusted sabkhas have almost no vegetation owing to their extremely high salinities. Only about 2–5% of the area are wet sabkhas vegetated with a community dominated by *Halocnemum strobilaceum*, and with cover of 70–95% associated with two halophytes; *Arthrocnemum macrostachyum* and *Suaeda vera*. In the dry sabkhas, vegetation cover is about 5–10%; *H. strobilaceum* is also the dominant, associated with *A. macrostachyum, Cressa cretica, Juncus rigidus, Limoniastrum monopetalum, Phargmites australis, Suaeda vera* and *S. vermiculata*. In sabkhas covered with a sheet of drift sand, plant cover varies from <5 to 15%. These areas are vegetated by two communities dominated by *Zygophyllum album* and *Anabasis articulata*, with cover of 5–10% and 10–15% respectively. Other associates are *Cressa cretica, Cyperus laevigatus* and *Salsola kali*.

The littoral dunes are mostly in two parallel lines with lows (pans) between. These lows act as drainage basins where halophytes dominate. The vegetation of the littoral dunes is largely of patches of *Ammophila arenaria*. Although the cover may reach 50% within the patches, not more than 5% of the total area of the dunes is plant-covered (Kassas, 1955). Beside the dominant grass, the flora of littoral dunes commonly includes: *Eremobium aegyptiacum, Lotus arabicus, Moltkiopsis ciliata, Polygonum equisetiforme* and *Salsola kali*. Among other associates are *Atriplex leucoclada, Cressa cretica, Cyperus laevigatus* and *Juncus acutus* (as halophytes in the lows). Species occasionally present are *Artemisia monosperma, Astragalus tomentosus, Cyperus capitatus, Echinops spinosus, Elymus farctus, Euphorbia paralias, Pancratium maritimum, Silene succulenta* and *Thymelaea hirsuta*.

"Plain" dunes are sand drifts that are lower and less mobile than the littoral dunes. In these dunes *Artemisia monosperma* is dominant. This community is subject to intense human interference by cutting and grazing. In places far from human settlements, the cover may reach 70% or more. The contrast between the vegetation inside a barbed-wire fence and outside is very striking (Kassas, 1955). The flora of an *A. monosperma* community includes several characteristic species, e.g. *Haplophyllum tuberculatum, Neurada procumbens, Panicum turgidum, Pituranthos tortuosus* (= *Deverra tortusa*) and *Urginea maritima*. The last species is subject to selective cutting for its medicinal value. *Cynodon dactylon* is a common species everywhere. Stabilized mounds covered by *Lycium europaeum* are local. In one locality *Lagonychium farctum* is very abundant, growing on the leeward side of these dunes. Other associates are *Alhagi maurorum, Astragalus spinosus, A. tomentosus, Atractylis prolifera, Bassia muricata, Chrozophora verbascifolia, Citrullus colocynthis, Cleome arabica, Echinops galalensis, Eremobium aegyptiacum, Euphorbia terracina, Heliotropium luteum, Ifloga spicata, Launaea glomerata, Lotus creticus, Mentha* sp., *Stipagrostis plumosa, Tamarix aphylla* and *Ziziphus spina-christi*.

The vegetation of inland sand dunes depends on their history which influences the composition of the plant cover. Dunes formed on sabkhas are dominated by *Zygophyllum album* at one stage of development and may contain, at an advanced stage, some other halophytes, e.g. *Nitraria retusa*. The stabilized inland dunes of this type are, in general, dominated by *Panicum turgidum* with cover up to 60%. Associate species are *Anabasis articulata, Artemisia monosperma, Convolvulus lanatus, Cornulaca monacantha, Echiochilon fruticosum, Eremobium aegyptiacum, Fagonia arabica, Moltkiopsis ciliata, Noaea mucronata, Stipagrostis scoparia* (abundant) and *Thymelaea hirsuta* (abundant).

The second type of inland sand dune of the Mediterranean coastal area of Sinai is formed by the accumulation of sand on desert mountains (Kassas, 1955). At the final stage there are huge sand dunes with rocky centres. In this type *Panicum turgidum* is the dominant on the lower dunes whereas *Artemisia monosperma* dominates on the higher parts. *Anabasis articulata, Convolvulus lanatus, Noaea mucronata* and *Thymelaea hirsuta* are common associate species. Other associates include: *Aerva javanica, Asparagus stipularis, Asthenatherum forsskaolii, Echiochilon fruticosum, Fagonia arabica, Haplophyllum tuberculatum, Pituranthos tortuosus, Stipagrostis plumosa* and *Teucrium polium*.

The third type of inland sand dunes are those formed by the accumulation of sand on desert plains. In these dunes three communities have been recognized: one dominated by *Panicum turgidum*, one by *Stipagrostis scoparia* and a third co-dominated by *P. turgidum* and *S. scoparia*. *Panicum* is a good fodder plant and is, thus, subject to heavy grazing. *S. scoparia* is less grazed, partly protected by its occurrence on the higher parts of the dunes not in easy reach of animals. The average cover of the *P. turgidum* community is 40% and that of *S. scoparia* 50% (Kassas, 1955). Common plants are *Artemisia monosperma, Convolvulus lanatus, Fagonia arabica, Gymnocarpos decander* and *Thymelaea hirsuta*. The flora of the third community also contains *Citrullus colocynthis, Hyoscyamus muticus, Reaumuria hirtella* and *Salsola volkensii*.

(ii) *Vegetation of Shoreline-landward transect*

The landward successive communities of the Sinai Mediterranean coastal land are described in three representative transects set up at three sites: Rafah coast, Sheikh Zuwayid coast and Wadi El-Arish coast (Zahran and Willis, 1992).

*Transect 1.* This transect is in the most eastern part of the Sinai Mediterranean coast at Rafah. In the first zone, the sand dunes are vegetated by two communities dominated by *Ammophila arenaria* and *Silene succulenta*. The cover of the *Ammophila* stands is thin (5–10%); associates are *Pancractium arabicum, Silene succulenta* and *Tamarix nilotica*. In stands dominated by *S. succulenta*, the cover is 20–30%. *Acacia saligna* (semi-wild), *Ammophila arenaria, Pancratium arabicum* and *Tamarix nilotica* are the associate species. In the second zone of the transect (1 km landward) are huge sand dunes richly vegetated by the semi-wild shrub *Acacia saligna*. The ground layer is almost covered (about 75%) with the succulent xerophytic halophyte *Mesembryanthemum forsskaolii (= Opophytum forsskaolii)*. Other associates are *Casuarina stricta* (cultivated), *Cyperus capitatus, Nicotiana glauca, Phoenix dactylifera* (semi-wild). *Polygonum bellardii, Silene succulenta* and *Xanthium pungens*.

The third zone of the transect (6 km landward) comprises dunes dominated by *Artemisia monosperma*. This vegetation extends landward for a considerable distance (about 22 km: from 6 to 28 km south of the shore-line). In the northern stands, the cover is high (up to 80%), contributed mainly by the dominant xerophyte (about 50%) and partly by the abundant associate species: *Senecio desfontainei* (about 20–25%) and *Neurada procumbens* (about 5%). Other associates are, *Astragalus alexandrinus, Cyperus capitatus, Onopordum alexandrinum* and *Urginea maritima*, but these have neglibible cover. The cover of the *A. monosperma* community decreases gradually southwards. At km 25 south of the coast, the cover is about 70% contributed by the dominant (40–45%), *Stipagrostis scoparia* (about 20%) and *Neurada procumbens* (about 5%). *Senecio desfontainei*, abundant in the northern stand, is absent further south where *Silene succulenta* is commonly present. At 28 km landward, the cover of *A. monosperma* on the dunes is reduced to only 5%. *Eremobium aegyptiacum* is abundant. Other associates are: *Ononis serrata, Onopordum ambiguum* and *Stripagrostis scoparia*. In this zone *Thymelaea hirsuta* dominates in scattered patches in the low areas between the dunes.

*Transect 2.* This transect is at the coast of Sheikh Zuwayid (about 20 km west of Rafah) and extended for about 22 km from sea landward. Along the whole stretch of the transect are extensive sand dunes. In the beach zone the dunes are dominated by *Ammophila arenaria*, with cover ranging between 30 and 50%. *Salsola kali* and *Silene succulenta* are only associates. The dominance by *A. arenaria* continues landward for about 3 km. The cover decreases in the landward stands to 20–25% and the number increases to four (*Acacia saligna, Artemisia monosperma, Calligonum comosum* and *Silene succulenta*). Gradually *A. monosperma* increases landwards and becomes dominant, associated with an increase in the number of xerophytes. The cover of the *A. monosperma* community ranges from 25 is 35%, contributed mainly by the dominant. Associated species include: *Cynodon dactylon, Haplophyllum tube-rculatum, Lycium shawii, Lygos raetam, Onopordum ambiguum, Panicum turgidum, Tamarix nilotica* and *Thymelaea hirsuta*. In the inland sections of the transect, *Thymelaea hirsuta* is abundant in low areas between the dunes.

*Transect 3.* This transect is in the downstream section of Wadi El-Arish (about 45 km west of Rafah) extending for about 24 km from the sea landward. In the first zone the semi-wild *Phoenix dactylifera* vegetation has a cover of 10–20%. Common plants are *Echinops spinosus*, *Mesembryanthemum crystallinum, Pseudorlaya pumila* and *Silene longipetala*. The dune formation after about 1 km landward is vegetated with *Ammophila arenaria* (cover <5%). Associates include: *Artemisia monosperma, Cutandia dichotoma, Echinops spinosus, Erodium oxyrhynchum subsp. bryoniifolium, Lygos raetam* and *Moltkiopsis ciliata*. The dunes in the third landward zone of the transect are dominated by *Tamarix nilotica* with cover of up to 40%. Asssociates are *Artemisia monosperma, Cornulaca monacantha, Lobularia libyca, Ononis serrata, Silene succulenta, Stipagrostis ciliata* and *Thymelaea hirsuta*. As in the other two transects in the southern section, the dunes are dominated by *Artemisia monosperma*. The cover of this community thins gradually from about 40% in the northern stands to <20% in the southern ones. The number of associated species is, relatively, high and includes: *Asparagus stipularis, Carthamus tenuis, Cornulaca monacantha, Cyperus capitatus, Echinops spinosus, Heliotropium digynum, Herniaria hemistemon, Hyoscyamus muticus, Ifloga spicata, Lycium shawii, Neurada procumbens, Panicum turgidum* and *Thymelaea hirsuta*. The dominance of *A. monosperma* on the dunes continues southward but the cover is thinner. The areas between the dunes support a community dominated by *Thymelaea hirsuta* with 35–40% cover contributed mainly by the dominant shrub and partly (about 5%) by the common associate *Hammada elegans*. Other associates include *Artemisia monosperma, Cornulaca monacantha, Fagonia arabica, Hyoscyamus muticus, Panicum turgidum, Peganum harmala, Zilla spinosa* and *Zygophyllum album*. In some patches of the wadi bed, *Fagonia arabica* dominates, with thin cover (<5%). *Anabasis articulata, Cleome africana, Farsetia aegyptia* and *Pancratium sickenbergeri* are the associate xerophytes of this community.

## Palestine-Israel Mediterranean Coastal Area

### a. *The physical environment*

Palestine-Israel is the historical geographical area occurring in the SW corner of the Mediterranean Basin. It borders to the east and south by the steppes and deserts of the Near East, which are partly included within its boundaries. It is an area of striking contrasts in its topography and climate (Zohary, 1962). Its highest mountain peak, the Edom (Transjordan), rises to 1,700; its lower part, in the Dead Sea depression falls to 396 below sea level. The northern district enjoys a typical Mediterranean climate while its southern parts are dry desert.

Topographically, Palestine-Israel area has four belts: the coastal plain, the western mountains, the Jordan Valley with its southern continuation to the Gulf of Aqaba and the Transjordan plateau. These topographic belts show particular geomorphological, climatic and habitat features which have their impacts on their natural vegetation types.

The coastal plain broadens towards the south and attains its maximum (60 km) in the Negeve. In its greater part, this coastal plain is subject to a true Mediterranean climate and harbors a flora and vegetation peculiar to that climate. It is here that the main citrus groves are centered. Even since Biblical times this plain has been transversally subdivided into four districts: the Negev Plain (in the south), and the Philistia, Sharon, and Acre plains (in the north). With a gradual deterioration of climatic conditions from north to south, each of these districts exhibits certain biogeo-graphical features of its own. Edaphically, one can distinguish four well-defined longitudinal zones in the coastal plain: the zone of mobile sand dunes, that of interrupted calcareous-sandstone hills, the belt of sandy clays ("red sands"), and the zone of alluvial-colluvial heavy soils. Impeded in their way to the sea by dune and sandstone ridges, some of the latitudinal watercourses which cross the coastal plain have turned considerable stretches of land into swamps and marshes. These have dried up only recently and are now being utilized for agriculture.

The western mountain region constitutes a belt of considerable width. It extends from the southern foot of Mount Lebanon to the Desert of Sinai and reaches its peak (1,208 m) in the north (Jebel Jarmak in upper Galilee). While the western, gently sloping side of these mountain ranges has a Mediterranean climate and vegetation, the abrupt eastern slopes, which face the Jordan Valley, are mostly desert or semi-desert. Of the latitudinal valleys that interrupt the continuity of the western mountain region, the largest are the Esdraelon Plain in the north and the Plain of Beersheba in the south. The Negev, Judea, Samaria, and Galilee are the four main districts into which that region is commonly divided. The Negev is the largest and most desolate part of Israel, a desert with a rough topography. The other districts, though fairly well distinguishable biogeographically, share many features. The rocky, heavily eroded landscape, with its many ravines and shallow valleys, is common to all three. The so-called Judean Desert occupies the eastern parts of Judea and Samaria.

The Jordan Valley, is the most significant topographic feature. It is part of a rift valley, extending from the Orontes River in Syria to the Gulf of Aqaba and continues further into the Red Sea and the continent of Africa. In Israel it ranges in elevation from –396 m in the Dead Sea region to +200 m in the Dan Valley (Huleh Plain); mountains border it on both sides. Its southern part comprises deserts, salinas and tropical oases, while in its northern part there are swamps and stretches of fertile land situated within a Mediterranean wood climax area.

The Transjordan Plateau is higher in its northern and southern edges than in its central part. To the east this plateau merges gradually into the Syrian Desert. Its steep western escarpments are crossed by a series of latitudinal rivers, which empty into the Jordan Valley and the Dead Sea and divide the comparatively narrow western strip of the plateau into transverse districts known since Biblical times as Edom, Moav, Ammon, Gilead and Golan. Edom, the southernmost district of this plateau, is marked by vast sand deserts that border on the highest mountain ranges, which reach a height of 1,700 m. These afford conditions for the extension of Mediterranean woods as far south as the latitude of Cairo (Zohary, 1962).

The climate of the Plestine-Israel area as a whole is of the Mediterranean type, marked by a mild, rainy winter and a prolonged dry and hot summer. Geographical latitude, altitude, the blocking effect of mountain ranges, and distance from the Mediterranean Sea are among the factors which modify this climate. The effect of latitude manifests itself in the abrupt north-to-south decrease of the annual amount of rainfall, so that within a range of about four latitudinal degrees rainfall drops from about 1,000 to 25 mm/year. Temperatures, on the other hand, increase from north to south; the mean annual temperature rises from just below 16°C in the north to approximately 23°C in the extreme south. In a west-to-east direction, annual rainfall and mean temperatures undergo similar but less regular changes. This is because of the interference of the Israel and Jordan mountain ranges. As a result of their interception of rains, part of the Jordan Valley is a rain-shadow desert. In addition, the mountain ranges limit the tempering influence of the Mediterranean Sea to a narrow strip, leaving the greater part open to a wider range of seasonal and diurnal temperatures (Zohary, 1962).

Palestin-Israel climatic zones, according to the amount of annual rainfall are: a sub-humid zone with 1,000–400 mm, a semi-arid zone with 400–200 mm and an arid zone with only 200–25 mm.

Although plant life is not eliminated anywhere in the country by extremes of temperature, plant distribution is greatly influenced by thermal conditions. For instance, a series of tropical plants, that thrive in the southern part of the Jordan Valley, do not advance northward to where winters are colder. Among them are *Maerua crassifolia, Moringa peregrina, Acacia raddiana, Balanites aegyptiaca, Salvadora persica, Abutilon pannosum, Calotropis procera* and *Sebestena gharaf.* Very puzzling is the fact that most of these plants are limited to very low altitudes (300–400 m below sea level), whereas in the central Sahara they reach a considerable height on the mountains (Maire, 1933, 1940). *Balanites aegyptiaca* has been recorded at 13,000 m, *Salvadora persica* at 1,500 m, *Calotropis procera* at 1,800 m, and *Maerua crassifolia* at 1,900 m. Other examples of the part played by minimum temperatures on plant distribution include: *Quercus ithaburensis*, not found above 500 m; *Ziziphus lotus*, which occurs only up to a height of 250 m; and a whole series of species characteristic of the coastal plain, which do not occur in the mountain belt because of the lower temperatures during the winter. Other typical Mediterranean plants (*Laurus nobilis, Spartium junceum, Pistacia lentiscus* etc.) are altogether lacking in the Mediterranean territory of Transjordan, probably also because of the minimum winter temperatures.

The gradient of decreasing rainfall in the region is much steeper in its northern than in its southern half, yet in the south the decrease is much more decisive on plant life than in the north. In the south, even the smallest deviation from the average makes itself felt in density and development of vegetation, while in the north differences amounting to as much as 200 mm/year are hardly reflected in the general appearance of the vegetation. This is self-evident from the fact that depending on soil properties the 100–50 mm isohyetes constitute the lower rain limit of plant life.

The Palestine-Israel area, with the great deserts of Asia and North Africa on its borders, offers the plant geographer an ample opportunity to follow distribution features in the flora and vegetation along rainfall gradients. Hundreds of Mediterranean species are arrested in their southward move by diminishing rainfall. Some of them reach the southern boundary of the Mediterranean territory, while others are detained long before it. The same is true for many plant communities: for example, the *Quercus calliprinos-Pistacia palaestina* maquis association reaches its terminus at the 400–350 mm isohyet the widespread typical Mediterranean dwarf shrub association, *Poterium spinosum*, is detained at the 350–300 mm isohyet; the *Artemista herba-alba* at the 200–150 mm isohyet; and *Zygophyllum dumosum* at the 100–75 mm line. Obviously, local topography and nature of soil causes these boundaries to move slightly on either side of the isohyets.

Among the isohyets outstanding in their biogeographical importance are the 400–350 mm isohyet, the lower limit of Mediterranean forest and, of stable dry farming, and the 100–500 mm isohyet, which constitutes the climatic threshold of plant life in these deserts. Beyond this limit, however, plant life does not cease entirely but becomes wholly dependent on topography. Wadis and depressions, which collect water draining off from the surrounding area, support a fairly well-developed vegetation. Even depressions hardly perceptible to the naked eye form favorable sites for vegetation. This is also why deserts with rough topography are less bare than those dominated by smooth unbroken plains. Equal importance must be ascribed to physical soil properties in deserts. Moisture-absorbing surface layers, such as sand and fine gravel are much more favorable to vegetation than exposed, heavy-textured soils.

Another important feature of rainfall affecting plant life is the instability of its total annual amounts. These fluctuations may be so considerable that 1 year's rain may be no more than a fraction of another year's total. This is especially true for the arid and subarid parts of Palestine-Israel area, as seen from the following figures recorded for the Negev (Ashbel, 1945); Gaza (33 years of measurements) – minimum 238.1 mm, maximum 810 mm; Beersheba (28 years) – minimum 129.8 mm, maximum 336.1 mm; Auja (Nitzana) (13 years) – minimum 10.4 mm, maximum 284.5 mm. In these arid areas a falling short of the average by as little as 20–40 mm may cripple vegetation and be fatal also to the dry farming scattered in lowlands and depressions. The instability of the annual rainfall in the arid zones makes the desert "blossom" in 1 year, restricts the annual vegetation to wadis and depressions in others, and almost annihilates plant life in years of extreme drought. Moreover, where a number of dry years follow each other consecutively, certain desert dwarf shrubs desiccate and die off. Thus, the floristic composition of a plant community may alter drastically from year to year.

The rainy season lasts from October to May, but about 75% of the total amount of rain falls between December and February. The amounts of rain in particular months varies greatly from year to year. For instance, the January total for Ramle was 300 mm in 1 year and 3.2 mm in another; equivalent reading for Jerusalem are 367 and 3.0 mm, the northern shore of the Dead Sea had a January total of 58 mm in 1 year and 4.0 mm in another. Such fluctuations favor the development of certain plants in particular years. The results is the so-called "flower year", a very striking phenomenon in which certain annual species dominate the landscape at irregular intervals of a few to many years. Annual fluctuations in the rainfall totals of particular months may also bring about prolonged periods of drought within the rainy season. This can be harmful to vegetation and is also apt to alter the species composition of plant communities from 1 year to another.

There is a gradual decrease in the annual, monthly, and diurnal averages of relative humidity from north to south and from west to east throughout the whole area, excluding the coastal plain. This plain also has a high rate of humidity in the south, which presumably accounts for the high growth rate of many tropical and subtropical plants and certainly also contributes to the comparative fertility of the coastal Negev. The average relative humidity of a summer midday amounts to 65% in Natanya (Sharon Plain) and to 75% in the coastal Negev, while in Jerusalem it is only 35 and 40% in the Dead Sea Valley (Ashbel, 1951).

Dew plays an important role in plant life. According to Ashbel (1951), there are over 200 dew nights in the coastal plains and much fewer in the interior parts. Duvdevani (1953) stated that the amount of the yearly dew precipitation varied from 4.7 mm/year in Jericho (Lower Jordan Valley) to 29 mm/year on Mount Carmel.

### b. *Plant life*

- *Floristic analysis*

The flora of Palestine-Israel Mediterranean Coastal lands comprises about 2,250 species belonging to 718 genera out of these (1,506 species, 66.9%) are uni-regional elements with bi- and pluri-regional elements (743 species, 32.6%) making up most of the rest. The uni-regional elements are distributed between the five floristic (phytogeographical) groups, as follows: 863 species are Mediterranean elements, 299 Saharo-Sindian, 309 Irano-Turanean, 15 Euro-Siberian and 20 Sudanian. The taxa of the bi-and pluri-regional groups are classified under five subgroups: subtropical (415 species), subtropical-boreal (112 species) subtropical – tropical (69 species), boreal – tropical (85 species) and tropical, 53 species (Zohary, 1962).

### A. The uniregional groups

It is worth that, taxa belonging to the Mediterranean, the Irano-Turanian and Saharo-Sindian regions are confined to particular territories of Palestine-Israel area. The other elements are scattered throughout the area or occupy special habitats within those territories.

The following is a short account on these groups.

1. *The Mediterranean element.* Among the Mediterranean element of Palestine-Israel flora, the following six groups have been distinguished by Eig

(1931–1932): sub-Mediterranean, omni-Mediterranean, east Mediterranean, west Mediterranean, north Mediterranean, and south Mediterranean.

(i) *The sub-Mediterranean group.* About 160 Mediterranean species also penetrate into some countries adjacent to the Mediterranean region. Particularly interesting are the Mediterranean littoral species which recur in the Atlantic sector of the Eurosiberian region, e.g. *Glaucium flavum, Cakile maritima, Euphorbia peplis, E. paralias, Eryngium maritimum, Crithmum maritimum, Statice limonium, Centaurium maritimum, Inula crithmoides, Diotis maritima, Juncus subulatus,* and *Pancratium maritimum.* Other examples include: *Thesium humile, Adonis autumnalis, Raphanus raphanistrum, Fumaria capreolata,* and *Ridolphia segetum.* Noteworthy among the sub-Mediterranean species, recurring in some Irano-Turanian countries, are segetals and components of primary vegetation preserved in Mediterranean enclaves of the Irano-Turanian region, eg., *Pinus brutia, Juniperus oxycedrus, Psoralea bituminosa, Putoria calabrica, Stipa aristella,* and *Oryzopsis coerulescens.*

(ii) *The omni-Mediterranean.* This group, which also comprises about 160 species, is widespread over the Mediterranean region. Many of them are leading plants in local plant communities, e.g. *Pinus halepensis, Juniperus phoenicea, Calycotome villosa, Ceratonia siliqua, Pistacia lentiscus, Rhamnus alaternus, Cistus villosus, Lavandula stoechas, Thymus capitatus,* and *Viburnum tinus.* Many species of this group are confined to the coastal plain. Examples of these are *Matthiola tricuspidata, Ononis variegata, Orlaya maritima, Statice sinuata, S. virgata, Alkanna tinctoria, Ajuga iva, Crucianella maritima, Scleropoa maritima, Avena longiglumis, Sporobolus arenarius, Cyperus mucronatus, Leopoldia maritima*, and *Narcissus serotinus*. There are no deciduous trees among the Omni-Mediterranean group.

(iii) *The east Mediterranean (including the sub-east Mediterranean).* This group is the most prevalent element in the flora of Palestine and Israel and includes the majority of the endemic species. It comprises 485 species, many of them are important components of the vegetation, for example the lead species of the maquis: *Quercus calliprons, Q. boissieri, Q. ithaburensis, Platanus orientalis, Prunus ursina, Cercis siliquastrum, Pistacia palaestina, Rhamnus palaestina, Acer syriacum, Arbutus andrachne*, and *Styrax officinalis*, as well as the dominant species of the garigue and batha formations: *Poterium spinosum, Euphorbia thamnoides, Hippomarathrum boissieri, Anchusa strigosa, Alkanna strigosa, Salvia triloba, Phlomis viscosa. Thymbra spicata, Teucrium creticum, Majorana syriaca* etc. Within this group also are the dominant species of the litho- and chasmophytic communities: *Arenaria graceolens, Dianthus pendulus, Micromeria serpyllifolia, Ballota rugosa. Hyoscyamus aureus, Michauxia campanuloides, Varthemia iphionoides, Centaurea speciosa,* and others. In addition, many local weeds belong to this group, e.g. *Silene crassipes, Medicago*

*galilaea, Euphorbia cybirensis, Eryngium creticum, Tordylium aegyptiacum, Exoacantha heterophylla, Cachrys goniocarpa, Astoma seselifolium, Convolvulus hirsutus, Heliotropium villosum, H. bovei, Molucella laevis, Carthamus tenuis, Centaurea verutum*, and *Cynara syriaca*.
A characteristic feature of this group is the high number of geophytes (over 12%), among them *Lilium candidum, Hyacinthus orientalis*, eight species of *Iris*, four of *Colchicum*, and four of *Crocus* are worthy of mention.

(iv) *The west Mediterranean*. This group comprises only 14 species and plays almost no part in the vegetation cover of the country (Eig, 1931–1932). One may consider some of them, such as *Ophioglossum lusitanicum* and *Euphorbia dendroides*, as being relics of a more humid period, during which local climatic conditions were more favorable for mesic species than they are today.

(v) *The north Mediterranean group* comprising 30 species, and *(vi) the South Mediterranean group*, with 14 species, are vegetaionally unimportant.

2. *The Saharo-Sindian element*. In the Palestin-Israel area, the Saharo-Sindian element comprises only 299 species, i.e. 13% of the total flora. Eig (1931–1932) has distinguished four groups among this element: the Omni-Saharo-Sindian, and the West, East, and Middle-Saharo-Sindian. He refers the bulk of the Palestine Saharo-Sindian element to the latter one. Although the tripartition of the Saharo-Sindian region is still to be re-examined with regard to its lines of demarcation, it is obvious that there are three distinct floristic centers within this region.
According to Eig (1931–1932), Maire (1933, 1940), the Saharo-Sindian flora is less autonomous in its Origin than are those of the neighboring regions, and many of its species and genera are derived from the Mediterranean, Sudanian, and partly also from the Irano-Turranian stock. Examples of Mediterranean derivatives are species of the genera: *Paronychia, Silene, Matthiola, Medicago, Lotus, Ononis, Erodium, Daucus, Origanum, Thymus, Anthemis, and Picris*. The genera *Capparis, Cleome, Caralluma, Gomphocarpus, Trichodesma, Iphiona, Varthemia*, and *Tetrapogon*, are examples of genera of Sudanian or otherwise tropical origin. The Irano-Turanian derivatives are represented by species of the genera *Calligonum, Suaeda, Salsola, Haloxylon (Hammada), Anabasis, Astragalus, Tamarix, Heliotropium, Onopordum, Echinops*, and *Carthamus*. The South African flora is suggested by the occurrence of the genera *Aizoon, Mesembryanthemum, Notoceras, Caylusea, Neurada, Citrullus, Ifloga* and *Aristida*. However, there are a fair number of genera autochthonous in the Saharo-Sindian region, e.g. *Gymnarrhena, Pteranthus, Gymnocarpos, Sclerocephalus, Anastatica, Zilla*, and *Savignya*.
The most important species in the Saharo-Sindian flora of Palestine-Israel area are: *Calligonum comosum, Anabasis articulata, Suaeda asphaltica, S. vermiculata, Salsola tetrandra, S. rosmarinus, Haloxylon persicum, H. salicornicum, Chenolea arabica, Traganum nudatum, Halogeton alopecuroides, Gymnocarpos fruticosum, Zilla spinosa, Retama raetam (= Lygos raetam), Fagonia mollis, Zygophyllum dumosum, Nitraria retusa, Convolvulus lanatus,*

*Danthonia forsskaolii, Aristida scoparia,* and many others. The low percentage of geophytes, amounting to about 4% among the Saharo-Sindian flora, is remarkable.

3. *The Irano-Turanian element.* The Irano-Turanian element, including the sub-Irano-Turanian group, comprises 309 species, i.e. more than 13% of the total flora of the area. When the 406 bi-or triregional species are added to that number, the affinity of the local flora to that of the Irano-Turanian region becomes strikingly apparent. In contrast to Mediterranean and Saharo-Sindian species, the Irano-Turanian taxa are rather restricted and there are scarcely any instances of Omni-Irano-Turanian expansion. The bulk of this element in the Palestine-Israel area belongs to the Mesopotamian, the Irano-Anatolian, and the Mauritanian Steppes groups.

(i) *The Mauritanian steppe subregion.* This occupies a belt of considerable width in North Africa, notably in its western part. The flora of this subregion is also marked by a series of generic and specific endemics. In Palestine the Mauritanian steppes element is represented by about 30 species, many of them dominants in the local vegetation. Examples are *Ephedra alte, Noea mucronata, Haloxylon articulatum, Salsola vermiculata, Pistacia atlantica, Rhus tripartita, Zizyphus lotus, Salvia lanigera, Marrubium alysson, Linaria aegyptiaca, Artemisia herba-alba,* and *Achillea santolina.* There is also a striking resemblance between the vegetation of the local Irano-Turanian territory and that of certain parts of the Mauritanian Steppes.

(ii) *The Mesopotamian subregion.* This comprises the Irano-Turanian territory of the Syrian Desert, the whole Jezireh (upper Meso-potamia), and parts of the plains in southern Anatolia and south-western Iran. This subregion has a considerable number of endemics and some particular plant communities by which it is fairly well distinguished from the above-mentioned subrgion. In comparison with the Irano-Anatolian and even the Turanian subregions, it is poor in species.
There are 100 or more species which represent the Mesopotamian element and the majority of the Irano-Turanian endemics in Palestine-Israel area show Mesopotamian affinities. The flora of this subregion includes a remarkable proportion of herbaceous segetals which have also penetrated into the Mediterranean territory. Others of its species associates of various herbaceous steppe communities, but there are scarcely any leading species among them. There are, however, a considerable number of species common to both above subregions, so that their separation into two subregions is still tentative.

(iii) *The Irano-Anatolian Subregion.* This comprises mainly the mountainous area of interior Anatolia, Armenia, and the Iranian Plateau (exclusive certain parts of eastern Iran and western Afghanistan). Parts of southern Transcaucasia may also be included within the Irano-Anatalian subregion.

Its outlines coincide with those defined by Boissier (1867–1888) as "Sous-region de Plateaux".
Characteristics of the subregion are its high mountain ranges alternating with plains and plateaus of varying altitudes and badly drained valleys and basins. It harbors several different elements of the flora of the Irano-Turanian region and is one of its largest centers of speciation in the region. Many endemic species belonging to the genera *Astragalus, Acanthophyllum, Onobrychis, Hedysarum, Acantholimon, Cousinia, Centaurea, Jurinea,* and *Helichrysum* are recorded from here. This subregion is represented in Mount Hermon with species such as: *Atraphaxis spinosa, Pterocephalus pulverulentus, Buffonia virgata, Delphinium antheroideum, Erysimum crassipes. Astragalus bethleemiticus, A. deinacanthus, Euphorbia macroclada, Daphne linearifolia, Scutellaria fruticosa, Thymus syriacus, Phlomis orientalis, Pyrethrum santolinoides (= Tanacetum santolinoides),* and *Cousinia hermonis*. Most of these are confined to the higher altitudes of Transjordan and the Negev.
The remaining subregions of the Irano-Turanian region are also poorly represented. The Turanian subregion (Aralo-Caspian sensu stricto) shows some floristic affinity to Palestine-Israel area but rather to its Saharo-Sindian than to its Irano-Turanian flora. Thus, a series of genera, which have their highest concentration of species in the Turanian sub-region, are characteristic of halic (saline) and sandy habitats in the Saharo-Sindian territory of Palestine-Israel area, e.g. *Calligonum, Haloxylon, Suaeda, Salsola, Anabasis, Nitraria, Zygophyllum, Tamarix*, and others.

4. *The sudanian elements*. This is represented by 20 species. Many of them are trees and shrubs, such as *Maerua crassifolia, Moringa peregrina, Acacia tortilis, Balanites aegyptiaca*, and *Ziziphus spina-christi*. Most of them are conspicuous in the tropical oases of the Jordan Valley. Other species include: *Acacia albida, Loranthus acaciae, Indigofera argentea, Tephrosia apollinea, Acacia laeta*, and *Capparis cartilagina*.

**B. The bi- and pluriregional groups**

The previously discussed groups are uniregional, i.e. each group is more or less confined to a single plant geographical region. The sum total of uniregional species enumerated so far amounts to 1,506, i.e. 67% of the flora. The remaining 33% of species of the local flora have a wider range of distribution and extend over two or more plant geographical regions. The presence of bi- and pluriregional groups in Palestine-Israel area is not necessarily the result of its particular phytogeographical position. Each plant geographical region harbors, in addition to the uniregional species which constitute its floristic element, a considerable number of bi- and pluriregionals, or interregionals, as they may generally be termed.

There are roughly 200 segetal and ruderal species among the bi- and pluriregional. The Borealo-Tropical group comprises many of them, e.g. *Chenopodium album, C. murale, Urtica urens, Polygonum aviculare, Amaranthus retroflexus,*

*Portulaca oleracea, Stellaria media, Capsella bursa-pastoris, Anagallis caerulea, Convolvulus arvensis, Cynodon dactylon*, and others. Many more segetals and ruderals are included in the Borealo-Subtropical group, of which the following may be mentioned: *Sisymbrium officinale, Lepidium latifolium, Geranium dissectum, G. molle, Erodium cicutarium, Euphorbia helioscopia, E. peplus, Malva sylvestris, M. neglecta, Conium maculatum, Galium aparine, Senecio vulgaris, Eragrostis minor, Lolium temulentum,* and *Hordeum murinum.*

Among the bioregional, and especially among the Mediterrano-Irano-Turanian group, are found the largest number of segetals and ruderals. Examples are: *Polygonum equisetiforme, Silene conoidea, S. longipetala. Glaucium corniculatum, Biscutella didyma, Diplotaxis erucoides. Neslia apiculata, Eruca sativa, Rapistrum rugosum, Calepina irregularis, Medicago orbicularis, M. hispida, Trifolium resupinatum, Hymenocarpus circinnatus, Vicia narbonensis, V. peregrina, Lathyrus marmoratus, L. annuus, L. erectus, Erodium malacoides, Malva nicaeensis, M. parviflora, Ammi visnaga, Heliotropium europaeum, Marrubium vulgare, Galium tricorne, Notobasis syriaca, Silybum marianum, Cichorium pumilum, Bromus scoparius, B. macrostachys,* and *Gladiolus segetum.*

Among the halophytic interregionals are species of *Suaeda, Salsola, Frankenia* and *Salicornia*, as well as of *Cressa cretica, Sonchus maritimus, Sphenopus divaricatus*, and *Aeluropus littoralis* (Zohary, 1962).

The 160 endemic species represent about 7.1% of the total number of the flora of Palestine-Israel area (Zohary, 1962) which is rather high when compared with that of Egypt (about 2.9%, Boulos, 1995) but very low when compared with that of the Balkan Peninsula (about 26%). According to Zohary (1962), there is no proportionality between the size of plant families and the number of endemic species: for example the total number of species in Leguminosae, Gramineae, Cruciferae, Caryophylleceae, Chenopodiaceae, Iridaceae are: 268, 198, 124, 97, 60, and 23 respectively, whereas the endemics of these families are: 21, 10, 9, 5, 4 and 8 species, respectively. Among the species giving rise to endemic varieties are: *Quercus calliprinos, Q. ithaburensis, Polygonum equisetiform, Erucaria myagroides, E. boreana, Matthiola longipetala, Lygos raetam, Tamarix gallica, Aegilops peregrina* etc. According to Zohary (1962), some endemics of the Palestine-Israel area show clear taxonomic relations to sister species occurring in or outside the area, e.g. *Reheum palaestinum, Dianthus judaicus, Psoralea flaccida, Pimpinella petraea, Eremostachys transjordanica, Campanula hierosolymitana, C. aaronsohnii, Calendula pachysperma, Cousinia moabitica, Scorzonera judaica, Poa hackelii* and *P. eigii*. There are also endemics that can be considered coastal vicariads (representatives) of corresponding taxa of the interior, e.g. *Polygonum palaestinum, Paronychia palaestina, Scrophularia telavivensis, Galium pasianthum, Senecio joppensis, Echinops philistaeus* and *Allium telavirensis*.

The larger three centers of endemics in Palestine-Israel area are: Coastal plain (20 species), Jordan Valley (25) and the high lands of Edom and the Negev (40). The remaining endemics (75 species) are distributed throughout other parts of Palestine-Israel area. The endemics that occur as obligatory weeds include: *Alkanna galilea, Galium chaetopodum, Salvia eigii, Stachys zohariane,* and *Lachnophyllum hierosolymitanum.*

- ***Vegetation types***

The Palestine-Israel Mediterranean area comprises variety of vegetation types ranging from dense forests to thin patches of desert. These are classified under five types (Zohary, 1962):

a. Psammophytic;
b. Halophytic;
c. Steppe and Desert;
d. Wood and shrub;
e. Aquatic.

The following is an ecological account on the first four types, the aquatic vegetation will be included in the chapter of wetlands.

**a. *Psammophytic vegetation***

Psammophytic vegetation of the Palestine-Israel Mediterranean coastal area inhabits the light soil belt of the coastal plain which comprises: sand dunes, sandy loess, calcareous sandstone soils (locally known as kurkar) and the sandy-clay soil. All of these habitats are poor in nutrients and have – low moisture-retaining capacity.

(i) *Sand dune vegetation.* According to Eig (1931–1932), Zohary (1962, 1973), the sand dunes of the Palestine-Israel area can be categorized into: northern and southern coastal dunes. For the northern dunes, there is a difference between the vegetation of the low and high shores. On the sand dunes of the low shores, the vegetation is arranged in more or less parallel belts, e.g. sand dunes of the Acre Plain where the narrow, plantless tidal zone is lined further inland by shallow, most sand strip predominated by *Ipomoea littoralis and Salsola kali.* The associates are *Cakile maritima, Euphorbia peplis* and a few other species able to withstand sea spray. The first low front dunes are occupied by a community dominated by *Sporobolus arenarius and Lotus creticus* with common associates: *Agropyron junceum, Pancratium maritimum, Euphorbia paralias* and *Oenothera drammaondii.* All are highly resistant to sea spray and most have roots that reach the saline water level. Further landward there is a sandy broad belt occupied by an *Ammophila arenaria-Cyperus conglomerates* community associated with: *Silene succulenta, Medicago marina, Scripus holoschoenus, Tamarix gallica, Oenothera drummondii, Lotus creticus* and *Sporobolus arenarius. A. arenaria* is a psammophytic grass resistant to sea water spray and is an excellent sand binder with an extensive branching root system. It is capable of sending new sprouts (shoots) after being buried by sand. Landward, there is an almost plantless strip of mobile sands, varying in depth. Further east there is a broad belt of partially stabilized sand dunes occupied by a community co-dominated by *Artemisia monosperma* and *Cyperus mucronatus* (*C. capitatus*, Boulos, 1995). It is a community rich in species composition, e.g. *Polygonum palaestinum, Tamarix gallica, Lygos raetam* var. *sarcocarpa, Convolvulus secundus, Medicago marina, Echium angustifolium, Orlaya maritima, Oenothera drummondii, Senecio joppensis, Plantago sarcophylla, Rumex occultans, Trisetum lineare,* and *Cutandia memphitica.*

The zonetion pattern of the high and steep shore has a somewhat different arrangement. The tidal zone is practically bare whereas the steep slope facing the sea is occupied by a community dominated by *Sporobolus spicatus,* and *Lotus creticus* associated with: *Diotis maritima, Crithmum maritimum, Inula crithmoides* and *Statice virgata.* The higher sand dunes are occupied mainly by a community dominated by *Artemisia monosperma* and *Cyperus capitatus.*

The belt of the southern coastal sand dunes, however, is, relatively, wider and higher and the dunes in full movement. Here, the vegetation is limited mainly to the sheltered depression between the mobile sand dunes, it is represented by only one community dominated by *Artemisia monosperma – Aristida scoparia* (= *Stipagrostis scoparia*, Boulos, 1995). The associates comprise some desert psammophytes lacking in the northern dunes, e.g. *Panicum turgidum, Pennisetum dichotomum* and *Convolvulus lanatus.*

(ii) *The Sandy loess plains* are the most fertile areas in the Negev, almost the whole area has been under farming cultivation since early times. However, the natural vegetation is represented by only one community dominated by *Artemisia monosperma* and *Lolium gaudini.* The wild grass (*Lolium*) gives the landscape a prairie-appearance. The associate species include: *Leopoldia eburnea, Linaria asclonica, Aegilops bicornis, Ononis serrata, Astragalus annularis, Hippocrepis bicontorta, Colchicum ritchii, Dipcadi erythreum,* and *Trisetum glumaceum (= Trisetaria glumacea).*

(iii) *The Kurkar hills* are the consolidated old dunes occur at some distance from the present shore-line as discontinuous longitudinal ridges of calcareous sandstones. This is inhabited chiefly by coastal dwarf shrub communities dominated by, e.g. *Poterium spinosum – Thymelaea hirsuta, Thymus capitatus – Hyparrhenia hirta, and Ceratonia siliqua – Pistacia lentiscus.*

(iv) *The sandy clay* belt is limited to the Sharon Plain and it is occupied by the prairie – like vegetation dominated by *Desmostachya bipinnata – Centaurea procurrens*. The associate species are many and include: *Aegilops longissima, A. sharonensis Tulpa sharonensis, Nigella arvensis ssp. tuberculata, Reseda orientalis, Anchusa aggregata, Brassica tournefortii, Daucus littoralis, Trifolium dichroanthum, Lupinus palaestinus, Medicago littoralis, M. obscura, Leopoldia maritima, Erodium telavivense, Allium curtum, Silene nodosa, S. gallica, Crucianella herbacea, Lotus villosus, Cutandia philistaea, Trigonella cylindricea, Anthemis leucanthemifolia, and Ifloga spicata.*

### b. *Halophytic vegetation*

The halophytic (salt marsh) vegetation in centered mainly in the Saharo-Sindian territory, i.e. areas surrounding the Dead Sea and in the Arava valley, a smaller part occurs in the vicinity of the Mediterranean coast. The salt marshes (salinas) occur in the coastal plain as well as in the interior desert. The first type has been formed through flooding of brackish or saline springs and water courses particularly those of the Acre Plain fed by the saline water of the Kishon and Na'aman rivers. The inland salinas are most abundant in the Dead Sea area and in the Arava Valley. These are called high water table salinas.

Three communities are recognized in the coastal plain salinas dominated by: *Salicornia herbacea, Tamarix meyeri* and *Arthrocnemum macrostachyum – Sphenopus divaricatus*. The halophytic vegetation of the inland salinas is organized into 10 communities named after the dominant or co-dominant species as follow: *Phragmites australis, Juncus rigidus, Arthrocnemum macrostachyum, Tamarix tetragyna, Tamarix gallica, Suaeda monoica, Suaeda forsskaolii, Nitraria retusa, Suaeda palaestina* and *Salsola tetrandra*. The following are short notes on these 13 communities (Zohary, 1962).

*Salicornia herbaceae* (= *S. europaea*, Boulos, 1995) is an annual herb, sometimes rather stiff at base. Its community occupies the swampy spots. It is associated with *Suaeda fruticosa, Cressa cretica, Pholiurus filiformis* and *Hordeum marinum.*

The community of *Tamarix mayeri* inhabits the banks of the Kishon and Na'aman rivers as well as the borders of the pools and puddles in their vicinity. The associates are mainly halophytic species including: *Arthrocnemum macrostachyum, Obione portulacoides* (= *Atriplex portulacoides*, Boulos, 1995), *Juncus rigidus*, and *Salicornia* spp.

The *Arthrocnemum macrostachyum – Sphenopus divaricatus* community is the most common in the area. It occupies relatively highly saline parts associated with: *Statice limonium, Plantgo crassifolia, Aeluropus repens, Juncus rigidus* and *Phragmites australis*.

The two communities dominated by *Phragmites australis* and *Juncus rigidus* are confined to banks of streams, runnels, springs and ponds containing muddy or brackish water throughout the whole or part of the year. Both are formed of almost pure stands arranged in parallel belts the seaward belt is that of *P. australis*. The *Juncus* community can be associated with rare individuals of *Statice limonium, Inula crithmoides, Aeluropus littoralis, Suaeda baccata* (= *S. aegyptiaca*, Boulos, 1995) and *S. fruticosa*.

The community of *Arthrocnemum macrostachyum – Tamarix tetragyna* occupies badly drained depressions and parts of the wadis inundated during winter. The soil usually remains wet and there is no wide fluctuation in soil moisture content (14–23.5%). The associated species are all halophytes including: *Atriplex halimus, Suaeda forsskaolii, S. palaestina, Mesembryanthemum nodiflorum, Phalaris minor, Cistanche lutea, Frankenia pulverulenta, Statice spicata* and *Beta vulgavis.*

*The Tamarix gallica, Suaeda monoica* community forms almost pure stands of woods in the outlet regions of the wadis as well as in the well-drained but permanently flooded depressions. Rare associate species may present, e.g. *Suaeda forsskaolii, S. palaestina, S. vermiculata* and *Atriplex halimus*.

The community dominated by *Suaeda forsskaolii* occupies broad belts along the bank of the Jordan River at the back of *Populus euphratica* and *Tamarix jordanis* woods and on the elevated foreshore of the Dead Sea. The associated species include: *Suaeda baccta, Rumex dentatus, Aeluropus littoralis, Phalaris minor, Sisymbrium irio, Lolium rigidum* and *Bromus scoparius*.

*The Nitraria retusa* community covers wide areas in the Arava Valley and is also common in the shores of the Dead Sea. The soil of this community has low moisture content. The stands is usually pure with only occasional individuals of *Prosopis farcta, Alhagi graecorum* and *Statice pruinosa* (= *Limonium pruinosum* subsp. *pruinosum*, Boulos, 1995).

The *Suaeda palaestina* community abounds in the Dead Sea region particularly in the plains between Jericho and its northern shore. The soil of this community is usually muddy in winter, dusty and structureless in summer. The common associates of this community are: *Spergularia diandra, Mesembryanthemum nodiflorum, Sphenopus divaricatus, Salsola tetrandra, Atriplex halimus, Prosopis farcta, Alhagi graecorum, Limonium pruinosum, Statica spicata, and Bassia eriophora.*

The community dominated by *Salsola tetrandra* is rich with associated species particularly annuals, e.g. *Stipa tortilis, Plantago ovata, Aizoon hispanicum, Mesembryanthemum nodiflorum, Reichardia tingitana, Crepis arabica, and Chlamydophora tridentata* etc.

Apart from these saline communities, there are a series of smaller ones in the Arava Valley region where *Desmostachya bipinnata* predominates associated with *Tamarix* spp. and *Nitraria retusa*.

#### c. ***Steppe and desert vegetation***

Steppes are areas lacking arboreal vegetation for the shortage of rainfall, i.e. a timberless landscape with an open but more or less continuous vegetation of xerophilous shrubs, undershrubs and herbs. Though under arid climate, the annual amount of rainfall on the steppes allows sporadic dry farming. Deserts can be divided into "rain-deserts", "run-off desert" and "absolute deserts". The rain-deserts are areas with patchy – very sparse – vegetation maintained by rainfall whereas the runoff deserts are areas in which the amount of precipitation is insufficient to support any kind of vegetation except in depression, and low-lying places where ground winter or runoff moisture accumulates. Absolute deserts, have no sign of vegetation and the limited rainfall prohibits farming except in oases or flooded lowlands.

In the Palestine-Israel Mediterranean area, the steppe and desert vegetation is applied to the vegetation of both the Irano-Turanian and Saharo-Sindian territories. The Irano-Turaniean territory is affected by annual rainfall of 200–350 mm and its habitats are generally non-saline gray calcareous steppe soil, loess soil and rocky hills. This vegetation consists of thorny and broom like brush woods and dwarf shrub communities (Eig, 1948). These widespread communities are dominated by: *Pistacia atlantica, Ziziphus lotus, Lygos raetam, Phlomis brachystylis,*

*Artemisia herba-alba, Haloxylon articulatum* and *Anabasis haussknechtii* (= *A. syriaca,* Boulos, 1995).

The *Pistacia atlantica* community occurs in the eastern slope of the Galilee Mountains, on the eastern slopes of the Saamarian and Judea Mountains, as well as in the Edom. The dominant species is the only tree of this community, it has been referred to an "eila" in the Bible. It is a handsome deciduous tree which may attain a considerable age with height up to 20 m (Zohary, 1962).

*The Ziziphus louts* community is distributed on hillsides facing the central Jordan Valley between Wadi Far'a and Beit Shean. This community is rich in its floristic composition, and among its associated species are: *Ballota undulata, Salvia graveolens, Echinops blancheanus, Carlina corymbosa, Asphodelus ramosus, Ferula communis, Alkanna strigosa, Astragalus macrocarpus, Gypsophila rokejeka, Anchusa strigosa, Salvia horminum, Scabiosa prolifera, Ajuga chia (A. chamaepitys,* Boulos, *1995 ), Majorana syriaca, Onosma fruticosa, Pterocephalus involucratus, Elymus geniculatus* etc.

The community dominated by *Lygos raetam* var. *raetam* inhabit wide areas of the rocky mountain side facing the lower and central Jordan Valley. The most common associates are: *Rhus tripartita, Podonosma syriaca, Centaurea eryngioides, Gymnocarpos fruticosum* (*G. decander*, Boulos, 1995), *Varthemia iphionoides, Carlina corymbosa, Gypsophila rokejeka* and *Carthamus nitidus*.

*The Phlomis brachystylis* community occurs in the Judean desert, particularly the rocky habitats. Its associates are: *Poa eigii, Larex pachystylis, Noea mucronata, Salsola vermiculata* ssp. *villosa*, *Artemisia herba-alba, Lactuca orientalis, and Varthemia iphionoides*.

The community dominated by *Artemisia herba-alba* inhabits the gray calcareous soils of rocky hillsides in the central Negev (south of Avdat at an altitude 800 m). Its associate species are: *Reaumuria palaestina, Gymnocarpos decander, Helianthemus vesicarium, Convolvulus oleifolius, Morecandia nitens, Noea mucronata, Astragalus sanctus, Echinops spinosus, Poa sinaica, Phagnalon rupestre, Erodium hirtum, Scorzonera judaica, Stipa szowitsiana, Tulipa amblyophylla* etc.

*The Haloxylon articulatum* (= *H. scoparium*, Boulos, 1995) community is largely confined to the heavy fluviatile loess soil of wadi terraces and lowlands slightly saline soils. *H. scoparium* is associated with *Stipa tortilis, Malva aegyptia, Trigonella arabica, Avena wiestii* (= *A. barbata* ssp. *wiestii*, Boulos, 1995), *Astragalus callichrous, Plantago ovata, Enarthrocarpus strangulatus, Ifloga spicata, and Aizoon hispanicum*.

*Anabasis syriaca* community occurs on loess or loess like soil of the Negev with a few associates, such as: *Poa sinaica, Carex pachystylis, Scrophularia deserti, Colchichum ritchii, and Leopoldia longipes*.

Nine main communities are recognized in the Saharo-Sindian territory dominated by: *Zygophyllum dumosum, Chenolea arabica* (= *Bassia arabica*, Boulos, 1995), *Salsola tetrandra, S. villosa, Anabasis articulata – Lygos raetam, Anabasis articulata – Zilla spinosa, Anabasis articulata, Haloxylon persicum* and *H. salicornicum*.

*The Zygophyllum dumosum* community inhabits two habitats. One of them is confined to the northern and central Negeve, to hammada hills covered with flint stones, where the subsoil consists of a fine gypseaous matter intermingled with gravel. In this habitat, the associated species include: *Gymnocarpos decander, Reaumuria palaestina, Noea mucronata, Astragalus spinosus, A. tribuloides Traganum nudatum, Atractylis serratuloides*, *Halogeton alopecuroides, Salsola tetrandra, S. inermis, Helianthemum kahiricum, Diplotaxis harra, Herniaria hemistemon, Scorzonera judaica, Aegilops kotschyi, Erodium laciniatum, E. hirtum, Rumex roseus, Plantago ovata, Gymnarrhena micrantha, Asteriscus pygmaeus,* and *Stipa tortilis.*

In the second habitat, *Z. dumosum* community is confined to fissures and interspaces of a rocky substratum in which saline soil accumulates. This habitat occurs in the eastern portion of the central Negeve where *Farsetia aegyptia, Asparagus stipularis* and *Limonium pruinosum* are the most common associates. *Chenolea arabica* (*Bassia arabica*) predominates in parts of the soft gypseous soil, whereas *Suaeda asphaltica* is a leading species in the gypsiferous rocky substratum of the Judean desert and Transjordan. On marls and other soft limestone substrata of the Judean Desert and Dead Sea regions the leading plants are *Salsola villosa* and *S. tetrandra.*

The community dominated by *Anabasis articulata* and *Lygos raetam* occupies areas in the sandy plain of Tureiba (Meishor Yemin) of the central Negev. The associate species include: *Thymelaea hirsuta, Artemisia monosperma, Verbascum fruticosum, Atractylis flava, Helicophyllum crassipes, Ornithogalum trichophyllum, Asphodelus microcarpus, Aristida obtusa, A. ciliata, Pancratium sickendbergii, Matthiola livida, Lotus villosus, Schimpera arabica,* and *Adonis dentata.*

The community dominated by *Anabasis articulata* and *Zilla spinosa* is very common in southern Transjordan in the vast stretches of sand fields derived from Nubian sandstone. The associates are: *Noea mucronata, Gymnocarpos decander, Lygos raetam, Helianthemum arabicum,* and *Carex pachystylis.*

There is another community dominated by *Anabasis articulata* confined to wadis, runnels and depressions cut across the sterile hammadas of the southern Negev and Edom. The soil is less saline supporting a richer flora and denser vegetation. The associate species are: *Zilla spinosa, Gymnocarpos decander, Anvillea gracini, Crotalaria aegyptiaca, Salvia aegyptiaca, Fagonia grandiflora, F. kahirina, Farsetia aegyptia, Pulicaria undulata* (= *P. incise*, Boulos, 1995) *Diplotaxis harra, Reichardia tingitana, Notoceras bicorne* and *Erodium hirtum.*

There are two communities dominated by *Haloxylon persicum* and *H. salicornicum* inhabiting the sand dune and sand plain habitats of the interior desert. *H. persicum* is at the southern limit of its range and is associated with: *Calligonum comosum, Lygos raetam, Zilla spinosa, Salsola foetida, Farsetia aegyptia, Pennisetum ciliare* and *Eromobium linear. The H. salicornicum* community, is confined to sandy soils or sandy hammadas derived from Nubian sandstone and igneous rocks. It never occurs in coastal sand dunes but in more thermophilous sites of the interior desert. The associate species of this community include: *Panicum turgidum, Crotalaria aegyptiaca, Aristida obtusa, Monsonia nivea, Fagonia bruguieri, Salvia aegyptiaca,*

*Citrullus colocynthis, Aerva tomentosa* (= *A. javanica* var. *bovei*, Boulos, 1995) *Robbairea prostrata, Salsola foetida,* and *Schismus barbatus.*

Apart from the above mentioned vegetation types, the steppe and desert vegetation of this region comprise about 60 species belonging to various tropical floras, most importantly the Sudanian element represented by two communities. The first is dominated by *Ziziphus spina – christi* and *Balanites aegyptiaca* in Ghov es Safi and Nimrin with rich associates, e.g. *Acacia tortilis* subsp. *tortilis, A. tortilis* subsp. *raddiana, Salvadora persica, Moringa peregrina, Grewia villosa, Abutilon muticum (= A. pannosum), Lavandula coronopifolia, Cassia obovata, Solanum incanum, Calotropis procera, Boerhavia plumbaginacea, Pennisetum ciliare, Suaeda forsskaolii,* and *Atriplex halimus.* The second community is dominated by *Acacia tortilis* subsp. *raddiana*, widespread in the Arava Valley and the surrounding tributaries. The common associates are: *Acacia tortilis,* subsp. *Tortilis, A. negevensis, Anabasis articulata, Haloxylon salicornicum, Gymnocarpos decander, Zilla spinosa, Ochradenus baccatus, Cleome droserifolia, Cassia obovata, Capparis cartilaginea, C. spinosa* var. *negvensis, Lycium arabicum* (= *L. shawii*, Boulos, 1995), *Daemia cordata, Anvillea garcini, Heliotropuim arabainense, Lavandula coronopifolia, Aerva tomentosa, Hyphaene thebaica,* and *Acacia laeta.*

### d. *Wood and shrub vegetation*

This vegetation type is identified also as "class of *Quercetea calliprini*". It comprises Mediterranean forests, maquis,[8] garigues[9] and bathas[10] subjected to a Mediterranean climate with a minimum annual rainfall limit of 350 mm. The vegetation of rocky and stony grounds also belongs to this group (Zohary, 1962).

### 1. The Maquis

The local maquis and forests can be grouped under four types: pine forests, deciduous tabor oak forests, everygrean oak maquis and forests and carob-lentisk maquis.

(i) Pine forests is the southernmost pine forest of the Mediterranean region characterized by three communities. The first is dominated by *Pinus halepensis* and *Hypericum serpyllifolium*, the second is dominated by *Pinus halepensis* and *Juniperus oxycedrus*; the third by *Pinus halepensis-Cupressus sempervirens.* Among the associated species are: *Quercus calliprinos, Q. ithaburensis, Pistacia palaestina, P. lentiscus, Arbutus andrachne, Smilax aspera, Rubia olivieri, Genista sphacelata, Calycotome villosa, Cistus villosus, Cytisopsis*

[8]Maquis = Mediterranean woodland dominated by sclerophyllous evergreen low trees and shrubs up to 4 m high.

[9]Garigue = French term for sclerophyllous scrub that consists generally of chamaephytes and nano-phanerophytes 1 m or 50 cm high.

[10]Batha = Dwarf shrub formation.

*dorycnifolia, Teucrium divaricatum, Fumana thymifolia, Ceratonia silique, Phlomis viscosa, Saturega thymbra, Salvia triloba, Poteruim spinosum, Thymus capitatus, Ononis natrix, Echinops viscosus, Ballota undulata and Carlina involucrata.*

(ii) The Tabor Oak belongs to a large group of broad-leaved deciduous forests, fairly common in some east Mediterranean countries notably Turkey, Iraq and Iran in addition to Israel. Three plant communities are recognized, the first dominated by *Desmostachya bipinata*, the second by *Quercus ithaburensis* and *Styrax officinalis* and the third by *Quercus ithaburensis* and *Pistacia atlantica*. The grass community of *D. bipinnata* inhabits the Sharon Plain; but was destroyed by expanding agriculture. The *Q. ithaburensis – Styrax officinalis* community is the most widespread particularly on the South Western slopes of the Lower Galilee, and its remnants are also met with in the Ephraun Hills of Samaria. The most common associates are: *Pistacia palaestina, Crataegus azarolus, Cercis siliquastrum,* and *Phillyrea media.* The other less common associated are: *Quercus ithaburensis, Q. calliprinos, Clematis cirrhosa, Calycotome villosa, Rhamnus palaestina, Smilax aspera, Rubia olivieri, Bryonia multiflora, Asparagus aphyllus, Thymus communis, Ruscus aculatus, Anemone coronaria, Mandragora officinarum, Cyclamen persicum, Arum palaestinum,* and *Thrincia tuberosa.*

The third community of the Tabor Oak forests is dominated by *Q. ithaburensis* and *Pistacia atlantica*, it usually inhabits the western slopes of Gilead and Bashan, and remnants of it occur on the Huleh Plain. The associated species include: *Crataegus azarolus, Amygdalus communis, Rhamnus palaestina, Styrax officinalis, Calycotome villosa, Ruta graveolens, Carlina corymbosa, Convolvulus dorycnium, Echinops viscosus,* and *Ferula communis.*

(iii) The Evergreen Oak Forest and Maquis is the most typical widely spread forest and maquis vegetation of Sout-West Asian part of the Mediterranean basin, occurring in Palestine, Israel, Syria, Lebanon, Turkey and the Balkans. It is subdivided into two communities dominated by *Quercus calliprinos – Pistacia palaestina* and *Q. calliprinos-Juniperus phoenicea.*

*The Q. calliprinos – P. palaestina* community occurs in various habitats such as northern Galilee under conditions of heavier rainfall and northern exposure. The associates are: *Q. boissieri, Acer syriacum, Pyrus syriaca, Prunus ursina* and *Sorbus triblobata*. This community is also common throughout the western mountain belt from the foot of the Lebanon up to the Judean mountains, in the south. The common associates are: *Cistus salvifolius, C. villosus, Thymbra spicata, Calycotome villosa, Thymus capitatus,* and *Poterium spinosum.*

*Q. calliprinos – J. phoenicea* community occurs in the South-Western part of Transjordan at 1,000–1,700 m altitude. This forest occupies a strip of about 80 km in length and with its southern limit near Patra. The common associates are: *Crataegus azarolus, Pistacia paleastina, P. atlantica, Rhamnus palaestina* and *Daphne linearifolia.*

(iv) The Carob-Lentisk Maquis forest is widespread in the western foothills of the mountain belt as well as in the eastern slopes of Galilee and Samaria. This vegetation type also occupies consolidated dunes and the Kurkar hills in the Sharon plain north of Natanya. Three communities are recognized in this vegetation type dominated by: *Ceratonia sliqua-Pistacia lentiscus typicum, C. siliqua – P. lentiscus arenarium* and *C. siliqua – P. lentiscus orientale.*

*The C. siliqua – P. lentiscus typicum* community predominates in the western – foothills along the mountain belt of Israel particularly in the Matsuba – Eilom area of Galilee, on Mount Carmel, in the Hartuv – Beit Guvrin area and further south to the latitutde of Hebron. The associated species in this community are mostly Mediterranean chamaephytes such as *Cistus villosus, C. salvifolius, Calycotome villosa, Salvia triloba, Ruta graveolens, Poterium spinosum, Thymbra spicata,* and *Phlomis viscosa*. The lianas (climbers) of this community include: *Asparagus aphyllus, Rubia olivieri, Clematis cirrhosa,* and *Smilax aspera*, and the associated perennial grasses include: *Hyparrhenia hirta, Andropogon distachyus, Phalaris bulbosa* and *Dactylis glomerata. Olea europaea* var. *deaster* shrubs occur in the northern section of this community but absent from the southern section.

*The Ceratonia siliqua – Pistia lentiscus arenarium* community inhabits the consolidated sand dunes notably in the northern part of the Sharon Plain, fragments of this community occurs also in the surroundings of Nahariya. The associated species are: *Lycium europaeum, Ephedra alte, Ballota philistaea, Lygos reatam* var. *sarcocarpa, Artemisia monosperma, Asparagus stipularis, Prasium majus, Ruta graveolens, Cistus villosus, Clematis cirrhosa, Rhamnus palaestina,* and *R. alaternus*.

*The Ceratonia siliqua – Pistacia lentiscus orientalis* community occupies the eastern escarpments of the lower Galilee and Samaria which are exposed to Irano-Turanian influence. This explains the presence of some species belonging to semi-steppeand Irano-Turania communities, e.g. *Ballota undulata, Ononis natrix, Ziziphus lotus, Pistacia atlantica, Amygdalus communis* and *Phlomis brachyodon*. The other associates comprises species of the other two communities of carob-lentisk maquis forest.

## 2. The Garigue and Batha Vegetation

This type of vegetation comprises six communities dominated by *Salvia triloba, Cistus villosus, Poterium spinosum, P. spinosum – Thymelaea hirsuta, Thymus capitatus – Hyparrhenia hirta, Salvia dominica* and *Ballota undulata*.

The garigue community dominated by *Salvia triloba* occurs in the western slope of Mount Carmel. The associated species include: *Calycotome villosa, Rhamnus palaestina, Poterium spinosum, Pistacia lentiscus, Teucrium divarictum, Fumana thymifolia, Rubia olivieri, Asparagus aphyllus, Smilax aspera, Ranunculus asiaticus, Anemone coronaria,* and *Arisarum vulgare*.

The second community representing the garigue vegetation is dominated by *Cistus villosus* and occurs near Kiryath Anavim west of Jerusalem. The list of the associates includes: *Cistus salvifolius, Phlomis viscosa, Poterium spinosum, Teucrium polium, T. divaricatum, Stachys cretica, Phagnalon rupestre, Scorzonera papposa, Lactuca cretica, Helichrysum sanguinum, Majorana syriaca, Serrata cerinthifolia, Bellis silvestris, Thrincia tuberosa, Salvia judaica, Oryzopsis holciformis, Smilax aspera* and *Rubia olivieri.*

The community dominated by *Poterium spinosum* is the commonest of the batha vegetation communities. It occurs on the northern slopes of the Judean Mountains near Jerusalem associated with many species, such as: *Cistus villosus, Fumana thymifolia, F. arabica, Calyctome villosa, Teucrium polium, T. divaricatum, Hyparrhenia hirta, Andropogon distachyus, Dactylis glomerata, Teucrium polium, T. divaricatum, Osyris alba, Micromeria nervosa, Eryngium creticum, Salvia judaica, Rubia olivieri, Orchis anatolicus, Ophrys fusca, Ranunculus asiaticus, Thrincia tuberosa, Bellis silvestris, Mediago orbicularis, Onobrychis squarrosa, Trifolium campestre, T. stellatum, T. purpureum, Lotus peregrinus, Lathyrus aphaca, Aegilops peregrina,* and *Bromus scoparius.* On the light soils of the coastal plain, *Poterium spinosum – Thymelaea hirsuta* predominates. The associate species are mostly psammophytes, e.g. *Hyperrhenia hirta* etc.

The Kurkir hill in the central Sharon Plain is characterized by the batha community co-domoinated by *Thymus capitatus* and *Hyparrhenia hirta.* The associated species include: *Teucrium polium, Fumana thymifolia, F. arabica, Asphodelus microcarpus (= A. ramosus), Thymelaea hirsuta, Helianthemum ellipticum, Alkanna tinctoria, Leopoldia maritima, Tulipa sharonensis, Plantago albicans, P. cretica, Astragalus callichrous, Medicago littralis, Anthemis leucanthemifolia, Erodium telavivense,* and *Lotus villosus.*

The fourth community of the batha vegetation dominated by *Salvia dominica* and *Ballota undulata*, inhabits the chalk rocks and the white-grayish soil of a hill near the Shoval area. It belongs to the semi-steppe batha vegetation. The following species were recorded by Zohary (1962) as associates: *Astragalus feinbruniae, Eremostachys laciniata, Noea mucronata, Phlomis brachyodon, Cachrys goniocarpa, Heliotropium rotundifolium, Lycium europaeum, Asphodelus microcarpus, Hyparrhenia hirta, Anchusa strigosa, Alkanna strigosa,* and *Gypsophila rokejeka.*

Apart from the associated species already mentioned, the communities of the garigue and batha formation, mostly chamaephytes and nano-phanerohytes, include many annuals, and geophytes belonging to the following genera: *Anemone, Ranunculus, Colchicum, Tulipa, Allium, Bellevalia, Scilla, Crocus, Iris, Orchis,* and *Ophrys.*

### 3. Vegetation of the Rocky and Stony Grounds

The Mediterranean wood and shrub vegetation of Palestine-Israel area comprise a type confined to stony habitat such as rocks, rock crevices, stone fences, walls, cave intrances and rubble heaps (Zohary, 1962). Each of these habitats has its distinctive set of plant life-forms, mainly lithophytes having roots able to penetrate deeply into solid hard rock, which crack as the roots thicken. The characteristic species of these stony habitats are: *Varthemia iphionoides, Stachys palaestina, Podonosma syriaca, Ballota rugosa, Pennisetum asperifolium, Michauxia campanuloides, Centaurea speciosa* and *Dianthus pendulus.* All are chamaephytes able to germinate in minute pits in the rock surface and then to send their roots into the unbroken limestone rock. Old stone walls have small pockets of soil between the layers of stones or in the crevices, and these allow a community dominated by *Umbilicus intermedius* and *Allium subhirsutum* to grow associated with: *Erophila minima, Crepis hierosolymitana, Veronica cymbalaria, Cyclamen persicum, Parietaria judaica* and *Taraxacum cyprium* in addition to several species of mosses and liverworts of these genera: *Targionia, Lunularia, Barbula, Tortula, Grimmia, Encalyptra,* and *Camptothecium.* The old walls in towns and cities are characterized by definit communities, the main species of which are: *Parietaria judaica, Hyoscyamus aureus Antirrhinum siculum,* and *Capparis spinosa* var. *aegyptiaca. Adiantum capillus – veneris* often also inhabits these moist walls and cave entrances. The habitat of rubble heaps on field margins is characterized by a community dominated by *Cynocramb prostrata.* Associate species include: *Galium articulatum, Cicer pinnatifidum, Veronica cynbalaria, Vicia sericocarpa, V. angustifolia, Crepis bulbosa,* and *Pisum fulvum.*

#### 1.8.2.5 Wetlands

The wetlands of the SW Asian Mediterranean coastal land are represented by two lakes, namely: Lake Bardawil on the Sinai coast of Egypt and Lake Huleh on the Palestine-Israel coast, including their associated coastal swamps. An ecological account of these wetlands is given below.

##### A. *Lake Bardawil (Sinai coast)*

Lake Bardawil, on the Mediterranean shore of Sinai, is considered a hypersaline lagoon bordering the sea. The narrow semi-circular barrier-beach that forms the northern boundary of the lake separates it from the sea. Artificially maintained inlets connect the lake with the sea. The lake has no fresh-water supply and no major source of enrichment other than the Mediterranean. The influx of water from the sea is very important to its ecology since this inflow and outflow maintain the salt concentration in the water at tolerable levels. The annual rainfall in the area of the lake is about 82 mm and the annual evaporation about 1,600 mm. This excess of evaporation over precipitation has a marked effect on the salinity of the lake when the inlets are closed and circulation cut off. The salinity of the lake is normally about 45–55 parts/1,000. The water temperature generally varies from 12°C in January to 30.5°C in June (Anonymous, 1982). The vegetation of the shore-line in the wet salt marshes of Lake Bardawil is dominated by *Halocnemum strobilaceum* and *Arthroconemum glaucum (=A. macrostachyum)* and an elevated saline belt by *Zygophyllum album* which forms a pure stand. Within the benthic region of the lake, dense growth of *Ruppia maritima* occurs.

The Mediterranean coastal strip adjacent to Lake Bardawil is about 10 km wide and about 50 km long. This strip of land is poor in species. Depressed areas near the lake are subject to periodic inundation by salt water from the lake; the water evaporates to leave a saline residue. Only halophytes grow in these depressions, the vegetation being dominated by *Halocnemum strobilaceum* and *Arthrocnemum glaucum*. On the elevated parts of the area *Zygophyllum album* forms a more or less pure community and often occupies the whole area, except for a few microhabitats, up to the sand dune zone. The littoral strip also contains several localities with slightly elevated terraces, some 3–5 m above water level, dominated by *Nitraria retusa* with *Lycium europaeum* as a common associate. The elevated slopes of the dunes support *Artemisia monosperma, Lygos raetam, Moltkiopsis ciliata* and *Thymelaea hirsuta*. Depressions between the dunes contain a characteristic salt marsh community of *Juncus subulatus* (dominant) and *Cynodon dactylon, Lycium europaeum, Nitraria retusa* and *Phragmites australis* as associates. In the slightly saline depressions are *Artemisia monosperma* and *Lygos raetam*. At the foot of these dunes are semi-wild date palms (*Phoenix dactylifera*).

Recent studies by Khedr and El-Gazzar (2006), El-Bana (2003), Khalil and Shaltout (2006) recognized six main habitats in Lake Bardawil and its surroundings, namely: 1. open water, 2. wet salt marshes, 3. saline sand flats and hummock (nebkas), 4. stabilized sand dunes, 5. interdune depressions and 6. mobile sand dunes. In the open water, only three flowering plants (*Cymodoceae nodosa, Ruppia cirrhosa* and *Halodule uninervis)* grow. The first two species are widely distributed in the lake, but *H. uninervis* has not been reported from the Mediterranean before, and seems to have invaded from the Red Sea through the Suez Canal (Täckholm, 1974; Boulos, 1995). The wet salt marshes of Lake Bardawil are characterized by 16 species of halophytes and helophytes. *Halocnemum strobilaceum* is the dominant species and its associates include: *Arthrocnemum macrostachyum, Frankenia pulverulenta, Phragmites australis, Suaeda maritima* and *Tamarix nilotica*.

A total of 34 species have been recorded from the saline flats and hummocks (nebkas), *Zygophyllum album* dominates the saline flats and *Nitraria retusa* the nebkas. The associated species include: *Bassia muricata, Cutandia dichotoma, Lotus halophylus, Mesembryanthemum creystallinum, Salsola tetragona, Spergularia marina* and *Schismus barbatus*.

The stabilized sand dunes are characterized by two groups of islets: eastern and western, with a floristic assemblage of about 78 species. The characteristic species are *Retama raetam* and *Stipagrostis plumosa* in the eastern islets and *Asparagus stipularis, Deverra tortusa* and *Echium angustjfolium* in the western islets. Associate species include: *Anabasis articulata, Bupleurum semicompositum, Daucas litoralis, Echiochilon fruticosum, Launaea capitata, Lycium shawii* and *Schismus barbatus*.

The inter-dune depression is a hostil habitat, divided into saline and non-saline facies on the basis of the level of ground water. In non-saline areas, *Artemisia monosperma, Panicum turgidum* and *Thymelaea hirsuta* are the characteristic species, with associates: *Atractylis carduus, Cynodon dactylon, Ifloga spicata, Pancratium maritimum* and *Salvia lanigera*. Halophytes, such as: *Arthrocnemum macrostachyum* and *Halocnemum strobilaceum* predominate in the saline areas.

A total of 49 species have been recorded on the mobile sand dune, *Stipagrostis scoparia* and *Calligonum comosum* are the characteristic species, with associates: *Artemisia monosperma, Cornulaca monacantha, Eremobium aegyptiacum,*

*Heliotrpium digynum. Lotus halophilus, Malva parviflora, Pancratium sickenbergeri, and Retama raetam.*

The flora of all habitats of Lake Bardawil contains a total of 136 species (109 genera and 42 families) representing 30% of the recorded species of the Mediterranean coastal plain of Sinai (Gibali, 1988). Gramineae is represented by 12.5% followed by Chenopodiaceae (11%), Compositae (9.6%), Leguminosae (8.1%), Caryophyllacaeae (6.6%), Cruciferae (3.7%) and Cyperaceae (1.5%). Most of the plant diversity of Lake Bardawil are therophytes (44.1%) followed by chamaephytes (25%), geophytes (11.8%), phanerophytes (5.1%), parasites (2.2%) and hydrophytes (0.7%).

Following the IUCN Red Data Book (El-Hadidi and Hosny, 2000), six threatened species are recorded from Lake Bardawil: *Astragalus camelorum, Bellevalia salaheidii, Biarum olivieri, Iris mariae, Lobularia arabica* and *Salsola tetragona.* The first four species are endemic or near endemic (Khalil and Shaltout, 2006).

**B. *Lake Huleh (Palestine-Israel area)***

Lake Huleh is formed by a natural dam of basalt on the NE side of Israel; the Jordan River exists from its southern end. It is a lake near sea-level which in ancient times was called the waters of Merom. Its area used to be about 12–14 $km^2$ (5.3 × 4.4 km), and about 50 cm deep in summer and 3 m deep in winter. In the 1950s, the lake and its surrounding swamps were subjected to an attempt to alter the environment for agricultural needs and began to dry up. Between 1950 and 1958 about 5,000 ha of the lake's swampy shore were drained and then cultivated for grains, fruit trees, vegetable and cotton. However, due to the very limited benefits obtained, it was recognized that successful development can endure only if a compromise between nature and development is reached. As a consequence, a small section of the former lake and swamp region was recently reflooded in an attempt to prevent further soil deterioration and to revive the nearly extinct wetland ecosystem (Anonymous, 2008b).

Apart from fishing activities, Huleh Lake is considered an important bottle neck for the migratory birds along the Syrian-African Rift Valley between Africa, Europe and Asia (Anonymous, 2008b).

The plant life of the wetlands can be grouped under two main categories:

(1) Water plants (hydrophytes).
(2) Canal bank plants.

The water plants (hydrophytes) are those species growing in the water bodies. They are either totally under water (submerged species) or partly under water (floating and emersed species). The canal bank plants (riparian plants) are those inhabiting the wet banks of the river and its tributaries.

According to Zohary (1962), while most of the hydrophytes of Huleh Lake and its surrounding swamps have their origin in northern temperate regions, there are also fair number of topical origin. The tropical elements include: *Marsilea diffusa, Cyperus papyrus, C. latifolia, C. articulatus, C. polystachyus, C. lanceus, C. alopecuroides, Dinebra retroflexa, Paspalidium geminatum, Nymphaea caeruleca, Polygonum accuminatum, P. senegalense* (= *Persicaria senegalensis*, Boulos, 1995) and *Jussiaea repens*.

Zohary (1962) classified the water plants of the Palestine-Israel sector under two plant sociological classes: Potamotea and Phragmitatea. The Potamotea class comprises the hydrophytes of lakes and slow water currents. It includes five plant communities named after the dominant species. The first is dominated by *Potamegeton lucens* and *Myripohyllum spicatum*, both submerged species growing in deeper water; the second is dominated by submerged: *Nuphar luteum* and *Ceratophyllum demersum* occupying shallow and rather stagnant water bodies. The community dominated by *Potamogeton fluitans* (= *P. nodosus*, Boulos, 1995) is found submerged in slow-flowing streams. Communities dominated by the two floating species *Lemna gibba* and *L. minor* are often encountered in shaded ditches. *Nymphaea alba* and *N. caerulea* have become exceedingly rare.

The class phragmitatea comprises emersed (= protruding above the surface of water) hydrophytes forming eight communities. The first community is dominated by *Phragmites communis* var. *isiacus* and *Typha angustata*. It occurs on the low banks of permanent rivers and streams and of periodically flooded swamps. The associate species include: *Cyperus longus, Panicum repens, Scirpus lacustris, S. littoralis, Polygonum nodosum,* and the very rare, *Caldium mariscus*.

The second community is dominated by *Cyperus papyrus* and *Polygonum acuminatum* and occupies large stretches of the flooded peat soils of the Huleh swamps. The associated species here include: *Galium elongatum, Roripa amphibia, Dryopteris thelypteris, Persicaria senegalensis, and Alternanthera sessilis*.

The community dominated by *Persicaria senegalensis* and *Sparganium neglectum* forms considerable belt in the Huleh swamps but also occurs on river banks. Its associate species include: *Cyperus lanceus, Iris pseudo-acorus, Marselea diffusa, Polygonum occuminatum, P. scabrum, Jussiaea repens, Panicum repens, Typha domingensis,* and *Butomus umbellatus*.

The grass *Panicum repens* is fairly common on wet river banks and in flooded low lands. It forms the outer belt of the swampy vegetation of Huleh Lake. Its associates are: *Scirpus maritimus, Alisma plantago-aquatica, Echinochlea crus-galli,* and *Jussiaea repens*.

The community dominated by *Rubus sanctus* and *Lythrum salicaria* is found on the elevated banks of permanent or ephemeral canals as well as on swampy borders. The common associates are: *Cynanchum acutum, Epilobium hirsutum, Saccharum aegyptiacum* (= S. *aegyptiacum subsp. aegyptiacum*, Boulos, 1995), *Convolvulus saepium, Cyperus longus, Dorycnium rectum, Pulicaria dysenterica, Verbena officinalis, Eupatorium cannabinum,* and *Salix acmophylla*.

The *Inula viscosa* community inhabits areas that are usually flooded during winter but drying up in early summer. Frequently encountered associated species include: *Juncus acutus, Pulicaria dysenterica, Teucrium scordioides, Stachys viticina, Festuca arundinaceae, Carex* spp., *Lotus tenuifolius, Oenanthe prolifera, Mentha pulegium, Centaurium spicatum,* and *Cynodon dactylon.*

The community dominated by *Crypsis minuartioides* is confined to slight depressions inundated during winter drying out in summer. Its associates are: *Heliotropuim supinum, Glinus lotoides, Pulicaria arabica, Verbena supina,* and *Chrozophora plicata.*

A minor community dominated by *Veronica anagallis – aquatia* occupies permanent rivulets, and the margins of springs. It is associated with: *Nasturtium officinale, Apium graveolens, Mentha incana, Helosciadium nodiflorum, Cyperus fuscus,* and *Fimbristylis ferruginea.*

The canal bank (or riparian) species form wood and scrub vegetation on the banks of the Jordan River and the Dead Sea and its tributaries. Two communities are recognized: the *Populus euphratica* and *Tamarix jordanis* communities. The vegetation forms dense and sometimes ever impenetrable woods. *The P. euphratica* community occupies the front part of the bank, whereas the *T. jordanis* community is mostly behind it. The associated species are: *Lycium europaeum, Atriplex halimus, Glycyrrhiza glabra, Asparagus palaestinus,* and *Prosopis farcta.*

# Chapter 2
# Afro-Asian Red Sea Coastal Lands

## 2.1 Introduction

The Red Sea has been an important route of trade throughout human recorded history, linking trade goods of India and the Far East with the historical markets in Egypt, the classical world and Europe. When Ferdinand de Lesseps completed the Suez Canal in 1869, the connection became direct, and now the Red Sea is one of the most important shipping routes in the world. The narrow southern strait of Bab Al-Mandab marks the boundary between the Red Sea and the Gulf of Aden.

Eight countries have shorelines on the Red Sea in Africa and Asia. In the African (western) Red Sea coast (about 2,860 km) there are: Egypt (about 1,100 km including the western coast of the Gulf of Suez, 340 km), the Sudan (about 740 km) and Eritrea (Ethiopia) (about 1,020 km). Djibouti has a shoreline that extends about 60 km, immediately outside the proper western Red Sea coast. The countries of the eastern shoreline (Asian Red Sea coast, about 2,600 km) are: Saudi Arabia (about 2,100 km including the eastern coast of the Gulf of Aqaba, 235 km) and Yemen (about 500 km). Both Israel and Jordan have a tiny strategically important footholds on the Red Sea at the northern tip of the Gulf of Aqaba (9 and 24 km, respectively, Shawar, 1989).

During the nineteenth century, the Red Sea with its characteristic environment and rich flora and fauna in both aquatic (marine) and terrestrial (coastal desert) habitats, being the nearest warm sea to Europe, was the center of attraction particularly to the European naturalists. The Austrian ship, the R/N Pola, was the base for the first scientific expedition especially devoted to the Red Sea. The northern Red Sea was explored in the winter 1895–1896 and the southern Red Sea in winter 1897–1898. The Austrian expedition was followed by the Italian Expedition (1923–1924), Dutch Expedition (1929–1930), British-Egyptian Expedition (1933–1934) known as John Murray Expedition on board of the R/V Mabahis and the Egyptian R/N Mabahis Expedition (1934–1935) (Morcos, 1990).

After the second world war there were four scientific expeditions in the Red Sea namely: (1) Swedish Expedition Abatrus (1948), (2) R/V Manihine Expedition (1949), (3) US R/V Atlantis Expedition (1953) and (4) R/V Vema Expedition (1958).

M.A. Zahran, *Climate–Vegetation*, Plant and Vegetation 4,
DOI 10.1007/978-90-481-8595-5_2, © Springer Science+Business Media B.V. 2010

Starting from the 1960s, awareness of the ecological and economic importance of the Red Sea environment and its natural resources has strongly developed. Countless number of publications (papers, reports, proceedings of conferences, books, etc.) dealing with the different environmental, biological and ecological aspects of the Red Sea water body and its coastal deserts including e.g. geology, geomorphology, geophysics, geochemistry, climatology, oceanography, ecology of the biota inhabiting the terrestrial habitats of the coasts, sociology, fisheries and aquaculture and tourism are now available. The author of this book compiled a detailed checklist of publications related to the Red Sea in a book published in 1990 entitled "*Red Sea, Gulf of Aden and Suez Canal: A Bibliography on Oceanography and Marine Environmental Research*" (PERSGA, UNESCO, Banaja et al., 1990).

Ecological studies on the coastal desert vegetation of the Red Sea include: Drar (1936), Andrews (1948, 1950–1956), Hassib (1951), Vesey-FitzGerald (1955, 1957), Draz (1956, 1965, 1969), Kassas (1956, 1957), Hemming (1961), Kassas and Zahran (1962, 1965, 1967, 1971), Zahran (1962, 1964, 1965, 1967, 1977, 1982a, 1982b, 1993a, 2002, 2004, 2007), Baeshin and Aleem (1987), Migahid and Hammouda (1978), Ahmed (1983), Younes et al. (1983), Zahran et al. (1983), Mandura et al. (1987), Zahran and Willis (1992), Zahran and Al-Kaf (1996), Le Houerou (2001), and Zahran et al. (2009).

## 2.2 Geology and Geomorphology

The Red Sea is a long narrow trough separating north-east Africa from the Arabian Peninsula. This body extends between longitudes 40° 80′E and 30° 20′E in NNE–SSE direction in an almost straight line from Suez in the north (northern part of the Gulf of Suez, latitude 30°N) to Bab Al-Mandab Strait in the south (latitude 12° 30′N), Fig. 2.1.

The formation of the Red Sea in geological times was discussed by Ball (1912). Between the Eocene and Oligocene periods there must have been a great rise in the land, and it seems probable that this rise was brought about by simple upward movements, there was considerable folding in the mountain ranges on both sides of the sea. This means that the present extent of the Red Sea has been caused by a great general subsidence of the land and not by trough-faulting.

The Red Sea is naturally separated from the Mediterranean Sea by the Isthmus of Suez which was cut in 1869 by the Suez Canal linking the waters of both seas. In the south, it is connected with the Indian Ocean through Bab Al-Mandab Strait (Banaja et al., 1990).

The Sinai Peninsula divides the northernmost Red Sea into the shallow (55–73 m deep) Gulf of Suez and the deep Gulf of Aqaba (1,100 m deep in the north and 1,420 m deep in the south, with a maximum depth of 1,829 m). The Gulf of Suez is about 250 km long and average width of about 32 km. The Gulf of Aqaba is about 150 km long and 16 km as average width. The sill depth in the strait of Tiran which separates the Gulf of Aqaba from the Red Sea is about 300 m (Morcos, 1990). Ras

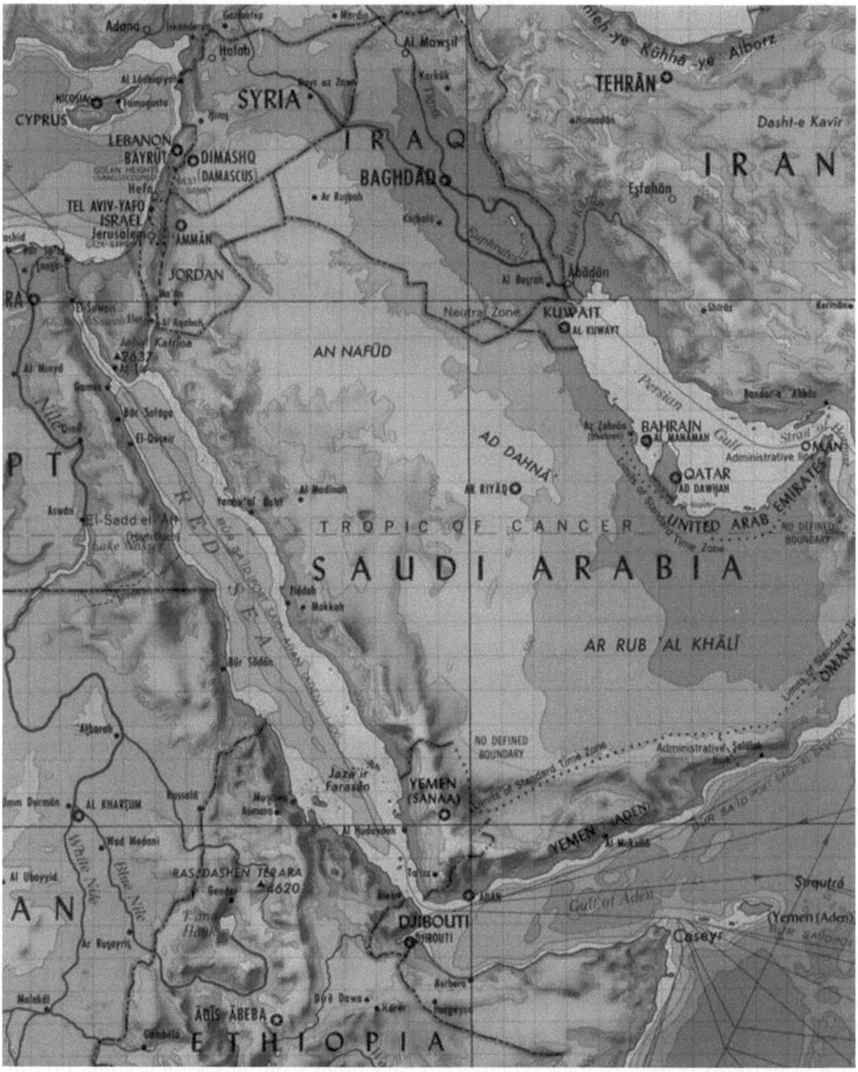


**Fig. 2.1** Map of the Red Sea and its Afro-Asian Coastal Lands

Muhammed (latitude 27° 45′N) is the southern apex of the Sinai Peninsula where the eastern coast of the Gulf of Suez meets the western coast of the Gulf of Aqaba (Zahran and Willis, 1992, 2008).

The strait of Bab Al-Mandab is divided by Perim Island into a small channel (4 km wide and 25 m deep) to the east and a larger channel (20 km wide and 300 m deep) to the southwest. On the southern side of the strait, the bottom slopes down to the Gulf of Aden, then branching after 35 km into two channels which are communicating with a deep furrow (1,700 m) extending westward into the Gulf of Jadjurn.

Throughout its length, the Red Sea opposing shorelines in Africa and Asia are remarkably parallel. In the north, the width of the sea is only 175 km but southwards it increases to a maximum of 350 km in the area between Jizan (Saudi Arabian coast) and Massawa (Eritrean coast), latitude 16° 55′N, longitude 42° 35′E. From there it decreases to a minimum of 24 km at Bab Al-Mandab Strait (latitude 12° 35′N, longitude 43° 30′E). Offshore of both shorelines are coral reef zones that descend to coastal shelves in a series of steps down to 300–600 m. This is followed by the irregular broken floor of the main trough which varies in depth from 600 m to more than 1,100 m, and then the axial trough which develops in a continuous deep within the main trough south of latitude 23°N. Ball (1912) stated that the Red Sea is bordered almost continuously by jagged coral reefs the bases of which extends downward to a depth more than 100 m. Such submerged coral reefs are exposed only at neap tides.

The land adjacent to the Red Sea is generally mountainous, flanked on the eastern side by high table-land of Arabia and Yemen, and on the western side by a range of mountains 1,705–2,187 m above sea level (Kassas and Zahran, 1971). A gently sloping plain extends in the deep trough between the shore and the high land. This coastal plain, which varies in width from <8 to >35 km, is covered with sand, over which the drainage system meanders by shallow courses. Along the Gulf of Aqaba (Sinai western side) and in certain parts of the western side of the Gulf of Suez (e.g. Khashm El-Galala, about 60 km south of Suez) the coastal plain is practically non-existing and the mountains rise almost directly from the water of the Gulf.

The eastern (Asian) Red Sea and Gulf of Aqaba coastal lands are called Tihama. It is a flat strip of land that extends from Aqaba (in Jordan) southwards to the Yemeni border and stretching inland to the foothills of the first range of mountains. It varies in width from almost zero in the north to more than 40 km in the south in Jizan (southern section of Saudi Arabian Red Sea coastal desert). North of Jeddah, Tihama is known as Hijaz.

Tihama coastal desert plain consists mainly of recent deposits while inland towards the foothills there are Tertiary sedimentary beds of crystalline rocks. The scarp of mountains runs parallel to the Tihama Plain and lies between the plain and the crest of the Red Sea Rift Escarpment. The mountain belt varies in width from about 30–40 km to a maximum of 140 km at the Yemeni border. Most of the outcropping rocks belong to the Precambrian Basement Complex which splits up into many fault blocks associated with the formation of the Red Sea Rift. Remnants of Tertiary sedimentary beds overlie some lower crystalline blocks. The drainage systems (wadis) of these scarp mountains run westwards towards the Red Sea and are controlled mainly by a screen of faults at right-angles of these mountains.

The general geomorphological features of the African Red Sea coastal desert are almost comparable to those of the Asian coast. In Egypt, the land adjacent to the Red Sea is generally mountainous, flanked on the western side by the range of coastal mountains (1,705–2,187 m above sea level, Kassas and Zahran, 1971). In the deep trough between the shoreline and the highlands extends a gently sloping plain. This coastal plain, which varies in width, is covered with sand, over which the drainage systems (wadis) meander with their shallow course. These wadis run eastward to drain their water into the Gulf of Suez and the Red Sea. In the Sudan, the Red Sea coastal plain ranges from 8 to 15 km throughout most of its length. It is often covered with drifted fine deposits (Kassas, 1957). In Eritrea (Ethiopia), Hemming (1961) stated that the Red Sea coastal plain averages 24–32 km in width, consisting of vast deposits of mixed sandy and stony alluvium derived from the ancient basement massif to the west of the stony pediment which lies along the foot of the hill-range in the form of an apron and which have been reduced by deflation to stone mantles.

The coastal desert plains of the Red Sea in the Sudan and Eritrea is dissected by set of wadis (drainage systems) cut through the massive of mountains (up to 1,523 m in the Sudan and 2,100 m in Eritrea) and run eastward to drain into the Red Sea.

## 2.3 Climate

The Red Sea provides a singular environment both meteorologically and oceanographically (Edwards and Head, 1987). This water body experiences some of the hottest and most arid conditions which occur in any marine area on Earth. To the west stretches the vast tract of the almost rainless North African Desert, over 4,800 km of sand and stone and gravel extending to the Atlantic, relieved only by the thin ribbon of vegetation in the Nile Valley which owes its existence to rainfall falling in two areas very far to the south. Eastwards and north-eastwards, desert and semi-desert extend even farther through Arabia, Iran and Afghanistan to central Asia and Mongolia. To the north lies the Mediterranean, almost rainless for 6 months or more in the year and where, only in winter, do travelling depressions bring periods of unsettled and rainy weather which may briefly affect the northern parts of the Red Sea. To the south, the copious summer rainfall of the Ethiopian Highlands remains distant and the completely "maritime" regime of the Indian Ocean, with its alternating summer and winter monsoon, rarely penetrates to the southern extremities of the Red Sea basin.

The climate of the Red Sea basin which is controlled by the distribution of, and changes in atmospheric pressure over a very wide area is considered in two types: the Mediterranean type in the north and the monsoon type in the south. The pressure centers involved are generally distant from the Red Sea and themselves undergo variations during the course of the year, which result in widespread and sometimes dramatic seasonal changes over extensive areas.

Generally, the climatic features that affect the Red Sea basin show uniformity throughout the year which may be divided into two main seasons: cooler and hotter seasons. Such classification is roughly corresponding to the conventional northern hemisphere winter and summer. Winter usually extends from mid-October to mid-April with summer from mid-April to mid-September. The mean daily maximum temperature in January (winter) ranges from about 20°C in the far north to about 29°C in the far south, the corresponding July (summer) figures being 35 and 40°C, respectively. In winter, lower night temperature resulting from the prevailing cool wind, provide relief from the heat of the day. However, summer may be most uncomfortable in all areas of the Red Sea basin due to the high temperature associated with high humidity. In the western (African) coast, the mean temperature of the hottest month ranges between 30 and 35°C. The high humidity along the margins of the sea renders the summer heat especially oppressive though a sea breeze often develops during the afternoon, and alleviates the heat. Even at night, in summer the temperature is usually above 26°C. The northern part of the Red Sea, including the Gulfs, is just cool enough in winter to fall outside the tropical category.

Rainfall is deficient throughout. The northern third of the Red Sea is extremely arid with mean annual rainfall as low as 3 mm (in Hurghada, 400 km south of Suez, Egypt). The least dry stations are along the southern part, e.g. in Suakin (in the Sudan) and Massawa (in Eritrea) with rainfall more than 150 mm/year.

The erratic rainfall, which characterizes the arid and semi-arid areas, is obvious along the Red Sea coasts. Along the African coast, the mean annual rainfall fluctuates from 422 mm (in 1925) to 9 mm (in 1910) in Port Sudan, from 617 mm (in 1896) to 33 mm (in 1953) in Suaken (Kassas, 1957) and from 2.0 mm (in 1949) to 56.8 mm (in 1952) in Suez (Kassas and Zahran, 1962). In Massawa, rainfall fluctuates between 15% above and 59% below the annual mean, and the monthly variation is even greater e.g. the December maximum is 188 mm and minimum is nil (Hemming, 1961). The annual Pluviothermic coefficients (Emberger, 1951; Zahran and Willis, 1992) for Suez, Port Sudan and Massawa are: 4.17, 8.2 and 27.5, respectively.

Along the Asian coast of the Red Sea the variability of rainfall is extremely high. Sometimes, very heavy torrential rains occur during short unpredicted intervals in very localized areas, and are followed by long dry periods (Zahran, 1982b). Accordingly, the value of the average annual precipitation is debatable unless data over a considerable number of years and from at least two meteorological stations are used. The highest annual precipitation (220 mm) falls in Al-Wagh and the lowest (10.8 mm) in Yanbu Al-Bahr. In Jeddah and Jizan, the total annual rainfall averages are 78 and 31.8 mm, respectively. According to Draz (1956), Zahran and Al-Kaf (1996), in Yemen, monsoon rains occur during August, and precipitation is regular during August, September and October. The second rainy season occurs during the cool season at the end of the year whereas the third rainy season comes in the early summer as thunderstorms.

Temperatures of the Asian Red Sea coast exceed 48°C in the shade during summer and never drop below zero during winter. The highest mean annual temperature (29.4°C) is recorded at Jizan; the highest maximum (48.4°C) and the lowest minimum (9°C) temperatures have been recorded in Jeddah and Yanbu Al-Bahr, respectively.

Wind direction is almost constant throughout the year in Jeddah (NNW-N), Yanbu Al-Bahr (NW) and Al-Wagh (WNW-NW). In Jizan, the wind direction is west during summer and autumn and southwest during winter and spring.

Climatic variation is relatively high in the northern part of the Red sea Basin particularly in winter when it may be influenced by disturbances in the Mediterranean. South of Massawa (latitude 25°N), the salient feature of the climate is monotony;

there is little change from day to day, perhaps the very occasional thunderstorms, and only the slow change from week to week and month to month as the seasons come and go.

The water of the Red Sea have unique characteristics when compared to the other seas and oceans of the world. The very high surface-water temperature and salinity makes the Red Sea one of the hottest and saltiest water bodies of seawater anywhere; only in the Arabian Gulf are temperature and salinities somewhat exceeded. But the Arabian Gulf is much smaller and, in comparison, is very shallow.

The average depth of the Red Sea, excluding the shoreline shallows, is about 700 m and the main trough everywhere exceeds 1,600 m. Several depressions, exceeding 2,000 m deep, are known. Unlike all of the open oceans and seas of the world where the water depth is close to freezing point, the waters of the Red Sea are warm throughout. Below the strongly heated surface layers, temperature falls in the top 300 m or so but below about 350 m a remarkably uniform temperature of a little less than 22°C is, with few exceptions, maintained to the bottom.

## 2.4 Habitat Types

The four major ecosystems of the Afro-Asian Red Sea coastal lands occur in two main habitat groups: saline and non-saline. The saline habitats harbour the mangrove swamps and the littoral salt marshes. The non-saline habitats are suitable for two ecosystems: coastal desert plains and coastal mountains.

### *2.4.1 Saline Habitats*

The universal valid classification of the littoral shoreline habitats is based on the relation between the ground level and the level of the tides. According to this classification, five zones are distinguished (Kassas and Zahran, 1967): (1) supra-littoral zone, (2) supra-littoral fringe, (3) mid-littoral zone, (4) infra-littoral fringe and (5) infra-littoral zone. Such zonation pattern, though universal, but may or may not occur in the shorelines of the oceans and seas. Along the shorelines of the Red Sea in Asia and Africa, the supra-littoral zone and supra-littoral fringe with adjoining parts of the mid-littoral zone are recognized (Zahran, 1962, 1964). The former is referred to as the littoral salt marshes and the second is the habitat of mangrove swamps.

The usual habitat of the mangrove swamps is the shallow water along the shore especially the unprotected areas: lagoons, bays, coral or sand bars parallel to the shore. However, in a few localities, mangrove plants may grow as isolated individuals on the terrestrial side of the shoreline. The tidal mud of the habitat of the mangrove swamp is usually salt-affected, grey or black in color and is often foul smelling from the high content of humus. However, salinity of the mangrove muddy substratum is usually less than that of the neighbouring soil of the littoral salt marsh habitat. Continuous washing by sea water reduces the salt content of the mangrove soil when compared to that of the salt marsh ecosystem.

The littoral salt marshes comprise the maritime habitats within the land bordering the shoreline. These habitats are affected by the neighbourhood of the sea. Though lying above the level of the tide-marks, the littoral salt marsh habitats are subjected to maritime influences: sea water spray, seepage of ground salt-water, tidal movement, and accidental inundation during exceptional gales. The vegetation is clearly zoned. The zonation of the salt marsh vegetation is obviously a universal phenomenon (Chapman, 1960) usually attributed to the: (1) width of the coast, (2) land levelling and (3) spatial gradient of the salinity of the soil. However, in the littoral salt marshes of the arid climate, e.g. Red Sea coastal areas, the gradient does not form a regular pattern of decreasing salinity further from the shoreline. Areas that are nearer to the shoreline and subjected to recurrent wash by sea-water are less saline than adjacent areas that are only accidentally inundated by salt-water. The zonation sequence is complete only where the shore rises gently and almost imperceptibly into the land. This is rarely the case. The topography of the salt-marsh is rarely uniform since it comprises different types of shoreline bars and its ground may be studded by mounds and hummocks of various dimensions. Again the ground of the littoral salt-marsh may rise landward such that it may reduce the salt-marsh or some of its zones.

The littoral salt-marsh ecosystem of the Asian and African Red Sea coasts comprises three habitats: two are wet and the third is dry, namely: saline flat habitat (wet), reed swamp habitat (wet) and piled sand (sand formation) habitat (dry).

The ground surface of the saline flat habitat is higher than the level of the tide, or areas that are protected against tidal inundation by shoreline bars piled by waves or tides. The soil of the reed swamp habitat is salt affected, usually covered with water. It occurs in channels and creeks at the mouth of the big wadis and areas which represent the combined influences of the brackish water springs (Kassas and Zahran, 1962). The sand formation (piled sand) habitat occur, in two forms: a. sand bars (coastal dune) along the shoreline formed of homogeneous slightly saline sandy material and b. sand mounds, hummocks and hillocks usually piled by wind (aeolian sand). These sand formations occur as elevated islands in the successive zones of the littoral salt-marsh ecosystem. Plants inhabiting the sandy habitat (psammophytes) are capable of trapping sand and may, sometimes, be completely covered with sand for long periods without being damaged.

### *2.4.2 Non-saline Habitats*

The non-saline habitats of the Red Sea coastal lands comprise two dry coastal ecosystems namely: coastal desert plain ecosystem and coastal mountains ecosystem.

The coastal desert plain ecosystem occupies the area between the littoral salt marsh ecosystem (on the seaward side) and the coastal range of hills and mountains (on the inland side). Unlike the littoral salt marsh ecosystem where the zonation pattern of vegetation is clear, the coastal desert plain ecosystem has a complicated pattern of topography, characteristics of surface deposits and relationships with mountains. For their intermediate ecological position, the coastal desert plains show, especially on their fringes, transitional characters between the two neighbouring seaward and landward ecosystems. However, the inland boundaries are not as clear as the seaward ones (Zahran, 1965).

The coastal desert plains are essentially covered with gravels, traversed by the mid-stream parts and down-stream extremities of the main wadis (dry water courses) and are dissected by smaller drainage runnels that may extend from the foothills to the coastal front or may not reach the coast. The downstream extremities of the main wadis may form deltaic basins with finer sediments when compared to the sediments of the mid-stream and up-stream parts. Superimposed of this pattern aeolian sand deposits may form desert sheets, mounds, hummocks and hills of various size and extent. This complex set-up produces likewise complex habitat conditions. The vegetation pattern associated with these habitat conditions is further complicated by differences in the intensity of human interference and wild-life grazing and also by differences in the geographical ranges of the floral elements.

The Red Sea coasts are bounded on the inland side by an almost continuous range of hills and mountains with various heights (<500 to >2,000 m). The presence of these coastal mountains has influenced the climate and the water resources of the nearby desert areas. For example, according to Murray (1951), in Egypt, though regular rainfall ceased about the close of the Plio-Pleistocene period, ¾ of a millions years ago, yet their torrential storms usually occur in the Red Sea mountains increasing the amounts of rainfall received in this ecosystem by 2–3 times. The mean annual rainfall of the Red Sea shoreline in Egypt ranges between 3.4 mm up to 30 mm. On the other hand, precipitation in Gebel Elba mountains range may reach 60–100 mm/year (Zahran, 1964). Kassas (1956) stated that Erkwit mountains of the Sudan Red Sea coast receive 218 mm annual rainfall which is greater than the amounts of rainfall on Suakin (181 mm/year) and Sinkat (127 mm/year). In Saudi Arabia (Anonymous, 1977; Zahran, 1982b), the average annual rainfall of the Red Sea shoreline ranges between 14.5 and 77.9 mm whereas the mountain ecosystem enjoys rainfall up to 655 mm/year.

The relatively ample orographic precipitation is referred to as: "occult precipitation", "horizontal precipitation" or "fog precipitation" (Kassas, 1956). The amount of this type of precipitation differs on the different slopes: the windward side is moist whereas the leeward side is very dry (rainshadow). The amount of this rain depends on: (1) the direction of wind in relation to land-water direction, (2) the expanse of the water body crossed by the wind, (3) altitude and (4) distance from the shoreline. This orographic precipitation may produce local oases of rich vegetation on the slopes of the mountains that are given the names "Nebelwald", "Nebeloasen", "Mist Oases", Kassas (1956).

Tothill (1948) and Kassas (1956) reported that the climate of Erkwit mountains in the Sudanese Red Sea coast may be due to the fortunate combination of three factors: latitude, situation and elevation. Its latitude (18° 46′N) is close to the latitude 19° which divides the Sudan into a desert region to the north and a tropical continental region to the south.

Generally, habitat variation of the coastal mountain ecosystem of the Red Sea is influenced by 5 main factors, namely:

1. Expanse of water body traversed by the wind before reaching the coastal land.
2. Distance from the shore-line to the mountains and the topographic features of the stretch involved.
3. Heights of the mountains.
4. Exposures of the slopes of the mountains.
5. Physical and chemical characteristics of the substratum.

All of these factors control, to great extent, the amount of water received in the different habitats of the mountains and the availability of the water to the plants.

## 2.5 Vegetation Forms

The main vegetation forms of the Afro-Asian Red Sea coastal lands are: mangrove (mangal) vegetation, reed swamp vegetation, littoral salt marsh vegetation and vegetation of the coastal desert plains and mountains.

Mangal (mangrove), reed swamp, and littoral salt marsh vegetation forms occupy the wet-saline habitats whereas the vegetation forms of the coastal desert plains and mountains occur in the dry non-saline habitats. In addition, psammophytic vegetation is present in the littoral-saline sand dunes as well as in the non-saline sandy hummocks and hillocks that characterize certain wadis of the Red Sea coastal desert plains.

### *2.5.1 Mangrove Vegetation*

#### 2.5.1.1 Ecological Features

Mangrove is a West Indian name given to halophytic trees and shrubs that normally grow in the intertidal zone of tropical and subtropical coast-lines. They usually occur where large deltas of running rivers (or desert wadis like those of the Red Sea coastal desert) empty their water into the seas (Chapman, 1977; Zahran, 2002). According to Walsh et al. (1975) the total area of approximately 170,000 km$^2$ of tropical and subtropical coast-lines is occupied by mangrove forests. Clough (1993) stated that the best developed mangroves occur along the humid equatorial coast-lines in south-east Asia where they often form extensive tidal forests with trees of more than 1 m diameter and up to 45 m in height.

The mangrove (mangal) vegetation, in general, occupies the supra-littoral fringes and the adjoining mid-littoral zone. In tropical regions, the upper zones of the shores of estuaries and the more sheltered marine shores are vegetated by mangrove trees

(Chapman, 1975). Zahran (2007) mentiond that mangals occur only in the tropical region, approximately between 35ºN and 38ºS. The restriction to these regions is conditioned by the sensitivity of mangroves to frost. Temperature conditions, thus, limit the spread of mangroves, whereas rainfall conditions are more decisive for the sequence of zones in the tidal range. On flat coasts protected from wave action and in estuaries, the duration of inundation with sea water during flood periods decreases from sea landward.

The latitudinal limits of mangroves are close to 31ºN (in southern Japan) and 38ºS (in southern Australia). At these latitudinal extremes mangal vegetation is floristically simple. In each case the community is monospecific, i.e. comprises a single species (*A. marina* in the southern hemisphere and *Kandelia candel* in the northern hemisphere). Mangrove swamps can be seen also along the coastal belts of the arid region of the world, notably the coasts of north-eastern Africa (e.g. Red Sea coast), Pakistan, the Arabian Peninsula (Asian Red Sea, Arabian Gulf, Arabian Sea, Gulf of Oman and Gulf of Aden coasts), and the western coasts of Australia (Clough, 1993). In these arid coastal belts, mangrove communities comprise relatively shorter trees and shrubs, lower plant cover and a smaller number of associate species than those of the humid coast-lines. According to Ball et al. (1983), the low floristic diversity of the arid region's mangrove vegetation may be attributed, at least partly, to differences in tolerance of salty water and radiation stress between species.

The main environmental factors influencing the growth and survival of mangrove plants have been discussed widely. Clough (1993) considers opinion that temperature, salinity and aridity appear to be the key factors along the environmental gradients. It is stated (Anonymous, 1987) stated that the main abiotic factors influencing the growth of mangrove plants are: (1) mixing of water from the landmass and the sea with consequent variation in salinity, (2) the accumulation of sand along the coastal belts, which make it easier for certain mangrove species to put down their pneumatophores, (3) heavy rainfall, (4) high temperature with scarcely any variation, and (5) a high degree of atmospheric humidity. In addition, mangrove swamps develop only where coastal physiography and energy conditions are favourable. They are more extensive where there is a low shore gradient, and a large tidal range. Davies (1972) believed that sheltered habitats are essential for mangrove development; coast-lines mangrove communities are localized in the lee of other coastal landforms, low wave energy shorelines are the most suitable habitat for mangrove development.

Biotic and abiotic factors of mangrove swamps have made it possible for many marine populations which have acquired significant economic value to establish their habitats in the mangrove ecosystems. The species of these populations are favoured by the mangrove swamps because, for example, of the abundant food supply and because they can find pockets of still water where they can lay their eggs and their larvae can grow and mature. "In addition to being fished commercially and for subsistence consumption, many of these species are suitable for fishing farming so that aqua-culture can be undertaken on a large scale, thus improving the food supply of the inhabitants" (Anonymous, 1987). The economic potentialities of mangrove ecosystems as habitat and detrital food sources for marine organisms as well as their direct commercial value as lumber, fire, wood, painting agents, animal fodder, raw materials for the drug industry as well as shoreline stabilizers against erosion are well documented by Chapman (1975). The net productivity of mangrove ecosystems has been estimated to be about 930 g $m^2$/year on $CO_2$ exchange state (Zahran, 2007).

#### 2.5.1.2 Are Mangroves Prehistoric Plants?

Raymond and Phillips (1983), stated that genera of extant mangrove communities can be traced back to the Early Tertiary, 55 million years ago, largely by means of palynology. There is a gradual expansion in the diversity and association of mangrove genera from earliest Eocene time. Pollen grains from the black mangrove genera, namely: *Nypa* and *Brownlowia,* occur together in the Lower Eocene of Borneo. Mangroves which appear in successive epochs include *Rhizophora* pollens in the Early Oligocene and *Sonneratia* pollens in the Early Miocene. In Australia, Churchill (1973) has described old pollen of *Nypa*, *Avicennia*, *Rhizophora* and *Sonneratia* types. However, the oldest putative mangrove community of angiosperms has been described from the Late Cretaceous Dakota Sandstone of Kansas (Ratallack and Dilcher, 1981).

#### 2.5.1.3 Mangroves of the Red Sea Coasts

The usual habitat of mangrove vegetation of the Red Sea coasts is the shallow water along the shorelines, specially in protected areas: lagoons, bays, corals or sand bars parallel to the shore (Kassas and Zahran, 1967; Zahran, 2007). According to IUCN/PERSGA (1987), the biodiversity and productivity of the mangrove communities in many parts of the world are generally higher than those found in the Red Sea coasts. There are no major deltaic regions having dense growth of mangroves neither in the eastern (Asian) nor in the western (African) coasts of the Red Sea.

Most of the Red Sea is located south of the Tropic of Cancer, i.e. within the climatic limits of the region known to be favourable for the occurrence of mangroves. Also, the shoreline morphology as well as the local habitats of both African and Asian Red Sea coasts are suitable for the growth and domination of certain species of mangroves. Along these two coasts there are a series of small bays that cut into the raised beach. These bays are partly land-locked by further coral reefs and the sea water in them is sheltered. Wave action is reduced to a minimum. Such protected and shallow lagoons and bays of the Red Sea coasts provide a favourable habitat for the growth of mangal vegetation.

##### African Red Sea Coast

Three mangrove species have been recorded along the African Red Sea coasts, namely: *Avicennia marina, Rhizophora mucronata* and *Bruguiera gymnorrhiza.* The cover, density, stratification and distribution of these species vary. They are absent from the shoreline of the coast of the Gulf of Suez but their presence at the mouth of the gulf at the Sinai side (27° 40′N) was confirmed by Ferrar (1914) and Anonymous (2006). *A. marina* grows in a narrow lagoon (marsa) with shallow water. Along the African Red Sea coast, the northern boundary of the mangal vegetation is represented by a few, scattered and depauperate individual bushes of *A. marina* in Myos Hormos Bay, about 22 km north of Hurghada (27° 14′N). From Hurghada southwards to Marsa Halaib (22°N), the *A. marina* mangal is notable and is one of the common features of the vegetation of the littoral landscape of Egypt. Pure stands of *A. marina* mangal vary in extent from limited patches of a few individuals to continuous belts of dense growth extending for several kilometers. This distribution seems to depend upon the local conditions of the shoreline morphology (Kassas and Zahran, 1967). On Abu Minqar Island (offshore of Hurghada) and on Safaga Island (offshore of Safaga, about 60 km south of Hurghada) there are, relatively, dense thickets of *A. marina* mangal.

*Rhizophora mucronata* is recorded in the most southern part of the Red Sea coast of Egypt. In the shoreline between Marsa el-Madfa (23°N, near Marsa Halaib on the Sudano-Egyptian border), *R. mucronata* forms an almost pure growth; in other places it is associated or co-dominant with *A. marina*.

*A. marina* thickets extend further southwards, dominating the shoreline of the Red Sea coast of the Sudan. In a localized area south of Suakin (latitude 19° 15′ to 19°N) early reports by (Brown and Massey, 1929; and Andrews, 1950–1956) recorded that *R. mucronata* and/or *Bruguiera gymnorrhiza* dominate or co-dominate with *A. marina*. However Professor M. Kassas (personal communication) in his visit to Suakin (1966) did not find a single tree of *B. gymnorrhiza*. The disappearance of this species may be attributed to man's interference, because the inhabitants used to cut the bark for use in dyeing.

The shoreline of the Ethiopian (Eritrean) coast and southward to the far end of the African Red Sea coast in French Somaliland is fringed by the growth of *A. marina* mangal. Other mangroves (*R. mucronata* and *B. gymnorrhiza*) are not recorded (Hemming, 1961; Zahran, 2007).

The structure of the mangal vegetation of the Red Sea coast is simple, usually one layer of *A. marina* (Zahran, 1977). In the localities where *R. mucronata* and/or *B. gymnorrhiza* are included, these plants form a layer towering over that of *A. marina*. The ground layer is formed of associated marine phanerogams, e.g., *Cymodocea ciliata, C. rotundata, C. serrulata, Diplanthera uninervis, Halophila stipulacea* and *H. ovalis*.

### Asian Red Sea Coast

The Asian Red Sea coast is characterized, also, by three mangrove species: *Avicennia marina, Rhizophora mucronata* and *Bruguiera gymnorrhiza*. The distribution and densities of these species and their communities are of ecological interest. Vesey-FitzGerald (1955) stated "*Avicennia marina* is not common north of Jeddah (20° 30′N) but a few small shrubs of this species do occur in sheltered creeks even as far north as Yanbu (24°N)." Zahran et al. (1983) found that the growth of *Avicennia marina* extends northwards as far as Al-Wagh (26°N) where it forms dense vegetation in Hanak Islands (facing Al-Wagh). This means that *Avicennia marina* mangrove is growing in areas north of the Tropic of Cancer. The same was recorded along the Western (African) shoreline of the Red Sea (Kassas and Zahran, 1967; Zahran, 1977). This confirms Chapman's statement (1977) that mangal reaches its optimal development in the tropics but it extends to the subtropics described as warm temperate.

The mangrove swamps of the Saudi Arabian Red Sea coast are generally dominated by *Avicennia marina* (Zahran, 2007). A few individual trees of *Rhizophora mucronata* were recorded associated with the *Avicennia marina* community in Jizan swamps (16ºN), the southernmost border of the Saudi Arabian Red Sea coast. However, Zahran (1982b) and Zahran et al. (1983) did not record *R. mucronata* neither in the Jizan coast nor northward. On the other hand, El-Demerdash (1996) found *R. mucronata* in the Farasan Islands, offshore of Jizan.

Density, stratification and vigor of *A. marina* stands vary in the different stands of the Saudi Arabian Red Sea coast. North of Jeddah (except in Hanak Islands), *A. marina* occurrs in scattered patches of low bushes, rarely trees. These coastal green spots are usually seen in the lagoons e.g., Thewel and Rabegh lagoons (100 and 160 km north of Jeddah, respectively). In these lagoons *A. marina* plants are protected from strong waves and winds. Unprotected shorelines north of Jeddah have almost no mangal vegetation. South of Jeddah, mangal vegetation flourishes where dense *A. marina* forests characterize this section of the Saudi Red Sea coast. In the Jizan and Sabya coasts, for example, there is luxuriant plant cover of *A. marina* vegetation. In some stands branches of *A. marina* are so closely intermingled that it is difficult to pass through (Zahran et al., 1983; Mandura et al., 1987).

In the Yemeni section of the Red Sea coast, *A. marina* constitutes the main bulk of mangrove vegetation where its community occurs in narrow, discontinuous and pure stands. In the past two more mangrove species have been recorded: *Rhizophora mucronata* and *Bruguiera gymnorrhiza*. In Khant Katik near Hodeidah fringing a small island. Draz (1956) recorded a single huge *B. gymnorrhiza* tree on the Hodeidah coast. Recent studies in the Yemeni Red Sea coast by IUCN/PERSGA (1987) and Zahran and Al-Kaf (1996) did not record *B. gymnorrhiza* in the mangrove swamps. It is obvious that the single tree has been cut down.

Extensive stands of *A. marina* along the northern part of the Yemeni Red Sea coast are reported by Al-Hubaishi and Müller Hohenstein (1984). It is the natural southward extension of the Saudi Arabian mangrove vegetation. Approximately 84 km of the Yemeni Red Sea coast supports this type of vegetation. But, less developed *A. marina* mangroves occur along the whole of the Yemeni Red Sea coast (about 550 km).

The soil associated with the Red Sea mangal vegetation generally consists of sandy mud, black in colour, rich in organic matter and decaying debris, and often foul-smelling. The great bulk of the soil (60–75%) is formed of two particle-size ingredients: 0.211–0.104 mm and 0.104–0.050 mm diameter, with a small proportion of coarse sand (10%, > 0.211 mm diameter). The very fine fraction

(< 0.05 mm diameter) increases from 10% at the landward edge to 20% in the waterlogged mud. The soil reaction is alkaline (pH = 8–8.95) and its organic carbon content ranges between 0.32 and 2.22%. The total soluble salts of the soil range between 1.2 and 4.3%, on an oven-dry weight basis, mostly chlorides (0.5–1.75%) and partly sulphates (0.03–4.2%). The calcium carbonate (total carbonates) of the mangrove soils varies from 80% in the soil of *R. mucronata* to 4.5–19.5% in the soil of *A. marina*. "This edaphic distinction requires confirmation" (Kassas and Zahran, 1967).

It is not unusual to find *A. marina* on the terrestrial side of the shoreline. In the delta of Wadi Gimal (ca. 50 km south of Marsa Alam, Egypt), there is a stand of *A. marina* in a terrestrial habitat where the bushes are covered by sandy hillocks. This is apparently due to silting up of the shoreline zone, which was occupied originally by a lagoon with mangal growth.

### 2.5.2 Reed Swamp Vegetation

The habitat of the reed-swamp vegetation of the African Red Sea coast is provided by the channels and creeks of the mouth of the big wadis, e.g. Wadi El-Ghweibba and areas which represent the combined influences of the brackish-water springs, e.g., the Ain Sokhna area in Egypt (Kassas and Zahran, 1962) The reed swamps are represented by *Phragmites australis* and *Typha domingensis*. *T. domingensis* usually grows in areas where the soil is relatively less saline and the water is not too shallow, e.g. estuaries of wadis that collect the occasional surface drainage. *P. australis*, on the other hand, grows in swamps close to the dry land, often with higher soil salt content and drier than those of *T. domingensis* (Zahran, 1977). The common associates of the reed-swamp vegetation of the African Red Sea coast include: *Berula erecta, Cyperus articulatus, C. dives (alopecuroides), C. (Pycreus) mundtii, Lemna gibba, Samolus valerandi, Scirpus mucronata, S. tuberosus, Spirodela polyrrhiza* and *Wolffia hyalinas*.

Semi-enclosed lagoons which allow salinities to drop during low water and water creeks are the suitable habitats of the reed swamp vegetation along the Saudi Arabian and Yemeni (Asian) Red Sea coasts. In some of these lagoons the water is only slightly brackish. Encrustations were evident in some other areas (Zahran and Al-Kaf, 1996). These reed swamps are dominated by *Typha domingensis* with *Phragmites australis* as abundant or co-dominant species. However, in the most downstream section of the delta of Wadi Jizan in the territories between Saudi Arabia and Yemen, *Typha elephantina* grows in the catchment areas of water coming from the coastal mountains (Zahran, 1982b). The other species recorded in the reed swamps of the Asian Red Sea coast in Saudi Arabia and Yemen include: *Cyperus conglomerates, Echinochloa colona, Fimbristylis* sp., and *Scirpus tuberosus*. The saline fringes of the reed swamp lagoons support vegetation with halophytes e.g. *Aeluropus lagopoides, Cressa cretica, Cyperus laevigatus, Tamarix nilotica,* and *Zygophyllum album*.

## 2.5.3 Littoral Salt Marsh Vegetation

### 2.5.3.1 African Red Sea Coast

The vegetation of the littoral salt marshes of the African Red Sea coast comprises sixteen main community types dominated by: *Aeluropus* spp., *Arthrocnemum macrostachyum, Halocnemum strobilaceum, Halopeplis perfoliata, Juncus rigidus, Limonium pruinosum, L. axillare, Nitraria retusa, Salicornia fruticosa, Sporobolus spicatus, Suaeda fruticosa, S. monoica, S. vermiculata, Tamarix mannifera,*[1] *T. passerinoides,* and *Zygophyllum album*. Other dominants are: *Alhagi maurorum, Cressa cretica,* and *Imperata cylindrica* (Zahran, 1977). The ecological amplitudes of these dominants vary and this results in differences in their distribution, density, stratification, zonation and floristic composition. A brief account of each of these community types follows.

#### (i) ***Halocnemum strobilaceum*** community

This community occurs on the inland side of the shoreline. The ground is either a tidal flat, recurrently washed by tidal flow, or a shoreline bar of sand on rock detritus heaped up by wave and tidal action.

The growth of *Halocnemum strobilaceum* occurs in two forms: (1) circular patches on flat tidal mud ground and (2) sheet of irregular shaped patches on the shoreline bars.

In Egypt, *H. strobilaceum* is common within the littoral salt marshes of the Gulf of Suez (from Suez to Hurghada, about 400 km stretch) but not in the region further south on the African Red Sea coast. The plant cover is often pure stands of the dominant species but it may be associated with *Arthrocnemum macrostachyum, Zygophyllum album, Cressa cretica* and *Alhagi maurorum.*

*H. strobilaceum* is not recorded in the littoral salt marshes of the Sudan (Andrews, 1950–1956; Kassas, 1957). On the Red Sea coast of Ethiopia, Hemming (1961) did not refer to this community type, but *H. strobilaceum* is recorded among the species of the flora of Ethiopia (Cufodontis, 1961–1966).

#### (ii) ***Arthrocnemum macrostachyum*** community

*A. macrostachyum* occupies the same shoreline zone as the *H. strobilaceum* community and shows a similar growth habit. In Egypt, *A. macrostachyum* occurs throughout the whole Red Sea coast but it is less common in the northern part (coast of the Gulf of Suez), where the *H. strobilaceum* community abounds. Its associated species are: *Atriplex farinosa, H. strobilaceum, Halopeplis perfoliata, Limonium axillare, Zygophyllum album,* etc. (Kassas and Zahran, 1967).

On the Red Sea coast of the Sudan, *A. macrostachyum* covers long but narrow strips fringing the shoreline. The plant cover is dense on the sandy beach ridges and thins off a few metres inland. The second type of habitat consists of flat patches of

[1] *T. nilotica* (Boulos, 1995).

newly formed littoral plains, where the land is just above the level of the high tide but may be washed by high seas (Kassas, 1957). *A. macrostachyum* forms circular patches that increase in size till they coalesce. In the meantime phytogenic mounds are gradually produced.

*A. macrostachyum* is very abundant on the Eritrean Red Sea coast and is dominant in the salt flats lying about 20 m from the sea. "The areas dominated by *A. macrostachyum* did not appear to differ noticeably from that dominated by *Halopeplis perfoliata* except in that it is nearer to the sea" (Hemming, 1961). This is also repeated in the Egyptian and Sudanese Red Sea coast where *A. glaucum* is the first community inland from the mangal vegetation followed, in certain localities, by *H. perfoliata.*

### (iii) *Salicornia fruticosa* community

*S. fruticosa* and *A. macrostachyum* are difficult to distinguish from each other in their vegetative form; it is only at their time of flowering that they can easily be separated, *A. macrostachyum* flowering in April while *S. fruticosa* flowers much later (Kassas and Zahran, 1967).

*S. fruticosa* is recorded in the floras of Egypt, the Sudan and Ethiopia, but its dominance is not recorded neither in the survey of the Sudan Red Sea coast (Kassas, 1957) nor in that of the Eritrean coast (Hemming, 1961). On the Red Sea coast of Egypt, *S. fruticosa* is dominant over a very limited area (Mallaha, 260–280 km south of Suez). "There, *S. fruticosa* forms patches of pure growth with plant cover ranges from 50 to 100%. The ground is covered with a thick crust of salts (total soluble salts = 80.05%)" (Kassas and Zahran, 1965).

### (iv) *Halopeplis perfoliata* community

*H. perfoliata* dominates a community type which occupies a zone that follows, on the landward side, the littoral zone of the *A. macrostachyum* community type. This littoral zone may be occupied also by sand mounds covered by *Zygophyllum album*. The ground zone of *H. perfoliata* is often lower in level than the littoral zone where the wave-heaped detritus or the wind-deposited sand may form slightly elevated bars. It is also lower in level than the higher ground of the zones further inland. This seems to cause conditions of poor drainage in most of the localities, so that the ground surface is usually wet and slippery. In a few localities the zone of *H. perfoliata* may be at the shoreline.

The biogeography of *H. perfoliata* along the African Red Sea coast is of ecological interest. Within the 950-km stretch from Suez to Marsa Kileis (in Egypt), *H. perfoliata* is recorded in one locality: ca. 185 km south of Suez (Kassas and Zahran, 1962) but is otherwise very rare or absent. Southward of Marsa Kileis, *H. perfoliata* and its community is a common feature of the littoral salt marshes. The plant cover ranges from 5 to 40% mainly contributed by the dominant. *A. macrostachyum, Z. album, Cressa cretica, and Suaeda vermiculata.* are common associates. On the Sudan coast, the *H. perfoliata* community type occupies, as in Egypt, a zone inland to that of *A. macrostachyum*. The plant cover is thin (5–10%) and the associated species include *A. macrostachyum, Aeluropus lagopoides, Suaeda fruticosa, S. monoica, S. vermiculata* and *S. volkensii* (Kassas, 1957).

On the Eritrean coast, *H. perfoliata* colonizes the fringes of the salt flats which have no superficial sand mantle. Large areas of this habitat exist 16 km south of Marsa Cuba (latitude 16º 15′N) and about 1 km from the sea. "*H. perfoliata* grows to about 23 cm and is able to tolerate more saline conditions than any other species found here. The plants are quite widely scattered" (Hemming, 1961).

### (v) *Limonium pruinosum* community

The *L. pruinosum* community type occupies a salt-marsh zone bordering the shoreline zone of the *Halocnemum strobilaceum* community. It is common within the northern 100-km stretch of the Gulf of Suez (in Egypt). The plant cover ranges from 10 to 20%, and the associated species include: *Arthrocnemum macrostachyum, H. strobilaceum, Nitraria retusa, Suaeda calcarata,* and *Zygophyllum album*. The presence of *L. pruinosum* is not recorded southward in the littoral salt marshes of Egypt and the Sudan nor in Ethiopia.

### (vi) *Limonium axillare* community

*L. axillare* is a non-succulent salt-marsh half shrub, which may extend inland to the fringes of the coastal desert plain. Its community abounds in the saline flats of the Red Sea littoral salt marshes.

Within the northern 650-km stretch of the Gulf of Suez and the Red Sea coasts (in Egypt), *L. axillare* is found in one or two localities that are widely spaced: 61–18 km south of Suez and 114 km south of Suez. In the former locality there are a few patches of *L. axillare* whilst in the latter locality there are a few individuals. Within the stretch extending southwards from 650 km south of Suez, *L. axillare* and the community type it dominates are among the common features of the salt-marsh formation. In this community the dominant species contributes the main part of the vegetation cover (5–50%). *Aeluropus* spp., *H. perfoliata, Salsola baryosma, Salsola vermiculata, Sevada schimperi, Sporobolus spicatus and Z. album* are among the common associate species of the stands of the *L. axillare* community in the Red Sea coast of Egypt (Zahran, 1964).

Kassas (1957) recorded *L. axillare* among the species commonly present in the coral limestone raised-beach habitat of the Red Sea coast of the Sudan. A surface crust of salts may render it sterile and the plant cover is usually 5–10%. The other species are: *Aeluropus lagopoides*, *Cyperus conglomeratus* var. *effusus* and *Suaeda vermiculata.* Also, on the Eritrean coast the raised-beach soil or the soil that overlies raised beaches is the favoured habitat for the growth and dominance of *L. axillare.* In the quartz sand over the raised-beach soil the vegetation is mixed. "*L. axillare* may extend on to the sand with some associate species: *Eremopogon foveolatus, Indigofera argentea, I. semitrijuga, Salsola sp., Zygophyllum coccineum* (*Z. album*), etc." (Hemming, 1961).

### (vii) *Zygophyllum album* community

*Z. album* is omnipresent in the Red Sea littoral salt marshes. On the Egyptian coast it may grow in the form of small individuals distantly spaced on the saline ground of the dried salt marsh. It may also build small phytogenic mounds of sand that stud the ground surface of the dried salt marsh. Being widely distributed, the community type dominated by *Z. album* includes numerous other species (44 species including 6 ephemerals) with differing ecological requirements (Kassas and Zahran, 1967). These include: *Aeluropus* spp., *H. perfoliata, L. axillare, L. pruinosum, Nitraria retusa,* and *Sporobolus spicatus* as salt-marsh species; and *Panicum turgidum* and *Salsola baryosma. Launaea spinosa,* as desert-plain species. The dominant contributes 5–50% of the plant cover.

On the Red Sea coast of the Sudan, *Z. album* is listed as among the most common associated species of the salt-marsh communities dominated by *Arthrocnemum glaucum,*[2] *Aeluropus lagopoides*, etc. (Kassas, 1957). In his second survey (1966) of the Red Sea coast of the Sudan, Professor M. Kassas (personal communication) found that *Z. album* is widespread along the whole stretch southward to the Sudano-Eritrean border. In Tokar (ca. 160 km south of Suakin and ca. 80 km north of the Eritrean border), for example, the *Z. album* community covers vast areas of the salt marshes.

In the Eritrean Red Sea coast, a *Z. coccineum* community has been included among the communities of the littoral salt marshes (Hemming, 1961). However, *Z. coccineum* is considered a xerophytic and salt-intolerant plant dominating a widespread area in the Egyptian inland desert (Täckholm, 1956, 1974; Zahran and Willis, 1992) and the Sudanese Desert (Andrews, 1950–1956; and herbarium specimens, Botany Department, Faculty of Science, Cairo University). The *Z. coccineum* mentioned by Hemming (1961) in the Eritrean littoral salt marshes may be confused, as far as identification is concerned, with *Z. album*, the true halophytic species. Cufodontis (1961–1966) recorded *Z. album* in the flora of Ethiopia. Hence, *Z. album* (*Z. coccineum*) is widely distributed in the Eritrean Red Sea coast. It occurs in several habitats, e.g., in the coral line sand dunes and mantles, in the quartz sand over raised-beach soil, and in the normal raised-beach soils, where it dominates a well developed community type. The soil supporting this community (*Z. coccineum*[3] of Hemming, 1961) is saline, the total soluble salts ranging between 3.1% in the superficial coral, 5.5% at 15 cm depth, and 2.9% at 30 cm depth. The associated species in the Eritrean coast include: *Cornulaca ehrenbergii, Cyperus conglomerates* and *Limonium axillare*.

### (viii) *Aeluropus* spp. community

The grassland community type dominated by *Aeluropus* spp. (*A. lagopoides* and *A. massauensis*) occupies one of the inland zones of the Red Sea littoral salt

[2] *A. macrostachyum* (Boulos, 1995).

[3] Vesey-FitzGerald (1955, 1957) and Younes et al. (1983) described a community dominated by *Z. coccineum* in the Saudi Arabian Red Sea Littoral salt marshes.

marshes, the dominant grass forming patches or mats of dense growth. These are sometimes covered by spray-like crusts of salt which denote that the plants may be temporarily covered by saline water, which, on receding and drying, leaves the salt on the shoots of the plants. The growth-form of the grass is usually that of the creeping type, but in one locality of the Red Sea coast of Egypt (Marsa Alam, 700 km south of Suez) it forms peculiar cone-like masses of interwoven roots, rhizomes and sand. The plant cover (10–80%) is mostly contributed by the dominant plant, and its associates are mainly halophytes, e.g., *A. glaucum (= A. macrostachyum), Cyperus laevigatus, Halopeplis perfoliata, Limonium axillare, Sevada schimperi, Sporobolus spicatus, Tamarix mannifera* (= *T. nilotica*) and *Zygophyllum album.*

The community type dominated by *A. lagopoides* in the Red Sea coast of the Sudan follows the zone of *H. perfoliata*, occupying areas where the surface layer of the soil is apparently wind-borne material. The plant cover ranges from 50 to 70% and it is mainly of *A. lagopoides* associated with *Suaeda fruticosa, Sporobolus spicatus*, etc.

In the Eritrean Red Sea coast, *A. lagopoides* prevails in impeded drainage sites, or often in areas behind coral sand dunes along the shore with alluvium deposits, where the surface of the soil is generally a fine brown silty sand with a cracked surface. In the sand-flat areas which lie at the edge of the sands of the coastal plain, *A. lagopoides* is the common grass, where it is found on sand 6 cm deep and also on low hummocks of wind-blown sand.

### (ix) *Sporobolus spicatus* community

The *S. spicatus* community occupies a zone inland to that of *Aeluropus* spp. where the sand deposits are much deeper than in the *Aeluropus* zone and the soil salinity is much less. In many localities it is seen that the two grassland community types grow mixed in the same zone forming a mosaic pattern, the *Sporobolus* grassland on the higher parts forming island-like patches amongst the sea of lower saline ground covered by *Aeluropus* grassland. In this grassland the ground-water level is usually shallow (40–100 cm) but in the *Sporobolus* grassland the ground-water level is usually deeper than 150 cm.

In the Red Sea coast of Egypt, the *S. spicatus* community is common in the southern part (south of 1,030 km south of Suez) but not in the northern stretch. In the Sudan coast, the *S. spicatus* community is common and it follows the *A. lagopoides* community in succession sequences. The sand drift is deeper and shows no profile feature. Apart from the salt-marsh associate species, e.g., *A. lagopoides, Limonium axillare* and *Suaeda fruticosa*, there are about twelve species which are common in the inland plain and not typical salt-marsh plants, e.g., *Panicum turgidum, Lycium arabicum* and *Euphorbia monacantha.*

The *S. spicatus* community abounds in several habitats in the Eritrean Red Sea coast, such as (a) the sandy beaches, in areas where the soil is brown and more silty than the coral soil above. Here, *S. spicatus* is mixed with *Cyperus conglomeratus, Aeluropus lagopoides, Panicum turgidum, Dactyloctenium aristatum, Urochondra setulosa*, etc.; (b) the alluvium over coral lime soil where *S. spicatus* and *Eleusine compressa* are abundant between the clumps of *A. lagopoides*; (c) the salt flats with a superficial sand mantle and in a neighbouring area of the *Aeluropus* zone, the transitional zone between these salt flats is occupied by *S. spicatus* and *Eleusine compressa*.

### (x) *Nitraria retusa* community

*N. retusa* is a salt-tolerant bush that grows in two types of habitat. In the first type *Nitraria* forms saline mounds or hillocks that stud the flat ground of the salt marsh. Commonly, *N. retusa* covers the north-facing part of the hillocks of this habitat type. The second type of habitat comprises sandy bars (actually chains of sandy hillocks) fringing the shoreline. These bars are less saline than the hillocks of the other type.

The *N. retusa* community type is widespread in the northern 600 km stretch of the Gulf of Suez and Red Sea littoral salt marshes of Egypt, but is absent in the coastal land south of Marsa Alam (700 km south of Suez). It is not recorded neither in the survey of the Sudan coast (Kassas, 1957) nor in the flora of the Sudan (Andrews, 1950–1956). In the survey of Hemming (1961) for the Eritrean Red Sea coast, *N. retusa* is absent, though Cufodontis (1961–1966) mentioned it in the flora of Ethiopia.

The floristic composition of the *N. retusa* community includes *Z. album, Cressa cretica, Alhagi maurorum, Suaeda monoica* and *Tamarix mannifera*.

### (xi) *Suaeda monoica* community

*S. monoica* is a frutescent plant of the dry salt marshes, comparable in habit and habitat to *Nitraria retusa*. The two species have an ecological range that extends beyond the limits of the salt marsh to the fringes of the coastal desert plain. "they may form and protect phytogenic mounds and hillocks of sand, though *Suaeda* hills may reach greater size" (Kassas and Zahran, 1967).

The biogeography of *S. monoica* and *N. retusa* differs. On the Red Sea coast of Egypt, *N. retusa* abounds in the northern 300 km stretch, *S. monoica* gradually replacing *N. retusa* within the 300–650 km stretch, until in the south (700 km south of Suez and southwards) *N. retusa* is absent and *S. monoica* is abundant. The stands of the *S. monoica* community type include 51 species (9 salt-marsh plants and 42 desert-plain plants) as well as 52 ephemerals. The diversity is attributed to the wide ecological range of this community.

On the Sudan Red Sea coast, the *S. monoica* community abounds in the inland margins, following in sequence of zonation the community type dominated by *Suaeda fruticosa* on the alluvial depressions. By virtue of its intermediate position between the salt marsh and the coastal plain, its floristic composition includes halophytic species, e.g., *Suaeda fruticosa. Cressa cretica, Eleusine compressa, Sporobolus spicatus* and *Cyperus conglomeratus*, and desert-plain species, e.g., *Panicum turgidum, Indigofera spinosa* and *Convolvulus hystrix*.

*S. monoica* is also present in the Red Sea coast of Ethiopia (Eritrea) and of French Somaliland (Hemming, 1961; Zahran et al., 2009).

### (xii) *Suaeda fruticosa* community

*S. fruticosa* is recorded along the whole stretch of the African Red Sea coast. In Egypt and Ethiopia, *S. fruticosa* is a common associate species with many community types, but it is not a dominant (Hemming, 1961; Kassas and Zahran, 1962, 1965, 1967). In the Sudan Red Sea coast *S. fruticosa* dominates a well-developed community type that follows, in sequence of zonation, the community type dominated by *Halopeplis perfoliata*. In certain localities vast stretches of land are covered by this community (plant cover 25–50%) which provides grazing ground for camels.

*S. monoica, Aeluropus lagopoides* and *Eleusine compressa* are the common associates (Kassas, 1957). The ground covered by *S. fruticosa* is raised into low hummocks 5–10 cm above the general level of the sterile land. The salinity in soil layers exploited by roots of *S. fruticosa* is less (3.04%) than in deeper layers (9–20%).

### (xiii) *Suaeda vermiculata* community

*S. vermiculata* is commonly recorded among the associate species of the Red Sea littoral community types. Its dominance is restricted to the delta of Wadi El-Ghweibba (Gulf of Suez, Egypt). The *S. vermiculata* community type occupies the zone of vegetation which lies between the *Halocnemum strobilaceum* littoral zone and the *Nitraria retusa* inland zone of the dry salt marsh. It is associated with *Halocnemum strobilaceum, Zygophyllum album, Nitraria retusa, Limonium pruinosum, Cressa cretica,* and *Salsola villosa* (halophytes) and *Haloxylon salicornicum, Zilla spinosa,* and *Farsetia aegyptia* (xerophytes).

### (xiv) *Juncus rigidus* community

*J. rigidus* is a halophyte highly tolerant of increased soil salinity, soil water stresses and aridity of the climate (Zahran, 1977). It is present in the littoral salt marshes of the Red Sea, but its densest growth and dominance are recorded in the salt marshes of the Gulf of Suez, e.g., the Ain Sokhna area (ca. 50 km south of Suez) where its cover is up to 90–100% (Kassas and Zahran, 1962). The associates are mainly halophytes including: *Halocnemum strobilaceum, Cressa cretica* and *Tamarix mannifera*.

The *J. rigidus* community represents the sedge-meadow stage of the halosere succession of the Red Sea that follows the reed-swamp stage of *Phragmites australis* and *Typha domingensis*.

### (xv) *Tamarix mannifera* community

*T. mannifera* (*T. nilotica*) is one of the common bushes in the Red Sea littoral salt marshes from Suez southwards to French Somaliland (Andrews, 1950–1956; Cufodontis, 1961–1966; Kassas and Zahran, 1962, 1967; Zahran et al., 2009); it grows in a variety of habitats and in various forms. In many parts of the dried salt marsh it forms thickets; and, in the sand-choked deltaic parts of wadis that drain inland country and pour onto the shoreline, it forms sand hillocks that reach considerable size. The floristic composition of the *T. mannifera* community type includes about 26 perennial species including 9 halophytes, e.g., *Z. album, N. retusa, J. rigidus, Cressa cretica* and *Suaeda monoica, Tamarix aphylla*, and 17 xerophytes, e.g., *Zygophyllum coccineum, Haloxylon salicornicum, Lygos raetam, Acacia tortilis* and *Leptadenia pyrotechnica.*

#### (xvi) *Tamarix passerinoides* community

*T. passerinoides* is morphologicaly and ecologically comparable to *T. mannifera*, but the former is rare along the Red Sea coast. It is recorded only once within the salina (El-Mallaha) 240 km south of Suez.

Apart from the previously mentioned communities of the African Red Sea littoral salt marshes, there are certain halophytes of ecological interest, e.g., *Cressa cretica, Alhagi maurorum* and *Imperata cylindrica* which are common associate species within a great number of the Red Sea littoral salt-marsh communities. They may also be dominant, but such communities are limited to the downstream part of Wadi Hommath that drains into the Gulf of Suez 30 km south of Suez (Kassas and Zahran, 1962). Their habitat is a littoral belt of sand extending parallel to the shoreline and varies in width from 150 to 300 m.

### 2.5.3.2 Asian Red Sea Coast

The littoral salt marsh vegetation of the Asian Red Sea coast has almost comparable ecological features to the African side.

In Saudi Arabia, Vesey-FitzGerald (1955, 1957) referred to this vegetation type as beach vegetation north and south of Jeddah. He stated "north of Jeddah the salt marsh vegetation is scanty and largely composed of salt bushes belong to Chenopodiaceae and Zygophyllaceae. Along the beach and just above high-water mark there may be an open fringe of *Atriplex farinosa* and on white coral sand of the coastal zone there is usually a fairly close stippling of *Zygophyllum coccineum.* Other halophytes associated with *Z. coccineum* are: *Suaeda schimperi, S. volkensii, Anabasis setifera* and *Halocnemum strobilaceum.*" The same author added that the coastal inlets are often bounded by salty flats or small dunes of coral sand. Some of these habitats are enclosed by headlands of coral rock which are drained by sandy gullies. The littoral vegetation usually occurs in community types dominated by one species, or even forming a pure stand. The dominance changes suddenly, due probably to variations in the physical and chemical properties of soil, such as salinity, compactness and nature of drainage, all of which frequently depend on very slight differences in elevation. Salty flats along the coast are often not vegetated, but in some places are stippled with dwarf shrubs. The presence of such vegetation causes building of windblown sandy hummocks which provide new habitats for other species. *Halopeplis perfoliata* dominates the lowest flats which are often flooded by the sea. *Cressa cretica* encroaches far into the *Halopeplis* zone and may even be abundant there. A slight elevation allows some other plants to become established, such as *Salsola* sp. which is abundant over wide areas near the coast. *Halocnemum strobilaceum*, *Aeluropus* sp. and *Limonium axillare* usually form closely spaced communities over wide areas in depressions between decomposed coral rock. The last has a long tap-root and is evergreen.

At Yanbu (38ºE, 24ºN), which is outside the tropics and very dry, salt-bushes form the characteristic coastal vegetation. The species present vary sharply according to slight variations in elevation and soil type. *Zygophyllum coccineum* dominates areas of coral rock and coral sand, and *Limonium axillare* and *Salicornia* sp. grow in low-lying places, where the sand is finer and stained a brownish colour. Habitats of tawny silt in depressions are dominated by *Salsola baryosma* and *Anabasis setifera*. Both of the last species have well developed tap-roots and are sand bindes. The wind-driven sand tends to pile up into hummocks round the plants. The soil of the barren coastal salt flats is distinctly stratified. About 5 cm of coral sand lies on the surface and below this there is a 10 cm horizon of "soda" crystals overlying a hard pan. *Halopeplis perfoliata* is the only plant that grows in such soils and even this species is widely spaced.

The dry estuaries of inland drainage lines, which are liable to very occasional flooding, are characterized by a fine-textured clay soil, which is usually dusty dry on the surface but moist below. In this habitat, the growth of halophytes is rich and the plants build mounds spaced at about 2-m intervals. The characteristic species in such places are: *Suaeda monoica, S. volkensii, Halocnemum strobilaceum, Limonium axillare* and *Aeluropus lagopoides*. Restriced groves of shrubby *Tamarix* sp. also occur.

In the littoral salt marshes south of Jeddah, three *Salsola* spp. (*S. Forsskaolii, S. bottae* and *S. foetida*) are very common. These plants grow on the compact alluvium where former floods have laid down a deposit which is above the level of recent floods. At intervals along the coastal plain, the land has become raised slightly in relation to sea level. At such places old reeds are found containing fossil corals in the position of growth. This is especially noticeable along the coast between Wadi Amk (18° 25′N, 41° 30′E) and Shuqaiq (17° 44′N, 42° 02′E) where the fossil reef is also partly overlain by tongues of lava boulders. The coral rock has been superficially decomposed into a fine tawny silt, which is soft and dusty when dry, and slimy when wet. The drainage of such areas is usually poor and surface flood water tends to accumulate in pans. There is some local transportation of surface-soil from raised areas, which often keeps the coral-rock exposed. There is also a considerable accumulation of debris in hollows and over extensive flats which are slightly raised above the present level of the sea. *Limonium axillare* dominates wide areas of such soil, in almost pure stands. This plant remains green throughout the year and some flowers are to be found over a protracted period, but the main flowering occurs during the last months of the year, without necessarily the incidence of rain. Nevertheless, it seems that this formation is dependent upon local precipitation, and after prolonged drought the plants tend to die out. The grasses *Aeluropus lagopoides* and *Desmostachya bipinnata* and salt bushes (Chenopodiaceae) closely fringe the *Limonium* community without necessarily being associated with it.

There is little tidal variation along the Saudi Red Sea coast but the wind often drives the sea-water far inland over level flats which become encrusted with salt as the surface dries. These areas are usually absolutely barren. All around the coast south of Jeddah, *Halopeplis perfoliata* colonizes the fringes of such flats, this plant being able to tolerate greater salinity than any other species; in fact its short roots may be found embedded in a substratum composed almost entirely of soda-like crystals. *Halopeplis* forms a pure stand fringing the barren flats, but the plants are always widely spaced. The width of the *Halopeplis* zone is variable, depending largely on levels. Where the coast is very flat it may be correspondingly wide, but where the land rises only very slightly above the level reached by sea-water, the *Halopeplis* belt may be narrow or quite absent. Like other salt bushes, this plant bears its inconspicuous flowers during the hot months of the year, but the flowering season is evidently prolonged because flowers have been noted as early as March and as late as October. The only plant which has been found associated with *Halopeplis* at all regularly is *Cressa cretica*, a small erect herb, with numerous small leaves, which is seldom found in flower, and is locally quite abundant on the saline flats.

The landward side of the *Halopeplis* zone is encroached upon by a variety of other plants mostly other halophytes. Where saline conditions are rather less extreme, and land floods rather than sea floods are a possibility, although evaporation rather than free-drainage occurs resulting in salt accumulation, robust mixed salt bushes on mounds become established, but the species have not been determined.

Zahran (1982b, 2002) recognized 18 plant communities in the littoral salt marshes of the Red Sea coast of Saudi Arabia. The following is a short account of these communities:

#### (i) *Halopeplis perfoliata* community

*H. perfoliata* is a succulent halophyte not common in the northern section of the Saudi Red Sea coast but it dominates a community type that occupies the first zone of the Red Sea salt marshes just inland from the mangrove swamps in the southern section of this coast. Owing to frequent inundation with low and high tides, the soil of this community is usually slippery with high water content. It is formed mainly of fine sand and silt.

The individual plants of *H. perfoliata* usually do not build sand formations (mounds or hummocks), but in certain open areas where there are possibilities of settlement and accumulation of wind-borne materials, small mounds can be seen around the individuals of *H. perfoliata*. The plant cover of the *H. perfoliata* community (30–40%) is contributed mainly by the dominant plant. Few associate halophytes e.g. *Aeluropus brevifolius, A. lagopoides A. massauensis, Cressa cretica, Halocnemum strobilaceum, Salicornia fruticosa, Atriplex farinose* and *Tamarix mannifera,* were recorded.

#### (ii) *Arthrocnemum macrostachyum* community

This succulent halophyte predominates in the littoral salt marshes of the southern part of the Saudi Red Sea coast.

### (iii+iv) *Halocnemum strobilaceum* and *Salicornia fruticosa* communities

These are two closely related succulent halophytes. Their ecological amplitudes are quite similar. Both species are recorded in the littoral salt marshes of the northern part of the Red Sea coast of Saudi Arabia.

The growth of *H. strobilaceum* occurs in two forms:

1. Circular patches on flat tidal mud ground, and
2. Sheets of irregular shaped patches on shoreline bars.

The community types dominated by these two halophytes (which sometimes are co-dominants) occupy the second zone of the littoral salt marshes of the Red Sea inland to that of *H. perfoliata*. In areas where *H. perfoliata* is absent, *H. strobilaceum* and *S. fruticosa* community types occupy the first zone. Soils of these two communities are highly saline and usually slippery. The associate species include: *Suaeda schimperi, S. volkensii, H. perfoliata, Limonium axillare, Cressa cretica, Aeluropus* spp. and *Salsola* spp.

### (v+vi+vii) *Aeluropus littoralis, A. lagopoides* and *A. massauensis* communities

These are salt excretive grasses with almost comparable ecological amplitudes. They usually occupy the second zone of the littoral salt marshes where the soil is moistened and highly saline. In areas where *Halocnemum strobilaceum* and/or *Salicornia fruticosa* predominate, *Aeluropus* spp. communities occupy the third zone. The dominant grasses grow in patchy mats of dense growth. In some stands, the individuals of *Aeluropus* spp. grow in the form of creeping grasses but they may build also the peculiar forms described by Kassas and Zahran (1967) as "cone like masses of interwoven roots, rhizomes and sand". The plant cover of these community types is usually high (up to 50% in most stands) and the dominants are associated with: *Suaeda monoica, Halopeplis perfoliata, Zygophyllum coccineum, Sporobolus spicatus, Tamarix mannifera* etc.

### (viii) *Juncus rigidus* community.

This cumulative rush halophyte is widespread in Saudi Arabia. It is not only tolerant to soil salinity but also to various climatic conditions. The growth and domination of *J. rigidus* occur in the littoral and inland saline habitats of Saudi Arabia, especially in moist saline areas. The associate species of this community are moist-loving halophytes e.g. *Cyperus* spp., *Carex* spp. and *Juncus bufonius*.

### (ix) *Limonium axillare* community

The community dominated by *L. axillare* covers wide areas of the northern part of the Saudi Red Sea littoral salt marshes. It occupies the third or the fourth zone depending upon the zonation pattern of the area. The soil of this community has been formed by the decomposition (superficial) of coral rocks into a fine tawny silt formed mainly of calcium carbonates, i.e. calcareous soil. The plant cover of this community is generally low (5–10%) and its associates are halophytes e.g. *Zygophyllum album, Cressa cretica, Salsola tetrandra, Halopeplis perfoliata* and *Suaeda pruinosa. Zygophyllum simplex* may grow during the rainy (winter) season.

#### (x) *Halopyrum mucronatum* community

This salt excretive grass is recorded in a very limited area of the most southern part of the Saudi Red Sea littoral salt marsh. The community dominated by *H. mucronatum* is formed of almost pure stands in which the grass builds low coastal sand dunes the material of which is white and homogeneous in texture (sand = 70%). *Panicum turgidum* is the only associated species recorded.

#### (xi) *Sporobolus spicatus* community

*S. spicatus* is a salt secretive grass of limited distribution in Saudi Arabia. Its presence is recorded in the littoral salt marshes of the Red Sea where it grows, either as an associate species of *Aeluropus* spp. communities in most stands or as co-dominant in a few stands. Societies of *S. spicatus* are not unusual in the zone occupied by *Aeluropus* spp.

#### (xii) *Anabasis setifera* community.

*A. setifera* is a succulent chenopod recorded everywhere in Saudi Arabia, not only in the halic but also in xeric habitats. Its community occupies a narrow area of the northern part of the Saudi Red Sea littoral salt marshes. The soil of this community is calcareous and the total plant cover is generally thin (<10%). The associate species include: *Limonium axillare, Heliotropium* spp., *Fagonia cretica, Salsola tetrandra, Zygophyllum coccineum* and *Suaeda pruinosa. Cistanche phelypaea* (a member of the family Orobanchaceae) usually parasitizes on *Salsola tetrandra.*

#### (xiii) *Zygophyllum album* community

The community dominated by this succulent halophyte has been noticed along the whole stretch of the Saudi Red Sea littoral salt marshes (Zahran, 2002). It usually occupies the zone inland to that of *Aeluropus* spp. The associate species of the *Z. album* community include: *Dipterygium glaucum, Panicum turgidum, Atriplex farinosa, Aeluropus massauensis* i.e. a mixture of xerophytes and halophytes.

#### (xiv) *Zygophyllum coccineum* community

The well developed community dominated by *Z. coccineum* is observed in the middle and southern part of the Red Sea littoral salt marshes of Saudi Arabia. Younes et al. (1983) shows that *Z. coccineum* occupies the fourth zone of the Red Sea salt marshes following mangrove, *Halopeplis perfoliata* and *Aeluropus massauensis* zones. The soil of this community is highly gypsiferous-calcareous. The associate species include: xerophytes e.g. *Salsola baryosma, Dipterygium glaucum,*

*Calotropis procera* and halophytes e.g. *Aeluropus massauensis, Tamarix mannifera, Suaeda monoica, S. pruinosa, Sporobolus spicatus* and *Anabasis setifera*.

#### (xv) *Nitraria retusa* community

*N. retusa* is a salt tolerant shrub with a narrow geographical distribution in Saudi Arabia. It is recorded in the northern part of the Red Sea as it starts to appear from Umm Lujj (c. 140 km. north of Yanbu and 450 km. north of Jeddah) northwards, but is never seen southward. The same geographical distribution was observed along the western Red Sea coast in Egypt (Kassas and Zahran, 1967; Zahran and Willis, 1992; Zahran, 2002).

The individual bushes and shrubs of *N. retusa* build huge hillocks of aeolian sand. The associate species are few and include: *Z. album, Tamarix mannifera* and *Atriplex farinosa*.

#### (xvi) *Atriplex farinosa* community

*A. farinosa* is a halophytic shrub that grows, like *Halopyrum mucronatum*, on the coastal sandy dunes. Its presence was recorded in the Saudi Red Sea coastal area. This community does not extend inland except for a few meters, i.e. it is restricted to the line of littoral sand bars. *Zygophyllum album, Halocnemum strobilaceum* and *Salicornia fruticosa* usually grow near these bushes.

#### (xvii) *Suaeda monoica* community

*S. monoica* is a succulent halophytic shrub that attains a height up to 4 m. It is a very common species in Saudi Arabia. The community dominated by *S. monoica* plays an important ecological role as it represents a recognizable part of the vegetation of Saudi Arabia. The plant cover of this community is relatively high (up to 70%) in most stands, contributed mainly by the dominant plant. *S. monoica* is capable of building huge hillocks of sand especially in the protected area of the inland salt marshes. In the littoral salt marshes, where wind action is relatively strong, *S. monoica* individuals are smaller, and build no sand formations.

Along the Red Sea coast, *S. monoica* is not recorded in the northern part, but is very common south of Jeddah. In the delta of Wadi Jizan (c. 670 km south of Jeddah) *S. monoica* forms dense thickets. The associate species include: *Tamarix mannifera, Zygophyllum coccineum, Dipterygium glaucum, Phoenix dactylifera, Hyphaene thebaica, Hammada elegans, Rhazya stricta, Leptadenia pyrotechnica* and *Acacia tortilis* i.e. mixed halophytes and xerophytes.

#### (xviii) *Tamarix* spp. community

According to Migahid and Hammouda (1978) five species of *Tamarix,* namely: *T. nilotica, T. aphylla, T. mannifera, T. passerinoides* and *T. amplexicaulis,* are recorded in Saudi Arabia. These salt excretive shrubs and trees represent the climax stage of the halophytic vegetation.

*Tamarix* plants are widely distributed in the Saudi Arabian Red Sea coast where they may build huge hillocks of fine materials.

### (xix) Other communities

Apart from the above mentioned 18 main plant communities of the halophytic vegetation of the Red Sea coast in Saudi Arabia, there are other halophytes that dominate communities of narrow ecological amplitude e.g. *Cressa cretica, Alhagi maurorum, Salsola tetrandra, Suaeda vermiculata, S. pruinosa, Atriplex leucoclada, Juncus acutus* and *Cyperus laevigatus*.

Date palm tree (*Phoenix dactylifera*) is, actually, very common in the Saudi Arabian Red Sea coast.

In Yemen, most of the Red Sea littoral belts is low lying and is seasonally inundated, leading to the formation of sabkhas. These sabkhas dry out seasonally to form abiotic salt pans. Some of these sabkhas may partly reflect a previous sea level rise about 4,000–6,000 years ago (Zahran and Al-Kaf, 1996). Sabkhas usually extend just into the lower intertidal to the point where salinity stress is minimized by regular tidal flushing. Here, it may be slightly overlapped with intertidal mangroves in soft-bottom habitats.

The extensive sabkhas of the Red Sea coastal belt of Yemen are often bare sand flats covering the shoreline to about 5 km inland. Extremely saline conditions prevail because the ground water is shallow and evaporation rate is very high, particularly in the surface layers. In these highly saline soils, only salt tolerant plants (halophytes) grow and predominate. The dominant halophytes in this coastal belt include: *Suaeda monoica, S. vera, S. vermiculata, Halopeplis perfoliata, Arthrocnemum macrostachyum, Salsola baryosma, Zygophyllum album,* (succulent halophytes), *Tamarix nilotica, T. aphylla, Cressa cretica, Limonium axillare, Aeluropus lagopoides, A. massauensis, Sporobolus spicatus* and *Atriplex* spp. (excretive halophytes).

A line of discontinuous sand dunes commonly occurs close to the shorelines of Yemen. These dunes are formed of homogeneous sandy, slightly saline, deposits. Salinity varies but little in the successive layers (0.56% at the surface, 0.5% at 15 cm and 0.6% at 30 cm). In the Red Sea coast, *Halopyrum mucronatum, Zygophyllum album* and *Panicum turgidum* are also present. In the sabkhas some dominant halophytes are sand binders. They are capable of building sand hillocks (e.g. *Suaeda monoica*) and sand hummocks (e.g. *Zygophyllum album, Halopeplis perfoliata* and *Arthrocnemum macrostachyum*) of different sizes. Also, the salt tolerant excretive grasses *Aeluropus* spp. may build small sand formations which are cone-like masses of interwoven roots, rhizomes and sands. The importance of the shoreline sand dunes lies in their considerable stabilizing abilities, limiting coastal erosion and allowing other less tolerant species (e.g. *Panicum turgidum*) to grow.

Most of the wadis in the Red Sea coastal belt of Yemen have outlets into the Red Sea. This results in a layer of freshwater floating over the sea-water near the coast. There is evidence of freshwater seepage along 98 km of Yemeni Red Sea coast where vegetation varies between dense and sparse (IUCN, 1986). Similar fresh water locations occur in the shoreline of the Egyptian Red Sea coast (Kassas and Zahran, 1967). In contrast to the thin vegetation of the sabkhas, the fresh-water vegetation is relatively luxurious. This low salinity tolerance vegetation requires a continuous supply of fresh-water which is not sustained by rain but by ground-water supply originating in the highlands inland of the coastal plains. In some places the supply may have originated during historically wetter periods and may weaken as the effect of the recent more arid conditions prevails. The Yemeni Red Sea coastal belt comprises considerable number of date plants (*Phoenix dactylifera*) which form extensive dense but isolated patches. Doum palm (*Hyphaene thebiaca*) is less common than date palm, possibly reflecting agricultural exclusion resulting from the specific cultivation of the date palms.

## 2.5.4 Vegetation of the Coastal Desert Plains and Mountains

### 2.5.4.1 Ecological Characteristics

The coastal desert plains of the Red Sea occupy the areas between the littoral salt marshes and the coastal ranges of mountains. They are characterized by a series of wadis with different lengths, widths, and areas, and run towards the Red Sea in two opposing directions. The wadis of the African (western) coast run eastward whereas those of the Asian (eastern) coast run westward. Kassas (1957) stated that, in the Sudan, khors are similar to wadis of the Egyptian deserts. The following are representative wadis in both African and Asian coastal deserts of the Red Sea: Wadi Aber, Wadi Hommath, Wadi Hagul, Wadi El-Bada, Wadi El-Gheweibba, Wadi Aideib, Wadi Araba, Wadi Di-ip, Wadi Gimal, Wadi Yahamib, Wadi Bali, Wadi Shillal, Wadi Lasitit, Wadi Naam, Wadi Hawaday, Wadi Serimtai, Wadi Kannsisrob, Wadi (Khor) Amat, Wadi Harashab, Wadi Arbaat, Wadi Dahand, Wadi Karora, Wadi Hum, Wadi Falcat, Wadi Al-dibabo, Wadi Misho, Wadi Mutabet and Wadi Nenin (in the African Red Sea coast), Wadi Fatma, Wadi Baysh, Wadi Sabya, Wadi Dameda, Wadi Jizan, Wadi Amlah, Wadi Magab, Wadi Khulab, Wadi Taashar, and Wadi Harad (in the Asian Red Sea coast) (Kassas, 1957; Hemming, 1961; Kassas and Zahran, 1962, 1965; Mandura et al., 1987; Zahran et al., 2009) By reason of the intermediate position of the coastal desert plain, its ecological condition show, especially on its fringe, transitional characters (ecotones). "But it may be noted that the inland boundaries of the coastal plain are not as clear as its seaward boundaries" (Zahran, 1964).

The Red Sea coastal desert is essentially a gravel-covered plain traversed by the downstream extremities of the main wadis and is dissected by smaller drainage runnels that may extend from the foot-hills to the coastal front or may not reach the coast. The downstream extremities of the main wadis may form deltaic basins. Superimposed on this pattern, aeolian deposits may form sheets, mounds or hills of various size and extent. This complex set-up produces a likewise complex of habitat conditions. The vegetational pattern associated with these habitat conditions is further complicated by differences in the intensity of human interference and wild-life grazing, and also by differences in the geographical ranges of the floral elements.

Some of the community types are mostly confined to the northern stretch of the coastal desert; others are mostly confined to the southern part and a third group are present all over the coastal desert plain.

The Red Sea coastal lands in Africa and Asia are bounded inland by an almost continuous range of mountains and hills. In Egypt (northern sector of the African Red Sea coast), this range has a number of high peaks. It forms the natural divide between the eastward drainage (to the Red Sea) and the westward drainage (across the Eastern Desert to the River Nile) (Zahran and Willis, 1992). The presence of this coastal range has a distinct effect on the climate and the water resources of the surrounding desert.

The influence of orographic precipitation is hardly noticeable within coastal mountains of the Gulf of Suez. By contrast, the coastal mountains of the Red Sea proper show the influence of orographic rain. The amount of this rain, judging by the type of vegetation, varies according to altitude and distance from the sea. Ball (1912) stated that the vegetation of the wadis draining from the Elba mountains of the Egyptian Red Sea coast was so abundant that it was impossible to approach the mountains very close with loaded camels, owing to the closeness of the trees. He added "of the 10 days I remained on the summit in April and May 1908 only 3 days were clear".

The Red Sea mountains in Egypt may be categorized into:

### 1. Mountains facing the Gulf of Suez

These are the coastal chain of mountains within the west coast of the Gulf of Suez, between Suez (30ºN) and Hurghada (400 km south of Suez). These mountains are grouped as Shayib Group and comprise: Gebel Ataqa (900 m), Gebel Al-Galala El-Bahariya (977 m), Gebel Al-Galala El-Qibliya (1,200 m), Gebel Gharib (1,751 m), Gebel Abu Dukhan (1,705 m), Gebel Qattar (1,963 m) and Gebel Shayib El-Banat (2,184 m). The highest is Gebel Shayib El-Banat and the lowest is Gebel Ataqa (Kassas and Zahran, 1962, 1965).

### 2. Mountains facing the Red Sea proper

These include the groups of mountains that extend between latitude 24º 50′N and 22ºN on the Sudano-Egyptian border and comprise three main groups: Gebel Nugrus Group, Gebel Samiuki Group and Gebel Elba Group (Zahran, 1964; Kassas and Zahran, 1971). Gebel Nugrus Group extends between latitude 24º 40′ and 24º 50′N and faces the full stretch of the Red Sea. It includes Gebel Hafafit (857 m), Gebel Migif (1,198 m), Gebel Zabara (1,360 m), Gebel Mudargag (1,086 m) in addition to Gebel Nugrus which is a great boss of red granite rising to a height of 1,505 m among schist and gneisses (Ball, 1912).

Gebel Samiuki Group comprises Gebel Abu Hamamid (1,745 m), Gebel Samiuki (1,283–1,486 m) and Gebel Hamata (1,977 m). The last is the nearest to the sea (40 km) whereas Gebel Abu Hamamid is furthest away (65 km).

Gebel Elba Group is an extensive group of granite mountains situated on the Sudano-Egyptian border (latitude 22ºN), that includes: Gebel Elba (1,428 m), Gebel

Shindeib (1,911 m), Gebel Shindodai (1,426 m), Gebel Shillal (1,409 m), Gebel Makim (1,871 m) and Gebel Asotriba (2,117 m). This group, especially Gebel Elba, is particularly favoured by its position near the sea.

In the Sudan, Kassas (1956) described the Erkwit Plateau as an example of the Sudanese Red Sea coastal mountains. He reported that Erkwit plateau lies at the edge of a steep escarpment dropping abruptly (600 m) to the Red Sea. At the northern boundary are Gebel Nakeet (1,176 m) and Gebel Essit (1,143 m). These two Gebels drop to Khor (wadi) Dahand which separates the Erkwit oasis from the barren hill on the other side of the Khor. At the eastern boundary is Gebel Sela (1,273 m) which is the highest evergreen mountain of the district. At the southern boundary are Gebel Tatasi (1,190 m), Gebel Lagagribab (1,209 m) and Gebel Auliai (1,191 m) which are separated by Khor Amat from the barren mountains further south (Gebel Erbab, 1,523 m). At the western boundary are Gebel Hadast (1,147 m) and Gebel Mashokriba (1,113 m) which drops to Khor Dahand that separates the Erkwit Plateau from the desert plains to its east.

The plateau and the hills arising on its top are built of basement complex rocks: gneiss, basalt, granites, shales, marble etc. Several khors (ephemeral water-ways) dissect the plateau. The khors usually contain alluvial deposits of different depth into which holes are dug to provide fresh water supplies. Permanently running water is found in Khor Harasab and Khor Amat. Calcareous deposits are found on the bottom of khors at several sites.

Erkwit receives a rainfall (218 mm) greater than that of the neighbouring areas represented by the two towns: Suakin (181 mm) to the north-east and Sinkat (127 mm) to the west. Suakin, a Red Sea port, has a coastal climate with winter rainfall. Sinkat, which lies on the inland plain, represents areas with summer rainfall. Erkwit lies between and receives both the summer and winter rainfalls.

Tothill (1948, p. 55, footnote) suggested, "... the climate of Erkwit may be due to the happy combination of three things: latitude, situation and elevation". Its latitude (18° 46′N) is close to latitude 19°N which divides the Sudan into a desert region to the north and a tropical continental region to the south (Ireland, 1948). In the north, prevail the dry northerlies and in the south the moist southerlies which cause the summer rainfall. Erkwit enjoys the maritime modification of the Red Sea: as the continental northerlies pass over the warm water of the sea they absorb a considerable amount of moisture. Being situated on a westerly bend of the Red Sea, Erkwit is exposed to about 650 km (400 miles) of open sea in the direction facing the northerlies (N.E. Trades). The moisture-laden wind meets no dissipating obstacle before impinging on the cool hills of Erkwit (3,000–4,000 ft O.D. to 900–1,200 m) and orographic rain consequently occurs. After passing over the Erkwit Plateau the northerlies resume their dry continental characteristics.

During the winter months, the Erkwit Plateau is frequently swathed in clouds for weeks. This entails considerable dew precipitation which is more marked the higher the elevation, and which supplies the vegetation with a valuable water resource.

In Eritrea, according to Hemming (1961) the Red Sea coastal plain averages 15–20 miles (24–32 km) in width and consists of a vast deposit of mixed sandy and stony alluvium derived from the ancient basement massif which lies to the west, The present situation on the plains is a complex mosaic of basement pediments, stone mantle areas produced by deflation of mixed alluvial deposits, consolidated sands appearing as plains and dune remnants, and loose sand occurring either as a mantle of variable depth or in the form of mobile dunes. The term "stone mantle soil" is used to describe any truncated soil, soil remnant or skeletal soil which has a layer of stones on the surface, similar to those previously described in an account of the soils of south Turkana in north-western Kenya. Stony pediments lie along the foot of the hill range in the form of an apron and have been reduced by deflation to stone mantles.

The Eritrean coastal plain is interrupted between the main hill mass and the coast by a broken line of isolated limestone hills running approximately parallel to and generally less than 10 miles (16 km) from the coast line. These hills are the dissected remnants of a more extensive Pliocene deposit. North of 15° 45′N the general pattern is relatively simple, with a single plain between the main hill mass and the coast, but in the area west of Massawa there is an intermediate range of hills, so that there is also a sub-coastal plain centred about 23 miles (37 km) from the coast and lying at an approximate altitude of 1,000 ft (300 m).

Between this sub-coastal plain and Massawa there is a wide area of rough stony country consisting of the dissected remnants of alluvial gravel banks with some residual outcrops, together with small mounds and banks of rotted lava. In these areas of rotted lava the soils are very variable in colour, but a distinct purple tone is characteristic. The Afta-Zuia silt plain (15° 15′N, 39° 42′E) is bounded on the west by lava overlain with coarse basement alluvial gravels. These pallid silty areas are found only near lava hills, both below them in the form of plains and within them as infilled depressions; thus this silt may be regarded as a lava-derived soil. The presence of minute mica flakes and other traces of basement alluvium may be explained by the presence of the more recent overlying basement gravels. About 3 miles (5 km) south of Zula there are two isolated lava hills rising out of the silt plain and these, together with the lava that largely surrounds the Gulf of Zula, represent outliers of the main Eritrean lava mass which lies to the south-west. The whole of the area, surveyed north of the lava at 15° 15′N, is bounded on the west by a highland hill mass composed of ancient rocks of the basement complex, principally quartz, gneisses and schists, rising to over 8,000 ft (2,400 m) to the west of Massawa and dipping slightly downwards towards the north. In the vicinity of latitude 17° 00′N the hills begin about 20 miles (32 km) from the sea. They rise to about 3,000 ft (900 m) within a few miles of the plain and reach to nearly 7,000 ft (2,100 m) about 15 miles (24 km) from the plain. The slopes are usually steep and stony and erosion processes generally outstrip those of soil formation. Storms on the mountains cause the wadis to flow very rapidly for short periods. These wadis are deeply incised into the plains and their flood waters seldom reach the sea. The water is gradually absorbed by the sandy substratum and in the lower reaches the rivers break up into numerous small streams watering extensive areas which are used for the cultivation of bulrush-millet and cotton.

The Wadi Falcat Sheri Ma'assrai flows out from the main hill mass and passes Cum Chewa (16° 27′N, 38° 45′E). In this area there are wide valleys which are partly infilled producing broad flat valley floors, where the soil consists of a sandy loam. Eight miles (13 km) north-east of Cum Chewa lacustrine soils are seen covering the floor of a small valley.

The coastal and sub-coastal regions of northern Eritrea are generally hot and dry and are considerably affected by topography. The plains are bounded on the west by the Ethiopian massif and the distance between this and the sea is an important factor affecting the rainfall. The main rains occur in the winter months, when Massawa, with a mean annual rainfall of 165.4 mm, receives 119.1 mm (72%) in the 5-month period October to February.

During the winter months the Eritrean coast lies in an area of convergence between the dry NW Monsoon and a south-easterly wind which brings moisture from the Indian Ocean. Rain on the Eritrean Red Sea coast reaches a maximum at Massawa, possibly owing to the proximity of the massif to the south-west and the added orographic effect of Gebel Ghedem, the foot of which lies only 9 miles (14 km) to the south-east of Massawa.

While most of the rain in this area falls in the winter months, rain has been recorded at Massawa and Tokar during every month of the year, but no more than a trace in June.

Inland areas of the Sudan and northern Eritrea receive their main rains during the summer months owing to the northern position of the inter-tropical front reaching its most northerly position in July and August, when it runs approximately from west to east across the Sudan reaching the Red Sea coast near Suakin (19° 05′N, Meteorological Office, 1951). It then bends southwards and approximately follows the eastern edge of the main hill mass. This means that while the areas to the south and west of the inter-tropical front receive summer rain from the SW Monsoon, the Red Sea coastal areas do not. In fact they lie in the rain-shadow of the highland mass. There is, however, some spill-over on to the plains, especially near the base of the hills, which tends to reach a maximum where the hills lie close to the shore, as in the area south of Massawa as far as the Buri Peninsula, but at Massawa itself it seldom produces measurable amounts. The considerable increase in rainfall across the coastal plain from the shore towards the massif is illustrated by the case of Damas (15° 28′N, 39° 13′E) which lies only 16 miles (26 km) inland to the south-west of Massawa, where the mean annual rainfall was 645.5 mm for the period between 1926 and 1948. For Karora (17° 42′N, 38° 22′E), situated about 21 miles (34 km) from the sea at an altitude of 800 ft (240 m), there exists a single year's rainfall record of 321 mm. For the sub-coastal plains there are no rainfall records, but in the Ailet (15° 33′N, 39° 11′E) and Sheb (15° 52′N, 39° 02′E) areas there is sufficient rain for some kind of crop cultivation to continue throughout most of a favourable year. It should be noted that rainfall in arid and semi-arid areas is very erratic and the figures probably do not apply to the country a few miles away from the rainfall station. In the 13-year period 1945–1957 the annual totals for Massawa fluctuated between 55% above and 59% below the mean value and the monthly variation is even greater, e.g. December maximum 188 mm and minimum nil.

At Massawa the annual mean temperature is 29.7°C, which is the highest for any regular meteorological station in the area. This high mean is due mainly to the high temperatures in the winter months and not to exceptionally hot summers. The hottest months are July and August when the mean daily temperature is 34.7°C and the coldest month is February with a mean daily temperature of 25°C. The average daily range of temperature throughout the year is only 7°C (Meteorological Office, 1951).

In Saudi Arabia and Yemen (Asian Red Sea coast), Tihama or the low, sandy and hot coastal desert is narrow in its northern and middle sections (about 25 km) but relatively wide in the southern section (about 45 km). Geomorphologically, Tihama may be divided into 3 main parts: 1. Higaz Tihama, Asir Tihama (in Saudi Arabia) and Yemen Tihama (in Yemen). It is characterized by 3 main morphological features: (1) Sabkhas and salt affected lands in the areas close to the sea, (2) Rocky outcrops and (3) Coral reefs and islands of different areas.

The range of mountains associated with the Asian Red Sea coast is called Sarawat (Al-Sarat or the high lands) and is formed of Jurassic and metamorphic rocks, covered in some places with sedimentary rocks. The width of these mountains ranges between 120 and 200 km with height ranging between 1,200 and >3,500 m.

The vegetation within the wet salt marsh ecosystem shows clear zonation which makes the vegetational pattern reasonably intelligible. That is not the situation concerning the plant growth in the dry non-saline desert plains and mountains which presents a complicated pattern owing to different conditions of topography, characters of the surface deposits, relationships between the desert plains and mountains, direction of slopes and altitudes of the different mountains, distance from the sea etc.

Generally the plant life of the Red Sea coastal desert and mountains comprises two growth forms: ephemeral and perennial. Ephemeral vegetation usually covers extensive areas of the coastal plain and mountains, especially on the parts covered by sheets of sand. This growth is dependent on the chance occurrence of rainfall which varies from 1 year to another. It also varies from one locality to another: the desert rainfall is usually due to accidental choudbursts which pour over limited areas. The appearance of local patches of ephemeral vegetation soon attracts herds of camels, goats and gazelles. The ephemeral growth may be associated with a framework of perennial plants or may develop on shallow sheets of sands that do not allow for the establishment of perennial vegetation.

The ephemeral vegetation indicates soil conditions that allow for no overyear storage of moisture: soil wetness is maintained during a part and not the whole of the year. This may be due to: (i) scantness of rainfall and/or (ii) the surface deposits being too shallow for subsurface storage of moisture. This vegetation comprises three forms: succulent-ephemeral forms, ephemeral-grassland form and herbaceous-ephemeral form.

The perennial vegetation of the Red Sea coastal desert and mountains may be categorized under two groups (4 subgroups) namely:

A. Suffrutescent perennial plants

   (a) Succulent halfshrub form.
   (b) Perennial grassland form.

B. Frutescent perennial vegetation

(a) Succulent shrub form.
(b) Scrubland form.

The suffrutescent perennial vegetation is most widespread. It is distinguished by a permanent framework of perennial plants. This vegetation comprises two main layers: a suffrutescent layer (30–120 cm high) and a ground layer. In the greatest majority of the desert communities the suffrutescent layer gives character to the vegetation; the ground layer comprises dwarf or trailing perennials and may be enriched by the growth of ephemerals during the rainy season.

The frutescent perennial vegetation includes the scrubland types of the desert vegetation where the plant cover comprises three main layers: a frutescent layer (120–500 cm high), a suffrutescent layer (30–120 cm high) and a ground layer. The first layer is here dense enough and gives the vegetation the characters that distinguish it from the suffrutescent type.

The frutescent vegetation comprises two main forms: the succulent-shrub form and the scrubland form. The scrubland form is widespread along both Asian and African coasts of the Red Sea. The succulent-shrub form, however, occurs only in the southern section of the Asian and African Red Sea coasts and is represented by *Euphorbia* spp. e.g. *E. thi* (*E. ammak*) in the Saudi Arabian and Yemeni deserts and *E. candelabra* in the Sudanese desert. Both species are absent from the Egyptian as well as the northern section of the Saudi Arabian Red Sea coasts (Vesey-FitzGerald, 1955; Zahran, 1964, 1982b; Al-Hubaishi and Müller-Hohenstein, 1984).

#### 2.5.4.2 Vegetation Types

The vegetation types of the dry non-saline ecosystems of the African and Asian Red Sea coastal desert plains and mountains have been studied by many authors, among these are: Vesey-FitzGerald (1955, 1957), Kassas (1957), Hemming (1961), Kassas and Zahran (1962, 1965), Zahran (1962, 1964, 1982b, 1993a), Baeshin and Aleem (1987), Batanouny and Baeshin (1982), Zahran et al. (1983), Zahran and Willis (1992, 2008) and Zahran et al. (2009).

##### African Red Sea Coast

The vegetation types of the African Red Sea coastal deserts and mountains in Egypt and Sudan are described below.

##### In Egypt

The vegetation of the dry non-saline ecosystems of the Red Sea coastal desert may be described under two titles: coastal desert wadis and coastal mountains.

## I. Coastal desert wadis

The vegetation of three representative wadis flow into the Gulf of Suez and the Red Sea, namely: Wadi Hagul, Wadi Araba and Wadi Serimtai, are described below.

### (a) Wadi Hagul

This is an extensive wadi occupying the valley depression between Gebel Ataqa to the north and Gebel Kahaliya to the south. Its main channel extends for about 35 km and collects drainage water on both sides. The upstream part cuts its shallow channel into ochreous-coloured marls and grits and carolin-beds of the Upper Eocene. The main channel proceeds in a southeast direction, traversing limestone beds of the Miocene. Further downstream, the wadi widens and cuts its way across recent alluvial gravels before it finally traverses the coastal plain towards the Gulf of Suez (Sadek, 1926).

With reference to the vegetation and geological features of Wadi Hagul, three main sectors may be distinguished: upstream, middle and downstream.

In the upstream sector the plant cover varies in apparent relation to the area drained. The finer runnels here are short, their cover is sparse and their beds are strewn with coarse boulders. The vegetation is usually a community of *Iphiona mucronata*. In the finer runnels the dominant species is *Fagonia mollis*. Other species include *Centaurea aegyptiaca, Gymnocarpos decander, Helianthemum lippii, Heliotropium arbainense, Linaria aegyptiaca, Scrophularia deserti* and *Zygophyllum decumbens*. Runnels with greater drainage are usually characterized by a community dominated by *Zygophyllum coccineum*. The bed is covered with a rock detritus including some soft materials. Associate species include *Artemisia judaica, Cleome droserifolia, Crotalaria aegyptiaca, Fagonia mollis, Heliotropium arbainense, Iphiona mucronata, Launaea spinosa, Lavandula stricta, Lygos raetam, Pennisetum dichotomum, Pituranthos tortuosus, Reaumuria hirtella, Scrophularia deserti* and *Trichodesma africanum*. Of particular interest here is *Cleome droserifolia*, a xerophyte common eastward but rare westward in the Eastern Desert of Egypt.

Within the upstream part of the main channel of Wadi Hagul, two communities may be recognized. One is dominated by *Zilla spinosa* and is well developed on elevated terraces of mixed deposits. The other is dominated by *Launaea spinosa* and represents a further stage in the building up of the wadi bed: the floor deposits are deeper and include a greater proportion of salt deposits admixed with coarse rock detritus. Associate species of the Z. *spinosa* community include: *Artemisia judaica, Asteriscus graveolens, Crotalaria aegyptiaca, Fagonia mollis, Launaea spinosa, Lycium arabicum, Pennisetum dichotomum, Pituranthos tortuosus, Pulicaria undulata, Scrophularia deserti, Trichodesma africanum, Zygophyllum coccineum* and *Z. decumbens*. The same associates have been recorded in the *L. spinosa* community with the addition of *Cleome droserifolia, Lavandula stricta* and *Reaumuria hirtella*. Futher downstream this community includes *Acacia raddiana, Lygos raetam* and *Tamarix nilotica*.

In the middle section of Wadi Hagul the vegetation consists of three communities. One is dominated by *Hammada elegans* and occupies the gravel beds that form the raised terraces. The second is co-dominated by *Launaea spinosa* and *Leptadenia pyrotechnica* and occupies the waterways – that is the areas that are flooded by the occasional torrents. *Leptadenia pyrotechnica* is common further eastwards but rarely found westwards. The third community is represented by scattered patches of *Tamarix aphylla* occuping relicts of terraces of soft deposits. There are few trees of good size and the vegetation is mostly patchy bushy growth. Associate species of this community include: *Artemisia judaica, Cleome droserifolia, Crotalaria aegyptiaca, Fagonia bruguieri, F. mollis, Gymnocarpos decander, Iphiona mucronata, Launaea spinosa, Lavandula stricta, Lycium arabicum, Lygos raetam, Pituranthos tortuosus, Zilla spinosa* and *Zygophyllum coccineum.*

In the downstream part of Wadi Hagul which cuts across the coastal gravel beds, the channel is wide, but the course of the ephemeral streams is ill-defined. The wadi obviously changes its course on different occasions. The result is a reticulum of thin, branching and coalescing water courses. The meshes of this reticulum are patches of raised gravel isles. All are included on both sides by the gravel bed of the coastal plain. The vegetation of the main wadi is dominated by *Hammada elegans* with individuals of *Launaea spinosa* and *Lygos raetam.* The vegetation of the affluent runnels is mainly a grassland with *Panicum turgidum* and/or *Pennisetum dichotomum.* Further downstream, the wadi meets the littoral salt marsh vegetation.

### (b) Wadi Araba

The limestone block of Gebel El-Galala El-Bahariya (north Galala) is separated from the comparable block of the Galala El-Qibliya (south Galala) by a valley about 30 km wide (N–S).This valley, mostly of Lower Cretaceous sandstone, is occupied by the Wadi Araba system which drains the southern scarps of north Galala mountain and the northern scarps of south Galala mountain. The main channel of Wadi Araba terminates near Zaafarana Lighthouse (120 km south of Suez) and extends westwards to the central limestone country of the Eastern Desert.

The coastal front of the wide valley is dissected by the main channel of Wadi Araba (about 70 km) and a number of smaller wadis which drain the southeast of north Galala mountain and the northwest of the south Galala mountain. On the northern and southern sides are several independent wadis.

Wadi Araba consists of a main channel and innumerable tributaries (some several kilometers long) draining the two Galala blocks on the northern and southern sides and the fringes of the limestone plateau on the west. The main channel has a well-defined trunk cut across the sandstone beds.

Ecologically, a distinction may be made between the littoral saline habitat and desert vegetation of the channels of the drainage system. Within the former group are the littoral zone of raised beach, the littoral zone of sand hillocks and the inland zone of dry saline plain. The whole seaward belt of the coastal plain is saline and its soil and vegetation are distinctly different from those of the inland desert. The ecology of the drainage system depends on the extent of the catchment areas. The flora, however, shows certain differences between the wadis (or parts of wadis) cutting across the limestone masses of the ridges and those cutting across the sandstone.

The littoral and coastal plains are characterized by simple vegetation types and a limited number of species. The difference between the various communities often reflects differences in relative abundance of species rather than of floristic composition.

The littoral beach of Wadi Araba is represented by Mersa Thelemit, a small bay 7–8 km south of Zaafarana Lighthouse. A flat zone fringing the shore-line is bounded inland by a second raised beach which forms the fringe of the inland, narrow coastal plain. The littoral beach is less than one metre above sea level whereas the inland raised beach is 3–4 m above it.

Mersa Thelemit has a plant cover consisting of *Nitraria retusa* and *Zygophyllum album*. The latter forms pure stands on the flat littoral ground; *Nitraria* is much more abundant on the littoral side.

The littoral sand hillocks extend to the north of Zaafarana Lighthouse. The surface of these dunes is covered by a saline crust and a rich growth of *Halocnemum strobilaceum* associated with a few plants of *Zygophyllum album* and *Nitraria retusa*.

The coastal plain is transitional between the littoral salt marsh and the inland desert plain. In this, *Nitraria retusa* is the most abundant species. In some patches *Cressa cretica* or *Juncus rigidus* is dominant.

In the main channel, the wadi is characterized by extensive growth of *Tamarix aphylla*. This species may develop into trees, but in Wadi Araba it forms sand hillocks (1–4 m high and 10–100 $m^2$ area) covered completely by the growth of *Tamarix*. The *Tamarix* hillocks occur all along the main channel of the wadi but their density varies. Human destruction is the apparent cause of the thinning of these hillocks. In certain localities the hillocks are so crowded that they look like thickets.

Associated with *T. aphylla* and forming hillocks of smaller sizes are: *Calligonum comosum* (locally dominant), *Ephedra alata* and *Leptadenia pyrotechnica*. Near the mouth of Wadi Araba *Tamarix amplexicaulis* is a common associate. The spaces between these hillocks and mounds are the habitat of a number of species: *Aerva javanica, Artemisia judaica, Centaurea aegyptiaca, Convolvulus hystrix, Farsetia aegyptia, Francoeuria crispa, Hammada elegans, Heliotropium luteum, H. pterocarpum, Hyoscyamus muticus* and *Taverniera aegyptiaca*. Near the mouth of the wadi *Nitraria retusa* and *Zygophyllum album* grow. Distantly, dispersed within the main channel of the wadi and its main tributaries, are *Acacia raddiana* trees and shrubs that represent relicts of a dense population, destroyed by lumbering.

The community dominated by *Hammada elegans* is obviously the most common. It seems to replace *T. aphylla* scrubland, wherever the latter is destroyed. The ground cover is often coarse sand and gravel. Growth of *H. elegans*, which may build small sand mounds, is also abundant within the tributaries and runnels of the drainage system. The associates of this community include: *Artemisia judaica, Calligonum comosum, Ephedra alata, Francoeuria crispa, Pergularia tomentosa* and *Zilla spinosa* (Sharaf El-Din and Shaltout, 1985).

In the upstream extremities of the main channel is an area where *Anabasis articulata* is locally dominant. The vegetation here is of a few species: *Ephedra alata, Hammada elegans* and *Lygos raetam.*

Within the western affluents of Wadi Araba which traverse the limestone plateau of the Eastern Desert, a community dominated by *Launaea spinosa* is recognized. Associate species include: *Acacia raddiana, Achillea fragrantissima, Artemisia judaica, Asteriscus graveolens, Atractylis flava, Echinops galalensis, Launaea nudicaulis, Lygos raetam, Matthiola livida, Ochradenus baccatus, Pituranthos tortuosus, Pulicaria undulata, Trichodesma africanum* and *Zilla spinosa.* Grassland vegetation dominated by *Lasiurus hirsutus* is also recognized within the upstream affluents. Associates include: *Acacia raddiana, Asteriscus graveolens, Calligonum comosum, Centaurea aegyptiaca, Cleome droserifolia, Echinops spinosissimus, Fagonia mollis, Heliotropium arbainense, Iphiona mucronata, Launaea nudicaulis, L. spinosa, Linaria aegyptiaca, Lygos raetam, Panicum turgidum, Pergularia tomentosa, Plantago ciliata* and *Pulicaria undulata.*

Within the affluents of the middle part of Wadi Araba, apart from the common *Hammada elegans* community, two other communities may be recognized. One is dominated by *Artemisia judaica*, the other is grassland dominated by *Panicum turgidum* which is severely affected by grazing. Associates of the *A. judaica* community are: *Calligonum comosum, Centaurea aegyptiaca, Cleome droserifolia, Farsetia aegyptia, Hammada elegans, Pergularia tomentosa, Taverniera aegyptiaca* and *Zilla spinosa.* In the *P. turgidum* community the associates are: *Artemisia judaica, Calligonum comosum, Ephedra alata, Farsetia aegyptia, Hammada elegans* and *Zilla spinosa.*

Within the upstream parts of the main tributaries that cut across the limestone blocks of the two Galala mountains the plant cover is characterized by the preponderance of *Zygophyllum coccineum* which is present within the main channel of Wadi Araba. Common associates include: *Asteriscus graveolens, Erodium glaucophyllum, Fagonia mollis, Iphiona mucronata* and *Zilla spinosa.*

The main channel contains a number of brackish-water springs, e.g. Bir Zaafarana and Bir Buerat. Around these springs are neglected groves of *Phoenix dactylifera* and patches of halophytes – *Juncus rigidus, Nitraria retusa* and *Zygophyllum album.*

#### (c) Wadi Serimtai

This is one of the biggest drainage systems in the southern section of the Egyptian Red Sea coast located at about 50 km north of Mersa Halaib. The plant cover of this wadi and of the other wadis of the southern section does not vary greatly.

The littoral downstream belt of the delta of Wadi Serimtai supports halophytic vegetation comprising two communities; one is dominated by *Aeluropus brevifolius* and the other by *Zygophyllum album.* The *Aeluropus* community occurs in the low ground of wet brown sand covered by a silty saline layer. The plant cover is 70%, contributed by the dominant grass. The associates, mainly halophytes, include: *Arthrocnemum macrostachyum, Halopeplis perfoliata, Limonium axillare, Suaeda pruinosa* and *Zygophyllum album*. The *Z. album* community grows on flat soft sand sheets on which the dominant builds small hummocks of sand. The plant cover ranges from 30 to 40%. Being nearer to the desert habitat of the wadi, associates of this community are a mixture of xerophytes and halophytes. The xerophytes (perennials and annuals) include: *Asphodelus tenuifolius, Euphorbia scordifolia, Launaea cassiniana, Monsonia nivea, Panicum turgidum, Polycarpaea repens, Stipagrostis hirtigluma* and *Zygophyllum simplex*. The halophytes are *Limonium pruinosum, Salsola vermiculata, Sporobolus spicatus* and *Suaeda monoica*.

The westward (upstream) change in the soil characteristics and land relief is associated with a change in the vegetation of the wadi. West of the area dominated by *Zygophyllum album* is a narrow zone (200 m) dominated by *Salsola vermiculata* (29–30% cover) that builds hummocks of $2\times1$ m area and 60 cm high. Associate species are mainly salt non-tolerant desert perennials and annuals, e.g. *Acacia tortilis, Asthenatherum forsskaolii* (= *Centropodia forsskaolii*), *Calotropis procera, Cyperus conglomeratus, Eragrostis ciliaris, Euphorbia scordifolia, Glossonema boveanum, Monsonia nivea, Panicum turgidum, Stipagrostis hirtigluma* and *Zygophyllum simplex*.

In the midstream part of Wadi Serimtai the main channel is dominated by *Panicum turgidum*. The substratum is of coarse sand mixed with scattered rock detritus. Plant cover is 10–15% and increases to 60–65% during the winter season with dense growth of therophytes. *Acacia raddiana* is the abundant associate perennial. *Aristida adscensionis* and *Stipagrostis hirtigluma* are the abundant ephemerals. Other associates include *Abutilon pannosum, Aerva javanica, Arnebia hispidissima, Calotropis procera, Caylusea hexagyna, Cucumis prophetarum, Leptadenia pyrotechnica, Monsonia nivea, Neurada procumbens, Polycarpaea repens, Rumex simpliciflorus* and *Stipagrostis ciliaris*.

The dominance of *P. turgidum* continues for a few kilometers in the main trunk of the wadi, but then its abundance decreases gradually until it is replaced by a community dominated by *Acacia tortilis*. The plant cover of this community is about 40–50%: 10–20% perennials and 30–40% annuals. The most common associates are: *Aristida adscensionis, Stipagrostis hirtigluma* and *Panicum turgidum*. Other associates include *Abutilon pannosum, Aerva persica, Aizoon canariense, Asphodelus tenuifolius, Caylusea hexagyna, Cenchrus pennisetiformis, Cleome brachycarpa, Cucumis prophetarum, Dipterygium glaucum, Farsetia longisiliqua, Heliotropium pterocarpum, H. strigosum, Ifloga spicata, Launaea massauensis, Neurada procumbens, Panicum turgidum, Polycarpaea repens, Salsola vermiculata, Tephrosia purpurea, Tragus berteronianus* and *Zygophyllum simplex*.

The vegetation of the upstream part of the wadi is an open scrubland dominated by *Acacia raddiana*. The ground is of coarse sand mixed with gravel and rock detritus. Between the two *Acacia* communities (*A. tortilis* in the east and *A. raddiana* in the west) is a transitional zone co-dominated by both *Acacia* spp. Apart from the above-mentioned associates, other species recorded in the upstream part of the wadi include: *Amaranthus graecizans, Antirrhinum orontium, Aristida meccana, Astragalus eremophilus, Chenopodium murale, Euphorbia granulata, Ochradenus baccatus, Robbairea delileana, Sisymbrium erysimoides,*

*Spergula fallax, Trianthema salsoloides, Tribulus pentandrus* and *Trichodesma ehrenbergii*.

**Communities**

The preceding presentation shows that the vegetation of the Egyptian Red Sea coastal deserts (non-saline dry habitat) may be grouped under two main types: ephemeral and perennial. These two types of vegetation and their communities are described below.

### (i) Ephemeral vegetation

Ephemeral vegetation indicates soil conditions that allow for no overyear storage moisture: soil wetness is maintained during only a part of the year. This may be due either to scantiness of rainfall or also to surface deposits being too shallow. The ephemeral nature is the distinctive feature of this vegetation and not the habit of the species since many perennials may acquire ephemeral growth form under these conditions.

During the rainy season, the desert plain may show green patches of ephemerals. These areas may be independent of the drainage system. A rainy season is not an annual recurring phenomenon within a particular area. The rainfall is mostly inconsistent in time and space; cloudbursts that bring the desert rain are often of limited extent.

Ephemeral vegetation usually forms a mosaic of patches each of which may be dominated by one or several species. One type of ephemeral growth may recur in the same patch for several years. This may be due to the availability of seeds where parent plants have existed. Depending on the growth form of the dominant (or most abundant) species, three types of ephemeral vegetation are recognized in the Red Sea coastal desert: one dominated by succulent plants, a second by grasses and a third by herbaceous species (Zahran, 1964).

The succulent type of ephemeral vegetation may be dominated by: *Aizoon canariense, Trianthema crystallina, Tribulus pentandrus* or *Zygophyllum simplex*. The tissue of these succulents can store water that may be used later in the growing season. They are characterized by shallow roots and survive throughout a season longer than other ephemerals. In exceptionally wet years or highly favoured localities, *Zygophyllum simplex* may extend its life-span for a whole year or more. Also, included in this group may be the ephemeral growth of *Spergula fallax* and *Spergularia marina* which appear in exceptionally wet seasons or in areas fed by springs.

Ephemeral grasses include several species of *Aristida, Bromus, Cenchrus, Eragrostis, Schismus, Stipagrostis,* etc. This vegetation is of special importance for the nomadic herdsmen for whom it is valuable pasture.

The herbaceous ephemeral type of vegetation may be dominated by one of a great variety of species or may be mixed growth with no obvious dominant. Within the Egyptian Red Sea coastal desert, patches of ephemerals dominated by one of the following species are recorded: *Arnebia hispidissima, Asphodelus tenuifolius, Astragalus eremophilus, A. vogelii, Filago spathulata, Ifloga spicata,*

*Malva parviflora, Neurada procumbens, Plantago ciliata, Schouwia thebaica, Senecio desfontainei, S. flavus, Tribulus longipetalus* and *T. orientalis*. The growth of these plants may also provide valuable grazing.

The patches of ephemerals are associated with soft deposits, usually shallow sheets of sand, or especially favoured localities. This type of habitat provides a briefly sustained water supply during the rainy season; the shallow surface deposits eventually dry almost completely. The vegetation may be quite dense and the number of individual plants several 100/km$^2$. Plant cover varies from 5 to 50% and the patches of ephemerals look like micro-oases amidst the dry desert plain.

### (ii) Perennial vegetation

The perennial – xerophytic vegetation of the Red Sea coastal desert comprises two main types: suffrutescent and frutescent. The vegetation of the suffrutescent type consists of two layers: an upper layer (30–120 cm) which includes the dominant species and a ground layer (<30 cm) including the associate annuals and cushion-forming perennials, e.g. *Cleome droserifolia* and *Fagonia mollis*. Woody plants of 120–150 cm are sparse. In the frutescent type there are three main layers – the two of the suffrutescent type and a higher one: the dominant. This vegetation includes some trees of more than 5 m, e.g. *Acacia raddiana* and *Balanites aegyptiaca*.

#### A. *Suffrutescent perennial vegetation*

The suffrutescent perennial vegetation is widespread in the Egyptian Red Sea coastal desert. It is distinguished by a permanent framework of perennial xerophytes. This vegetation is formed of two layers: a suffrutescent and a ground layer. In the majority of communities the suffrutescent layer characterizes the vegetation; the ground layer is of dwarf or trailing perennials, enriched by the growth of ephemerals during the rainy season.

The suffrutescent perennial vegetation is recognized into 3 forms each characterized by its communities.

(a) Succulent half-shrub form

1. *Zygophyllum coccineum* community
2. *Salsola baryosma* community
3. *Hammada elegans* community

(b) Grassland form

4. *Panicum turgidum* community
5. Other communities

(c) Woody form

6. *Zilla spinosa* community
7. *Launaea spinosa* community
8. *Cleome droserifolia* community

9. *Sphaerocoma hookeri* subsp. *intermedia* community
10. Other communities

The following is a short account of each of these communities.

### (a) Succulent half-shrub form

#### 1. *Zygophyllum coccineum* community

*Z. coccineum* is a leaf and stem succulent that remains green all through the year. Yet this xerophyte seems to have an age limit of several years and in this differs from *Z. album* which does not appear to have a limited life. The regeneration of *Z. coccineum* is confined to rainy years when numerous crowded seedlings may appear but are eventually thinned to a limited number (1–2 per $m^2$). In exceptional localities where the *Z. coccineum* population may have a higher density, the individuals are usually small, being far below the normal size of up to 1 m. The most abundant species is *Zilla spinosa; Cleome droserifolia* and *Indigofera spinosa* are commonly present.

Within the *Z. coccineum* community, the frutescent layer is very open and may be of one or more of the following: *Acacia raddiana, A. tortilis, Leptadenia pyrotechnica, Lycium arabicum, Lygos raetam* and *Tamarix nilotica*. The suffrutescent layer contains the main bulk of the perennials including the dominant and its most common associates. The ground layer may include such dwarf or prostrate species as *Fagonia mollis, Heliotropium arbainense, Robbairea delileana* and *Cyperus conglomeratus*. The species population may be enriched during the rainy season by the profuse growth of therophytes.

The presence of *Z. coccineum* in the Egyptian Red Sea coastal desert is not very widespread, being confined to the limestone country. It occurs from the seaward fringes of the coastal desert plain to the inland mountain country. This wide range of ecological conditions is reflected by the presence of associates, including halophytes, e.g. *Limonium axillare* and *Nitraria retusa*, and plants of the mountain habitat e.g. *Moringa peregrina*.

In the Red Sea coastal lands of Saudi Arabia (Asian coast) and the southern section of the African coast (Eritrean coast), *Z. coccineum* is one of the halophytic plants that forms a widespread community (Hemming, 1961; Younes et al., 1983). There might be two *Z. coccineum* ecotypes, one intolerant of soil salinity and the other a salt-tolerant form.

The *Z. coccineum* community of the Red Sea coast is present within the drainage systems. Within the main channels of the larger wadis its growth is usually confined to the parts flushed by torrents and the community dominated by *Z. coccineum* is rarely found on the terraces higher than the water course.

#### 2. *Salsola baryosma* community

Within the Egyptian Red Sea coastal desert the community dominated by *S. baryosma* is common within the section south of Mersa Alam (700 km south of

Suez). Within the perennial framework of this community, *S. baryosma* contributes the main cover (10–50%). *Panicum turgidum* is an abundant associate and *Salsola vermiculata* and *Acacia tortilis* are commonly present. Less common associates include *Sevada schimperi* and *Sporobolus spicatus*. The growth of therophytes is notably rich. *Aristida adscensionis, A. meccana, Astragalus eremophilus, Eragrostis ciliaris, Stipagrostis hirtigluma* and *Zygophyllum simplex* cover about 50% of the spaces between the perennials. The parasite *Cistanche tinctoria* is usually present on the bushes of *S. baryosma*.

The frutescent layer of this community is made up of *Acacia tortilis* and *Lycium arabicum* and a few individuals of *Calotropis procera*. The suffrutescent layer includes the prostrate perennials and is enriched during the rainy season by the ephemerals.

Unlike the *Zygophyllum coccineum* community which is usually confined to the water courses flushed by torrents, the *S. baryosma* community spreads within the whole of the wadi channel and extends over the sides of the wadi delta. It is particularly common in the area between latitude 23° 10′N and 22° 20′N. In this part the coastal plain is wide (15–25 km). The downstream extremities of the wadis flowing into the plain may have ill-defined courses that are often lost among sheets of sand. The *S. baryosma* community grows on these sheets of sand. However, like *Z. coccineum, S. baryosma* does not build mounds and hillocks.

Within the Saudi Red Sea coastal land, the community dominated by *S. baryosma* occupies the transitional zone between the littoral salt marsh ecosystem and the coastal desert (Younes et al., 1983).

### 3. *Hammada elegans* community

Within the Red Sea coastal plain of Egypt the community dominated by *H. elegans* is confined to the coastal plain of the Gulf of Suez, i.e. the 400 km stretch from Suez to Hurghada. From Hurghada southwards to the Sudano-Egyptian border, *H. elegans* is absent. A comparable community dominated by *H. elegans* has also been recorded within the Saudi Red Sea coast (Zahran, 1982b).

In this community, *H. elegans* is consistently the most abundant species though its cover may not exceed 20% and may be much lower (<5%). The most common associate is *Launaea spinosa. Artemisia judaica, Francoeuria crispa, Panicum turgidum* and *Zilla spinosa* are common associates. *Lygos raetam, Pennisetum dichotomum* and *Zygophyllum coccineum* are less common. Other perennial xerophytes include *Anabasis articulata* and *Cleome droserifolia. A. articulata* is a species similar in habit to *H. elegans*; both are desert succulent xerophytes of the family Chenopodiaceae. *Anabasis* dominates a particular community with close physiognomic similarities to the *Hammada* community. It is abundant in several parts of the Egyptian deserts: Cairo-Suez Desert (El-Abyad, 1962) and Mediterranean coastal desert (Tadros and Atta, 1958). Within the Egyptian Red Sea coastal land, *A. articulata* is restricted to a few localities, e.g. delta of Wadi Hommath (33 km south of Suez) and delta of Wadi Araba (Zahran, 1962).

*A. articulata* is not recorded in the Saudi Red Sea coastal land (Migahid and Hammouda, 1978).

The vegetation of the *H. elegans* community is represented by three layers. The frutescent layer is of little significance and includes the shrubs: *Acacia raddiana, Calotropis procera, Tamarix aphylla* and *T. mannifera*. The suffrutescent layer contains the dominant and numerous associated perennials. The ground layer includes dwarf and prostrate perennials, e.g. *Centaurea aegyptiaca, Erodium glaucophyllum, Fagonia mollis, Paronychia desertorum* and *Polycarpaea repens*. During the rainy season this layer is enriched with the growth of therophytes.

Within the coastal land of the Gulf of Suez, the *H. elegans* community is abundant on the gravel terraces of the wadis. In a few localities e.g. Wadi Hommath, *Hammada* builds hillocks that may reach considerable size but the formation of such sand hillocks by *Hammada* is exceptional. Usually, the growth of *H. elegans* in the Egyptian desert is taken to indicate conditions where the softer deposits are gradually removed, leaving the coarse lag materials (Zahran and Willis, 2008).

### (b) Grassland form

#### 4. *Panicum turgidum* community

The *P. turgidum* community is one of the most common within the desert of Egypt, especially on sandy formations. It is also one of the most extensively grazed. *P. turgidum* is a good fodder plant. There is also evidence that it is one of the wild grain-plants which may be used as a supplementary food resource for desert inhabitants. Zohary (1962) stated that the grain of *P. turgidum* is collected in the Sahara by the natives of southern Algeria, ground and baked into bread.

*P. turgidum* is a tussock-forming grass that may acquire an evergreen habit, especially in favourable environments. Under less favourable conditions it may have a strictly deciduous growth form and remain dry, look dead for a prolonged period (a few years) but may regain its green habit after the rain.

The vegetation, though distinctly a grassland type, shows the usual three layers. The frutescent layer includes bushes (*Acacia tortilis, Leptadenia pyrotechnica, Lycium arabicum* and *Maerua crassifolia*) and trees e.g. *Acacia raddiana, Balanites aegyptiaca* and *Calotropis procera*. The suffrutescent layer is made up of the dominant grass and a range of perennial associates, e.g. *Hammada elegans, Launaea spinosa, Pituranthos tortuosus, Salsola baryosma, Sporobolus spicatus* and *Zilla spinosa*. The ground layer includes several dwarf or prostrate perennials e.g. *Citrullus colocynthis, Fagonia mollis* and *Polycarpaea repens*. The therophytes: *Aristida adscensionis, Brachiaria leersioides, Caylusea hexagyna, Eragrostis ciliaris, Euphorbia scordifolia, Stipagrostis hirtigluma* and *Zygophyllum simplex* form dense populations in especially favoured localities or in years with high rainfall, with cover, between the perennials, that may reach 40–60%.

Widespread in the Red Sea coastal desert, the *P. turgidum* community is made up of a rather large number of associate species having different geographical ranges within this coastal desert. For instance, *Acacia tortilis, Maerua crassifolia* and *Salsola baryosma* are present in the southern section of the coastal plain only, whereas *Hammada elegans* is confined to the northern section.

*P. turgidum* is mostly a sand-dwelling grass. Its growth, if not excessively grazed, may build up sand mounds and hillocks. It is an effective sand-binding xerophyte which also grows on mixed sand and gravel deposits but is uncommon on limestone detritus.

The *P. turgidum* grassland is widespread throughout the African Sahara (Zohary, 1962). In the arid desert of the Arabian Peninsula *P. turgidum* is one of the most common grassland types (Zahran, 1982b).

### 5. Other grassland communities

Reference has been made to two grassland types: the *Pennisetum dichotomum* community and the *Lasiurus hirsutus* community. The former is abundant in the wadis of the limestone country and is associated with salt terraces. Within the Red Sea coastal desert, a few stands of this grassland type are recorded in the wadis dissecting limestone formations. *Pennisetum dichotomum* is a tussock-forming grass similar in habit to *Panicum turgidum*. Without the inflorescence it is often difficult to distinguish between them. Both are grazed though *Pennisetum* seems to be less palatable.

*Lasiurus hirsutus* is apparently more drought-tolerant than both *P. turgidum* and *P. dichotomum*. It grows in smaller runnels with smaller catchment areas and hence less water resources. Within the Red Sea coastal desert, there are a few stands dominated by *L. hirsutus* associated with small runnels dissecting the sand and gravel formations of the coastal plain.

There is a fourth grassland type dominated by *Hyparrhenia hirta* which is very rare in the Red Sea coastal desert of Egypt. The dominance of *H. hirta* is confined to some of the fine upstream runnels dissecting the limestone formations. Associated species are chasmophytic and lithophytic species such as *Fagonia mollis* and *Helianthemum kahiricum*. *H. hirta* is of considerable importance in the southwestern Red Sea coastal mountains of Saudi Arabia. Its dense growth and dominance have been recorded in most of the protectorates of these mountains (Zahran et al., 2009). It is also recorded in other parts of the Middle East (Zohary, 1962).

### (c) Woody forms

### 6. *Zilla spinosa* community

*Z. spinosa* is a spinescent woody undershrub. It is normally a perennial xerophyte which, under favourable conditions, grows as an evergreen that flowers throughout most of the year. Under less favourable conditions it acquires a deciduous growth form. It may behave as an annual under extreme conditions. Individuals of *Z. spinosa* may reach considerable size with an area of 1–1.5 $m^2$ and height up to 80–170 cm. It may also be a small woody plant not exceeding 20–30 cm. A species with a wide range of habit and size is a valuable indicator of habitat conditions.

Within the Red Sea coastal desert, the plant cover of the *Z. spinosa* community is usually low (5–20%) and its individuals are normally small. Within the southern section, *Z. spinosa* very often acquires a deciduous or an annual growth form. *Artemisia judaica* and *Cleome droserifolia* are the most common associates. The abundance of *C. droserifolia* is a character that distinguishes the *Z. spinosa* community of the Red Sea coastal land. *Aerva javanica, Francoeuria crispa, Leptadenia pyrotechnica, Pulicaria undulata* and *Zygophyllum coccineum* are common associates. Less common species include: *Acacia flava* (= *A. ehrenbergiana*), *A. raddiana, Hammada elegans, Iphiona mucronata, Lasiurus hirsutus, Launaea spinosa, Lycium arabicum, Lygos raetam, Panicum turgidum, Pennisetum dichotomum* and *Zygophyllum decumbens*.

The vegetation of the *Z. spinosa* community shows the usual three layers. The dominant species and a great variety of associated perennials form the suffrutescent layer which contributes the main bulk of the plant cover. The frutescent layer is often very thin and may include one or more of the following bushes: *Acacia flava, A. tortilis, Leptadenia pyrotechnica, Lycium arabicum, Lygos raetam, Maerua crassifolia* and *Moringa peregrina* and trees, e.g. *Acacia raddiana* and *Balanites aegyptiaca*. The ground layer includes prostrate perennials such as *Citrullus colocynthis, Cucumis prophetarum, Fagonia parviflora, Monsonia nivea* and *Robbariea delileana*. The ephemerals enrich this layer during the rainy season.

The community dominated by *Z. spinosa* is confined to the channels of the wadis and is absent from other smaller runnels. It usually occurs in parts of the wadi bed that are covered with alluvial deposits. The size and growth-form of the dominant vary in obvious relationships with texture and depth of these surface deposits. In contrast to some other communities described, this community is made up of xerophytes only with none of the halophytes.

### 7. *Launaea spinosa* community

*L. spinosa* (Asteraceae–Compositae) is a milky sap xerophyte having a growth form comparable to that of *Zilla spinosa* (Brassicaceae–Cruciferae); both are spinescent with individuals ranging from small woody plants to bushy undershrubs of considerable size. Also, both may develop an evergreen growth form under favourable conditions and a summer deciduous growth form under less favourable ones. The two communities dominated by *Z. spinosa* and *L. spinosa* are confined to the channels of the wadis and their main tributaries. However, the community dominated by *L. spinosa* is restricted to the northern part of the Red Sea coastal desert. i.e. the coastal desert of the Gulf of Suez. It occurs as an associate species in the northern groups of mountains, namely Shayib Group (Zahran, 1964).

The plant cover of the *L. spinosa* community ranges between 10 and 25% and is contributed mainly by the dominant and the abundant associate, *Zilla spinosa*, which is recorded in all stands (Zahran, 1962). Other most common associates are: *Artemisia judaica, Cleome droserifolia, Crotalaria aegyptiaca, Echinops galalensis, Iphiona mucronata, Lasiurus hirsutus, Lavandula stricta, Lygos raetam,*

*Matthiola liviola, Pennisetum dichotomum, Pituranthos tortuosus, Tamarix nilotica* and *Zygophyllum coccineum.*

The three layers typical of the desert ecosystem communities are represented here. The most important layer is the suffrutescent one which contains the dominant species and the most common associates. The ground layer includes a number of perennials, e.g. *Asteriscus graveolens, Citrullus colocynthis* and *Helianthemum lippii* and the therophytes. The frutescent layer is usually thin and includes one or more of the following trees and shrubs: *Acacia raddiana, Lycium arabicum, Lygos raetam* and *Tamarix nilotica.*

### 8. *Cleome droserifolia* community

*C. droserifolia* is a low, much branched, densely glandular-hispid undershrub with small orbicular three-nerved leaves. It forms low cushions of spreading growth that may not exceed 30 cm in height; it occurs in the ground layer.

In Egypt, *C. droserifolia* is present in all regions except the Mediterranean coastal desert (Täckholm, 1974). It is recorded in the southern section of the Saudi Red Sea coastal land (Zahran, 1982b).

The dominance of *C. droserifolia* in the Egyptian Red Sea coastal desert is confined to the larger runnels; it does not occur in the channels of the main wadis covered with very coarse detritus including large boulders. The dominant may accumulate pads of soft material and it grows in isolated patches confined to particular positions among the coarse boulders.

Ecologically, the community dominated by *C. droserifolia* is associated with the limestone country of the Red Sea coastal desert (Zahran, 1964). *Zilla spinosa* and *Zygophyllum coccineum* are the most common associates. *Fagonia mollis* and *Francoeuria crispa* are frequently present whereas less common associates include: *Acacia flava, A. raddiana, A. tortilis, Artemisia judaica, Hammada elegans, Iphiona mucronata, Launaea spinosa, Leptadenia pyrotechnica, Lycium arabicum, Lygos raetam, Panicum turgidum, Pituranthos tortuosus* and *Zygophyllum decumbens.*

Stratification of vegetation is clear in this community. The frutescent layer is extremely thin. The suffrutescent layer is made up of a variety of species including the most common associates mentioned above. The dominant is in the ground layer which includes also *Citrullus colocynthis, Corchorus depressus, Cucumis prophetarum* and *Fagonia mollis.* This layer is enriched by therophytes during the rainy season.

### 9. *Sphaerocoma hookeri* community

*S. hookeri* is a glabrous, blue-green, densely branched shrub (60–70 cm high), with knotty branches, and fleshy, terete, opposite or whorled leaves. Its presence in Egypt is confined to the Red Sea coastal desert: *S. hookeri* ssp. *intermedia* is rarely recorded in the southern region of the inland part of the Eastern Desert of Egypt (Täckholm, 1974).

The community type dominated by *S. hookeri* is present in the southern part of the Egyptian Red Sea coastal desert (south of the Tropic of Cancer). It grows in a special type of habitat – sand dunes formed at the boundaries between the littoral salt marshes and the coastal desert plain. These dunes may also form embankments covering the edges of the desert plain in places where a low plateau rises abruptly at the border of the low ground of the littoral belt. Thus, the associate species of this community include halophytes, e.g. *Cyperus conglomeratus, Limonium axillare, Salsola vermiculata, Sevada schimperi* and *Sporobolus spicatus* and xerophytes, e.g. *Acacia tortilis, Asthenatherum forsskaolii, Heliotropium pterocarpum, Lycium arabicum, Monsonia nivea, Panicum turgidum, Polycarpaea repens* and *Salsola baryosma* and also ephemerals, e.g. *Aristida* spp.

The three common layers of vegetation can easily be recognized in this community. The suffrutescent layer includes the dominant and most of the associates, but the frutescent layer is thin, including the shrubs e.g. *Acacia tortilis*. The ground layer comprises species of: *Aristida, Cyperus, Eragrostis, Monsonia, Sporobolus* etc.

### 10. Other communities

There are a number of communities dominated by woody perennials that are less widespread in the Red Sea coastal desert. These include communities dominated by each of: *Iphiona mucronata, Artemisia judaica, Pituranthos tortuosus* and *Calligonum comosum*.

The community dominated by *Iphiona mucronata* is confined to smaller affluent wadis or the upstream extremities of greater tributaries traversing the limestone country. The most abundant associates include: *Gymnocarpos decander, Launaea spinosa, Zygophyllum coccineum* and *Z. decumbens*.

The *Artemisia judaica* community is present in a few localities within the wadis. It is apparently confined to conditions where the surface deposits are a mixture of alluvial limestone detritus and aeolian sand. Such mixtures are particularly those which traverse and drain the limestone and sandstone formations. The most common associates of this community include: *Calligonum comosum, Ephedra alata, Farsetia aegyptia, Hammada elegans* and *Zilla spinosa*. This community is widespread in the mountainous country of southern Sinai (Migahid et al., 1959).

The community dominated by *Pituranthos tortuosus* occurs in a few of the runnels dissecting the gravel plain. Associate species include: *Artemisia judaica, Fagonia mollis, Iphiona mucronata, Lavandula stricta* and *Lindenbergia sinaica*.

The *Calligonum comosum* community is associated with sand aeolian deposits. The dominant forms sandy mounds and hillocks. Associate species include: *Artemisia judaica, Francoeuria crispa, Hammada elegans, Tamarix aphylla* and *Zilla spinosa*.

The frutescent perennial vegetation includes the scrubland type of the desert vegetation. The plant cover is in three layers: a frutescent (120–500 cm), suffrutescent (30–120 cm) and a ground layer (<30 cm). The frutescent layer is here dense enough to give the vegetation characters that distinguish it from the previous categories. Two main forms may be recognized in this type: the succulent shrub form and the scrubland form. The former is not represented in the Red Sea coastal desert but a type dominated by a succulent xerophytic bush, *Euphorbia candelabra*, is recorded from one part of the Red Sea coastal desert of the Sudan some 50–70 km south of the Egyptian coast (Zahran, 1964). The scrubland form is represented by the communities dominated by *Acacia raddiana, A. tortilis, Lycium arabicum* and *Tamarix aphylla* and several less common types. In these communities the vegetation is well developed as is evident from the long list of associate species. All these communities are present in the channels of the main wadis and their larger tributaries. The distribution of the *Acacia* scrubland types extends over wadis traversing the coastal plain and their upstream parts across the mountain area. The description of these communities, given below, depends on stands representing their growth within the two ecosystems. The vegetation of these communities, especially that dominated by *A. raddiana*, are (and for a long history have been) subject to extensive lumbering for fuel and charcoal manufacture. Camel breeding and charcoal manufacture are the main industries of the local inhabitants, especially in the Gebel Elba district (Zahran, 1964).

### B. Frutescent perennial vegetation

This vegetation includes the following communities:

11. *Lycium shawii* community
12. *Acacia tortilis* community
13. *A. raddiana* community
14. *Tamarix aphylla* community
15. Other communities.

### 11. *Lycium shawii* community

*L. shawii* is a spinescent shrub which may develop an evergreen growth-form under favourable conditions of water resource. Growing in less favourable habitats it is deciduous, shedding its leaves early in the season and the main part of its shoot may also dry up. Within the thick growth of Gebel Elba scrubland, *L. shawii* may show a climbing habit.

Within the Egyptian Red Sea coastal desert the *L. shawii* community is particularly common in the southern part. In the coastal plain extending from mountain ranges to the littoral salt marsh belt this community is one of the most common of the wadis and the deltas of the main channels. It may also dominate the channels of the large tributaries. The plant cover is in the range of 10–50%. The dense cover is due to the rich growth of the therophytes which almost completely cover the gaps between the perennial vegetation. It is estimated that the cover in some stands is 70–90% contributed by the ephemerals and 20% by the perennials. In other stands the cover of the ephemerals is slightly higher (40–50%) than that of the perennials (30–40%). In certain stands the cover is mainly perennials and the ephemerals contribute very little to this.

In the *L. shawii* community the abundant associates are: *Acacia tortilis, Panicum turgidum* and *Salsola vermiculata* (perennials) and *Euphorbia scordifolia* and

*Zygophyllum simplex* (annuals). Less common species include *Acacia raddiana, Aristida adscensionis, A. funiculata, A. meccana, Asphodelus tenuifolius, Balanites aegyptiaca, Calotropis procera, Launaea cassiniana, Leptadenia pyrotechnica, Lotononis platycarpos, Polycarpaea repens, Salsola baryosma* and *Stipagrostis hirtigluma.*

Stratification of vegetation is clear in this community. The frutescent layer includes the dominant and several shrubs and trees, e.g. *Acacia raddiana, A. tortilis, Balanites aegyptiaca* and *Lygos raetam*. This layer contributes the bulk of the perennial cover and forms the main part of the permanent framework which gives the character to the vegetation. The suffrutescent layer, though made up by the majority of the perennials, including some of the most common associates, contributes little to the cover. In the ground layer there are several perennials and almost all the ephemerals. In years of good rainfall ephemerals form patches of dense growth between the components of the other layers.

### 12. *Acacia tortilis* community

*A. tortilis* is a flat-topped or umbrella–shaped spinescent shrub. The size varies from dwarf shrubs to much larger ones. The *A. tortilis* community is a most common scrubland type within the desert area extending to the south of Mersa Alam but it is absent from the stretch from Mersa Alam northward to Suez (c. 700 km). There is, however, a limited locality along the Suez-Ismailia desert road (16–17 km north of Suez) on the Suez canal west bank where there are a few patches of *A. tortilis* scrubland.

The *A. tortilis* community occurs in a variety of habitats. It may be seen on the slopes of the low hills at the northern and eastern foot of the mountains, e.g. Gebel Elba, and the similar coastal mountains. The most common habitat is the channels of the main wadis of the larger tributaries. Unlike the *A. raddiana* community which occurs on the softer deposits of the wadi channels, the *A. tortilis* community is present on the coarse deposits.

The vegetation cover of the *A. tortilis* community is primarily formed by the canopy of the dominant. In a few instances, the most common perennial associates, e.g. *Lycium shawii* and *Panicum turgidum*, and therophytes, e.g. *Aizoon canariense, Aristida adscensionis* and *Zygophyllum simplex,* contribute substantially to the cover of the permanent framework. Other associates include *Acacia raddiana, Aerva javanica* (*A. persica*, Täckholm, 1974), *Balanites aegyptiaca, Indigofera spinosa, Maerua crassifolia, Moringa peregrina, Salsola vermiculata* and *Zilla spinosa*.

By virtue of the growth form of the dominant species, the frutescent layer is the most notable part of the vegetation. This layer includes trees, e.g. *Acacia raddiana* and *Balanites aegyptiaca*, and shrubs, e.g. *Leptadenia pyrotechnica, Lycium arabicum* and *Maerua crassifolia*. Most of the associates, e.g. *Aerva javanica, Panicum turgidum, Salsola vermiculata* and *Zilla spinosa*, are in the frutescent layer. The main bulk of the ephemerals are in the ground layer.

### 13. *Acacia raddiana* community

*A. raddiana* is one of the most common and widespread plants of the Eastern Desert of Egypt. It is recorded in almost all the main wadis of this desert, including the Red Sea coastal desert. Comparison of the distribution of the two *Acacia* species, *A. raddiana* and *A. tortilis*, in Egypt shows that the former is much more widespread all over the Eastern Desert whereas the latter is mostly confined to the southern part. But, where the two species grow together, there is evidence that *A. tortilis* is the more drought tolerant. A study of the wadis of the Abu Ghusson district (latitude 24° 20′N, longitude 35° 10′E) showed that most of the very abundant *A. raddiana* shrubs were dry, dead or almost dead whereas shrubs of *A. tortilis* were thriving almost normally. This was repeatedly observed in a number of wadis where the habitat was subject to a spell of rainless years (Zahran, 1964).

In the main wadis draining the Gebel Elba area, the two *Acacia* scrubland types may be present. *A. tortilis* shrubland covers the gravel terraces and the *A. raddiana* open forest is in the channels which contain ephemeral streams and where the deposits are much softer. In these localities *A. raddiana* shows some features of forest growth such as the presence of lianas (*Cocculus pendulus* and *Ochradenus baccatus*) and parasites, e.g. *Loranthus acaciae* and *L. curviflorus*.

It may be added that *A. raddiana* scrub and open forest is confined to the channels of the main wadis and is not present on the hills and mountain slopes where other *Acacia* spp. (*A. etbaica, A. laeta, A. mellifera* and *A. tortilis*) may grow. This has been also observed in the Saudi Red Sea coastal desert (Zahran, 1993a).

### 14. *Tamarix aphylla* community

*T. aphylla* is a species that normally grows as a tree that may reach considerable size. It may form a dense open forest that represents one of the main climax communities of the desert wadis. Under the influence of cutting, grazing and other destructive agencies it acquires a bushy growth-form that covers the ground in patches. In localities subject to the accumulation of windborne sand it may form dunes which it covers.

In the *T. aphylla* community of the Egyptian Red Sea coastal desert the cover ranges between 5 and 50%. It is mainly contributed by the dominant and the ephemeral vegetation during the rainy years. The most common associates include: *Acacia raddiana, Balanites aegyptiaca, Calligonum comosum, Farsetia aegyptia, Francoeuria crispa, Hammada elegans, Iphiona mucronata, Lycium shawii, Ochradenus baccatus, Salvadora persica, Zilla spinosa* and *Zygophyllum coccineum*.

The flora of this community is of a variety of species with different ecological amplitude. For instance, *Acacia raddiana, Balanites aegyptiaca* and *Tamarix aphylla* are trees that require ample water resources and deep valley-fill deposits and that represent the climax stage in the wadi-bed development. *Hammada elegans* and *Zygophyllum coccineum* are succulents that withstand different conditions. *Gymnocarpos decander* and *Iphiona mucronata* represent pioneer stages in the wadi-bed development and occur in localities where the surface deposits covering the bed rocks are very thin.

*T. aphylla* is present in the whole stretch of the Egyptian Red Sea coastal desert. Because its wood is softer than that of the other trees (*Acacia* and *Balanites*) of the area, its scrublands are apparently the first vegetation to be destroyed by cutting. This may be one of the main causes for the limited dominance of *T. aphylla* in the coastal desert. Relict patches of *T. aphylla* forest are usually seen in the main wadis. Climatic changes affect trees of *T. aphylla* more adversely than other trees.

**15. Other scrubland communities**

Reference may be made to four scrubland types that are occasionally found within the Red Sea coastal desert. One is dominated by *Acacia ehrenbergiana*. It is usually associated with valley-fill deposits that include considerable proportions of soft material that make the deposits compact and difficult to excavate.

A second community is dominated by *Capparis decidua*. It is recorded in the wadis of the southern region where the dominant may form large patches of bushy growth that may build sand hummocks. *C. decidua* may also have a tree growth-form. This community is usually associated with soft deposits that may be alluvial or aeolian.

The *Leptadenia pyrotechnica* community is more widespread in the Egyptian Red Sea coastal desert than the two previously mentioned communities. Like all the other scrubland types, this type is confined to the channels of the main wadis. It is usually associated with wadi terraces of mixed deposits.

The *Salvadora persica* community is recorded in a few of the main wadis, e.g. Wadi Bali. The dominant forms patches of growth that show the influence of repeated cutting. *S. persica* is the tooth-brush tree; if saved from cutting, it forms trees with upright trunks. In the mid-part of Wadi Bali, *S. persica* is associated with *Artemisia judaica, Cleome droserifolia, Pulicaria undulata, Zygophyllum coccineum* etc. (Zahran, 1964). In the Saudi Arabian Red Sea coastal desert, *S. persica* (Arak in Arabic) grows densely in Wadi Arak (Zahran, 1982b). The dense growth of *S. persica* in the Sudanese Red Sea coastal desert has also been mentioned by Kassas (1957).

A fifth community is an open forest dominated by *Balanites aegyptiaca*. Patches of this are present in a few of the main wadis especially within the mountain ranges. These patches are clearly relicts of a much more widespread growth. Trees of *B. aegyptiaca* are recorded in almost all wadis of the southern section. The fleshy fruits are collected and eaten by the Bedouins.

**II. Coastal mountains**

The coastal mountains of the Egyptian Red sea coast have been classified by Kassas and Zahran (1962, 1965, 1971) under the following groups: Shayib Group, Nugrus Group, Samiuki Group and Elba Group. Shayib Group comprises the range of mountains of the western coast of the Gulf of Suez whereas the other three groups are located in the proper Red Sea coast of Egypt. The following is an account on the vegetation of these groups.

### (a) Shayib group

Within the coastal hills of the Gulf of Suez (Shayib group) the vegetation is confined to the upstream part of the drainage system and to the slopes of these hills. The water courses are usually well defined in the hill country. Across the coastal plain, on the other hand, the courses are usually ill-defined runnels within the much wider courses of the wadi. Plant cover varies in relation to the extent of the area and the texture of the bed cover. Several plant communities may be recognized; some are common in both coastal ecosytems (desert and mountains) and some are confined to the hill ecosystem. *Acacia raddiana, Anabasis articulata, Artemisia judaica, Hammada elegans, Launaea spinosa, Leptadenia pyrotechnica, Lygos raetam* and *Panicum turgidum* belong to the first category. The *Zilla spinosa* and *Zygophyllum coccineum* communities are common in the wadis of the limestone hills.

There are a few plants that characterize the cliffs and dry waterfalls that intercept the courses of the wadis traversing the hills. These are *Capparis cartilaginea* (*C. sinaica*, Boulos, 1995), *C. spinosa, Cocculus pendulus* and *Ficus pseudosycomorus*. Reference may also be made to the slopes of the Cretaceous limestone hills which include: *Anabasis articulata, Halogeton alopecuroides, Heliotropium pterocarpum* and *Salsola tetrandra* together with such common species as *Hammada elegans, Ochradenus baccatus* and *Pergularia tomentosa*.

Within the drainage system of the mountains facing the Gulf of Suez (Shayib group) two main communities may be recognized. One is dominated by *Zilla spinosa* which is widespread within the channels of the wadis and the second is characterized by the preponderance of *Moringa peregrina* confined to the upstream parts of the wadis draining the slopes of the higher mountains.

#### 1. *Zilla spinosa* community

*Z. spinosa*, the most abundant plant in the majority of the wadis, acquires, in this district, a distinctly deciduous growth form. The shoot is dry and plants often appear to be dead. In rainy years, which are not of regular occurrence, plants are profusely regenerated. It is suspected that *Z. spinosa* is here a particular variety (*Z. spinosa* v. *microcarpa*, Täckholm, 1974) or an ecotype (potential annual, Zahran, 1964). This requires further ecological and taxonomic studies. Boulos (1995) reported it as *Z. spinosa* subsp. *spinosa.*

Common perennial associates are *Aerva javanica, Artemisia judaica, Calligonum comosum, Cleome droserifolia, Fagonia mollis, Leptadenia pyrotechnica, Solenostemma argel* and *Zygophyllum coccineum*. Abundant and common ephemerals are *Aizoon canariense, Arnebia hispidissima, Asphodelus tenuifolius, Ifloga spicata, Lotus arabicus, Reichardia orientalis, Robbairea delileana* and *Senecio flavus*.

In certain parts of these wadis *Acacia raddiana* is locally dominant: there are patches of *A. raddiana* scrubland which are apparently relicts of better growth that has been destroyed. *Acacia* scrubland is presumed to represent the natural climax vegetation of these wadis.

## 2. *Moringa peregrina* community

*M. peregrina* is one of the most interesting plants in the mountain ranges of the Red Sea coastal land. It is a "10–15 m high tree, [with white bark] usually destitute of leaves. These, when present, consisting of 3 pairs of long, slender, junciform pinnae, looking like opposite virgate branchlets. The pendulous pods ripen in October, the angled nut-like, white seeds (behen-nuts) are of a bitter-sweet nauseous taste and rich in oil (ben-oil)" (Täckholm, 1974). The behen-nuts are collected by the local natives and sold at a good price. The ben-oil of these seeds is used for special lubrication purposes. This particular attribute has saved this plant which is too valuable to be cut for fuel.

The *Moringa* scrub is represented by patches that cover limited areas of the upstream runnels of the drainage systems. These are runnels collecting water at the foot of the higher mountains.

A survey of *Moringa* within the Red Sea mountains of Egypt extending from latitude 27° 20′N to 22°N (Table 2.1) shows that this species is confined to the foot of the mountains that are higher than 1,300 m. Lower mountains and hills have almost no *Moringa* at their foot. The ground where *Moringa* grows is usually covered with coarse rock detritus, a character typical of the upstream runnels at the foot of the mountains.

The usual association of *Moringa* with coastal mountains more than 1,300 m high is not a sharp limit, since mountains nearer to the coast are better favoured than those further from it. Within the mountain range of Hurghada, the *Moringa* community contains the following xerophytic associates: *Acacia raddiana, Aerva javanica,*

**Table 2.1** Mountains within the Red Sea coastal desert of Egypt showing altitude and the presence (+) or absence (–) of *Moringa peregrina* (after Kassas and Zahran, 1971)

| Gebel | Altitude (m) and occurrence of *Moringa* | Gebel | Altitude (m) and occurrence of *Moringa* |
|---|---|---|---|
| Abu Harba | 1,705 (+) | Umm Laseifa | 1,210 (–) |
| Abu Dukhan | 1,661 (+) | Nugrus | 1,504 (+) |
| Abu Guruf | 1,099 (–) | Hafafit | 857 (–) |
| Qattar | 1,963 (+) | Zabara | 1,360 (+) |
| Shayib El-Banat | 2,187 (+) | Miqif | 1,198 (–) |
| Umm Anab | 1,782 (+) | Hashanib | 1,133 (–) |
| Abu Fura | 1,032 (–) | Abu Hamamid | 1,745 (+) |
| Weria | 1,035 (–) | Samiuki | 1,486 (+) |
| Mitiq | 1,112 (–) | Hamata | 1,977 (+) |
| El-Sibai | 1,484 (+) | Elba | 1,428 (+) |
| Abu Tiyur | 1,099 (–) | Shindodai | 1,426 (+) |

*Artemisia judaica, Capparis cartilaginea, C. decidua, Chrozophora plicata, Cleome droserifolia, Fagonia mollis, Francoeuria crispa, Hyoscyamus muticus, Launaea spinosa, Lavandula stricta, Leptadenia pyrotechnica, Lindenbergia sinaica, Lycium arabicum, Ochradenus baccatus, Periploca aphylla, Zilla spinosa* and *Zygophyllum coccineum.*

### The nakkat habitat

The convectional rainfall of the Gulf of Suez coastal area, according to the average rainfall of Hurghada, is 3 mm a year (Anonymous, 1960). But the vegetation and the water resources in the mountain area indicate greater precipitation. In the wadis running at the foot of the mountains there are several shallow wells of fresh water. On the slopes or cliffs of the mountains there are cracks from which a continuous trickle of water oozes ("nakkat" is Arabic for "dropper") and runs down the slopes, water collecting in a pot-hole at the foot of the slope forming a "bir" (well) or in some parts forming hollows (gelts) along the slope. The source of the nakkat is often a fissure in the solid basement complex rocks of the mountains, usually situated near the top. The courses of the runnels dissecting the slopes of the mountains may contain pot-holes that are periodically filled with water. These pot-holes are usually lined with calcareous skin material dissolved in the water that collects in them. In this peculiar habitat of these pot-holes, ferns e.g. *Adiantum capillus-veneris*, mosses and algae, alien strangers of the desert environment, grow. Associates are such water-loving plants as *Imperata cylindrica, Phragmites australis, Solanum nigrum* and *Veronica beccabunga.*

The nakkat habitat is also typical for *Ficus pseudosycomorus*. Stunted individuals of *Phoenix dactylifera* also occur in many nakkats, hanging from the top or near the source. The wet areas that fringe the birs (wells) and gelts are often covered with a rich growth of *Cynodon dactylon, Cyperus laevigatus, Imperata cylindrica* and *Juncus rigidus.*

The restriction of *Moringa* to the foot of the higher mountains and the presence of nakkats as features peculiar to such mountains indicate that the high altitude leads to greater water resources.

In one of the runnels across the eastern slope (facing the sea) of Gebel Shayib El-Banat, the following plants have been recorded: *Acacia raddiana, Aerva javanica, Artemisia inculta, A. judaica, Capparis cartilaginea* (*C. sinaica*, Boulos, 1995), *Chrozophora oblongifolia, Citrullus colocynthis, Cleome droserifolia, Fagonia mollis, Francoeuria crispa, Hyoscyamus muticus, Lavandula stricta, Lindenbergia abyssinica, L. sinaica, Moringa peregrina, Periploca aphylla, Pulicaria undulata, Solenostemma argel, Teucrium leucocladum, Zilla spinosa* and *Zygophyllum coccineum.*

### (b) Nugrus group

This group, which extends between latitudes 24° 40′N and 24° 50′N, faces the full stretch of the Red Sea. The highest of this group is Gebel Nugrus, "a great boss of red granite rising to a height of 1,505 m among schist and gneisses" (Ball, 1912).

This group includes Gebel Hafafit (857 m), Gebel Migif (1,198 m), Gebel Zabara (1,360 m) and Gebel Mudargag (1,086 m).

In the district of Gebel Nugrus the growth of ephemerals in the wadis and the rills across the mountain slopes is usually rich during the wet years.

The growth of *Moringa* characterizes the foot of the high mountains (above 1,300 m) but not the lower ones. The main wadis draining this group are the habitat of various types of open scrub dominated by one of the following: *Acacia raddiana, A. ehrenbergiana, Balanites aegyptiaca, Leptadenia pyrotechnica* and *Salvadora persica*. The *A. raddiana* scrub is confined to wadis draining westward. The *B. aegyptiaca, L. pyrotechnica* and *S. persica* scrublands are less common and are mostly confined to the channels of the main wadis.

### (c) Samiuki group

This group comprises Gebel Abu Hamamid (1,745 m), Gebel Samiuki (1,283–1,486 m) and Gebel Hamata (1,977 m). The last is the nearest to the sea (40 km) whereas Gebel Abu Hamamid is furthest away (65 km).

The vegetation in the Gebel Hamata is much richer in species and in plant cover than that in the Gebel Samiuki area. However, the flora of this group as a whole is richer than that of the Gebel Nugrus group to its north which is again richer than that of the Gebel Shayib group still further northward.

### (d) Elba group

This is an extensive group of granite mountains situated on the Sudano-Egyptian border (latitude 22ºN) and includes: Gebel Elba (1,428 m), Gebel Shindeib (1,911 m), Gebel Shindodai (1,426 m), Gebel Shillal (1,409 m), Gebel Makim (1,871 m) and Gebel Asotriba (2,117 m). This group, especially Gebel Elba, is particularly favoured by its position near the sea. The richness of the vegetation of the Gebel Elba area is so notable, compared to the other regions of Egypt, that this is considered as one of the main phytogeographical regions of the country (Drar, 1936; Hassib, 1951). The flora of the Gebel Elba group is much richer than that of the other Egyptian Red Sea coastal mountain groups. The number of species collected (Zahran, 1962, 1964; Kassas and Zahran, 1971) within the areas of the four mountains are: 33 species in the area of Gebel Shayib group, 92 species in the area of Gebel Nugrus group, 125 species in the area of Gebel Samiuki group and 458 species in the area of Gebel Elba group.

Though not the highest of its group, Gebel Elba is the nearest to the sea (20–25 km). The whole group faces a northeast bend of the shore such that Gebel Elba faces northward to an almost endless stretch of water. Gebel Elba is the most northerly of its group, the rest being in its shadow from the north winds.

Within the block of Gebel Elba, the vegetation on the north and northeast flanks is much richer than that on the south and southwest. The difference is equal to that between rich scrubland or open parkland on one side and desert vegetation on the other.

The northern and northwest slopes of Gebel Elba are drained by Wadi Yahameib and Wadi Aideib. These wadis are densely covered with *Acacia* thickets, the only place in the Egyptian coastal and inland deserts where the vegetation looks like a forest. The north and northeast slopes of Gebel Elba are richly vegetated. Three latitudinal zones of vegetation may be recognized: a lower zone of *Euphorbia cuneata*, a middle zone of *E. nubica* and a higher zone of moist habitat vegetation. In this higher zone are stands of *Acacia etbaica, Dodonaea viscosa, Ficus salicifolia, Pistacia khinjuk* and *Rhus abyssinica*. Within these higher zones ferns, mosses and liverworts are present.

Gebel Karm Elba is one of the main foot-hills of Gebel Elba which lies on its east. The north and northeast slopes of this block are characterized by the abundance of *Delonix elata*.

The southern slopes of Gebel Elba drain into Wadi Serimtai, one of the most extensive drainage systems within the whole district. The *Acacia* scrub of this wadi is much more open than that of Wadi Aideib. The southern slopes are notably drier: the vegetation is mostly confined to the rills and runnels of the drainage system. The most common type of vegetation within these runnels is a community dominated by *Commiphora opobalsamum*. On the higher altitudes, some shrubs of *Acacia etbaica* and *Moringa peregrina* may be found.

The western slopes of Gebel Elba are even drier than the southern ones. The vegetation is of therophytes (mostly *Zygophyllum simplex*) which appear in rainy years.

The differences in vegetation on different slopes are also notable in the eastern foot-hills of the Gebel Elba group. The higher hills rise to 175 m in the coastal plain to the east of Gebel Elba. On the north and northeast slopes the vegetation is characterized by the preponderance of *Euphorbia cuneata*. On the southern slopes *Aerva javanica* is dominant with only rare individuals of *E. cuneata*. On one of the foot-hills (nearer to the shore-line) the northern and eastern slopes are covered by a rich growth of *Acacia nubica* whereas on the southern and western slopes there is an open growth of *Aerva javanica*.

The vegetation of Gebel Shindodai and Gebel Shillal is also richer than that of Gebel Shindeib. The three mountains lie along the same latitude but Gebel Shindodai and Gebel Shillal are on the east, nearer to the sea than Gebel Shindeib. It will similarly be noted that the eastern side of Gebel Asotriba (Soturba, Schweinfurth, 1865) has also a richer flora than the inland slopes.

The eastern flanks of Gebel Shindodai and northern slopes of Gebel Shillal feed the upstream tributaries of Wadi Shillal which is covered by dense *Acacia* scrub. The vegetation of the northeast slopes of Gebel Shindodai show four main zones from base to top: (1) a zone characterized by the abundance of *Caralluma retrospiciens*, (2) a zone characterized by the abundance of *Delonix elata*, (3) a zone of *Moringa peregrina* and (4) a zone with bushes of *Dodonaea viscosa, Pistacia khinjuk* and *Euclea schimperi* and with numerous bryophytes and ferns including *Ophioglossum polyphyllum* and other moisture-loving species such as *Umbilicus botryoides*.

The northeast slopes of Gebel Shillal are richly vegetated with a great variety of species. A number of zones may be recognized from base to top: (1)

a zone of *Acacia tortilis* and *Commiphora opobalsamum*, (2) a zone of *Acacia etbaica* and *A. mellifera*, (3) a zone with patches of *Cordia gharaf, Dodonaea viscosa, Maytenus senegalensis, Rhus oxyacantha*, and a number of bryophytes and ferns.

**Notes on the Flora of the Red Sea Coastal Mountains in Egypt**

The vegetation on the slopes of the mountains, especially Gebel Elba, is delimited into altitudinal zones, the lower of which shows recognizable characters of community structure. The vegetation of the upper zone is obviously influenced by minor differences of habitat. The individual plants are crowded in patches, forming a mosaic pattern that makes the recognition of clearly defined communities difficult. With such an ill-defined pattern, the relationship between the habitat conditions and vegetation may be interpreted on the basis of moisture requirements of species. This interpretation is supported by studies carried out on the mountain groups further south in the Sudan (Kassas, 1956, 1957, 1960) and in the East African territories (Keay, 1959).

The flora of the Egyptian Red Sea coastal mountains includes over 400 species classified under three main growth-form categories: (a) trees, shrubs and undershrubs, (b) persistent herbs and (c) ferns and bryophytes. The species of categories (a) and (b) are subdivided into two classes according to their drought tolerance and moisture requirements.

Trees, shrubs and undershrubs are represented by the following species (Kassas and Zahran, 1971) and (Kassas, 1960): Seven species belong to the genus *Acacia*, namely: *A. ehrenbergiana, A. nubica, A. raddiana, A. tortilis, A. etbaica, A. mellifera* and *A. laeta*.

*A. tortilis* forms open thickets on the north slopes of the coastal hills. It is also very abundant on all slopes of the foot-hills of Gebel Elba and on the lower parts of the north slopes of the same mountain. In these habitats it is associated with *Euphorbia cuneata* which is often dominant. This is very similar to the growth of *A. tortilis* on the foot-hills of the Red Sea coastal mountains of the Sudan. *A. raddiana* is almost absent from the coastal hills and foot-hills. It is only occasionally found on the north slopes of Gebel Elba within the runnels dissecting the south slopes. *A. raddiana* is less drought tolerant than *A. tortilis*.

*A. nubica* has a limited distribution within the Red Sea coastal mountains of Egypt. This drought-deciduous species occurs in a few types of habitat within the Elba district. It dominates the north slopes of a few of the coastal hills that are covered by surface sheets of sand but is absent from the north slopes of the small hills. It is occasionally found within the runnels of the north slopes of the foot-hills and within the lower levels of the slopes of the Elba mountains.

The other group of *Acacia* species includes *A. etbaica, A. laeta* and *A. mellifera*. *A. etbaica* is a mountain species that forms rich growth within the higher zone of the mountain slopes of Elba. It is also present on the north slopes of the coastal mountains (Shindodai, Shillal and Asotriba) but not on the slopes of the inland ones. *A. mellifera* has a wider range within the northern slopes of Gebel Elba (from base to top). *A. laeta* is less common but does well on the highest zones of the north slopes of Gebel Elba. It is recorded from one locality within the foot-hills type. It is also present in the Gebel Hamata foothills (Samiuki group).

Other species within the group of trees, shrubs and undershrubs with low water requirements include several species which are widespread within the desert habitats of the Egyptian Red Sea coast. Most of these species are either absent or very rare on the southern slopes of the coastal hills and the foot-hills. *Leptadenia pyrotechnica* is much more abundant within the wadis draining the hills and mountains than on the slopes. *Ochradenus baccatus*, which is also common in the wadis, does better on the slopes. It is the most abundant bush within the runnels of the slopes of Gebel Elba where it may form groups of thickets. *Salvadora persica*, which forms patches of rich growth in parts of the main wadis, is rare on the mountain slopes. *Lycium arabicum* (= *L. shawii*, Boulos, 1995), a species that dominates a community on the coastal desert plain, is common on the coastal hills and the foot-hills. It is extremely rare on the north slopes of the Elba mountains but locally abundant within some of the runnels of the south slopes of the other mountain groups. *Ephedra alata* is present on the north slopes of the coastal hills but not in the other habitats of the mountains. *Grewia tenax* is common on the north slopes of the coastal hills, foot-hills and the mountains but absent from the south slopes. *Indigofera oblongifolia* is common in several localities of the north slopes of the foot-hills and the runnels of the south slopes of these hills. It is also present on the north slopes of the Elba mountains. *Balanites aegyptiaca* occurs on the higher zones of the north slopes of the mountains and the runnels of the south slopes of the foot-hills. It is also present on the Gebels of the Samiuki and Nugrus groups. *Maerua crassifolia, Cadaba farinosa* and *C. rotundifolia* are recorded within the foot-hills and the north slopes of the Elba mountains.

Trees and shrubs that are less drought tolerant include, apart from the *Acacia* spp. mentioned above, several species which are dominant within other communities. *Moringa peregrina* is present on the higher zones of the north-facing slopes of the mountains, especially Gebel Shindodai. It is also present within the mountains of the Samiuki, Nugrus and Shayeb groups. *Ficus pseudosycomorus* is associated with dry water-fall habitat types and is especially common in the northern mountain groups: Samiuki-Shayib.

*Dracaena ombet* is recorded in the highest zones of the north and east slopes of Gebel Elba. In several localities there are limited groves of this tree; otherwise there are isolated individuals. Reference may be made to the studies on the growth of *D. ombet* within the Sudanese coastal mountains including the Mist Oasis of Erkwit (Kassas, 1956, 1960). The occurrence of *Dracaena* in Gebel Elba is its most northern limit within the Red Sea coastal mountains (Kassas and Zahran, 1971; Zahran and Willis, 1992). In the high mountains of the Asian Red Sea coast *Dracaena* has been recorded by Migahid and Hammouda (1978) in Saudi Arabia and by Wood (1997) in Yemen.

*Euphorbia cuneata* is one of the most abundant species within the coastal hills, the foot-hills and the base zone of the Elba mountain. It dominates a species community on the north slopes but is only occasional on the south slopes of the coastal hills. On the south slopes of the foot-hills it may be common on the runnels but very rare on the slopes outside the runnels. It forms a rich growth that characterizes the lower zones of the north slopes of the Elba mountain, but its growth is gradually reduced up these slopes. *E. cuneata* is not recorded from the northern mountain groups: Samiuki-Shayib. *E. nubica*, apparently a species with higher water requirements, is only occasional on the north slopes of the coastal hills and the foot-hills, absent from their slopes and occasional within the lower zone of the north slopes of Gebel Elba where *E. cuneata* dominates. *E. nubica* dominates a middle zone of these slopes, thins up the slopes and is absent from the south slopes. The distribution of the two communities dominated by these two *Euphorbia* species is comparable to their distribution within the coastal hills and mountains of the Sudan (Kassas, 1960). Täckholm (1974) mentioned that *E. thi* is very rare in Gebel Elba mountain.

*Delonix elata* is abundant within certain localities – Gebel Karm Elba of the foot-hills and the zone NS of the Gebel Shindodai of the Elba group.

There is a group of species that are the least resistant to drought and are confined to the highest zones of the north slopes of the Elba group. These include: *Dodonaea viscosa, Ephedra foliata, Euclea schimperi, J. floribundum, Jasminum fluminense, Lantana viburnoides, Maytenus senegalensis, Olea chrysophylla, Pistacia khinjuk, Rhus abyssinica* and *Withania obtusifolia*. Most of these are species that are dominant or very abundant within the wettest zone of the mist oasis of Erkwit, Sudan (Kassas, 1956).

The persistent herbs and clearly herbaceous herbs may also be classified into two groups, one having lower water requirements than the other. The first group includes *Aerva javanica* which dominates the vegetation on the southern slopes of the coastal hills and the foot-hills and which is common in the highest (wettest) zone of the north slopes of the Elba mountains.

*Cleome droserifolia, Fagonia boveana, F. bruguieri* and *Launaea spinosa* seem to be confined to the northern mountain groups (Shayib – Nugrus – Samiuki) and are not recorded in the Elba group. Except for *Cleome droserifolia*, these species are not recorded in the Sudanese flora (Andrews, 1950–1956) and appear to be geographically confined to the northern parts of the Red Sea coast.

*Francoeuria* (*Pulicaria*) *crispa* occurs on the coastal hills, the lower zone of the south slopes of the mountains and also in the runnels of their south slopes. *Zilla spinosa* is rare within the Elba group but is one of the most abundant plants within the northern mountain groups. *Echinops galalensis* and *Solenostemma argel* are similarly rare within the Elba group and are common within the northern mountain groups. *Convolvulus hystrix, Salsola vermiculata, Seddera latifolia* and *Solanum dubium* are commonly found within the habitats of the coastal hills, foot-hills and the zones of the north slopes of the Elba mountains but are rare in the northern mountain groups. *Farsetia longisiliqua* is occasional on the north slopes of the foot-hills and Elba group. It is also present in the Samiuki and Nugrus groups. The second group of herbs is much more restricted in distribution as they are confined to the less arid localities. The 25 species of this group are all recorded in the higher zones of the north slopes of the Elba group. In this habitat they grow better than in any of the other habitats.

The ferns and bryophytes are mostly confined to less arid habitats and all grow best on the upper zones of the north slopes of the mountains. *Adiantum capillus-veneris*, the most widespread fern in the Egyptian desert, is present on the north slopes of the Elba mountains and their foot-hills. It is also common within the pot-hole and nakkat habitats of the northern groups. Other ferns and bryophytes are confined to the Elba group.

*In the Sudan*

### I. Coastal desert plain

In the Sudan, according to Kassas (1957), the vegetation of the Red Sea coastal plain is formed of widely open community where the average cover is less than 5% as well as scrubland types where the coverage may reach 60%. Sterile areas are among the obvious features of the plain. Here, the ground surface is covered with closely strewn gravels and boulders (desert armour) which provide little possibility for plant growth.

Six plant communities have been recognized dominated by: *Indigofera spinosa, Panicum turgidum, Acacia tortilis, Calotropis procera, Acacia nubica* and *Capparis decidua.*

#### (1) *Indigofera spinosa* community

This community type is found where the surface deposits are admixed gravel and sand and where the site receives little run-off water. The plant cover is usually less than 5% and the individuals are stunted by grazing. Associate common species are *Salsola vermiculata, Convolvulus hystrix, Cadaba farinosa* and *Blepharis edulis*. Other associate species occasionally found include: *Cadaba farinosa, Panicum turgidum.*

#### (2) *Panicum turgidum* community

This is a grassland type. *Panicum turgidum,* the dominant species, is a valuable fodder plant and a good sand binder. The plant coverage varies according to the water resources of the site which depend on its level in relation to adjacent land. *Acacia tortilis* and *Calotropis procera* are commonly found (Pr. = 60%); their density is, however, much thinner than in other community types where they dominate. *Salsola vermiculata* and *Indigofera spinosa* are also common associates. Other less common species include: *Capparis decidua, Leptadenia pyrotechnica, Convolvulus hystrix, Lycium shawii and Cassia senna.*

#### (3) *Acacia tortilis* community

This is the most common type and is especially well developed along the runnels. *Acacia tortilis* is a shrub found all over the arid part of the Sudan. It shows a wide range of size but nearly always grows into an umbrella-shaped form. Among the common bushes are *Acacia nubica, Capparis decidua, Cadaba farinosa* and *Lycium shawii.*

A second layer of the community includes *Salsola vermiculata* and *Panicum turgidum.* A few climbers are occasionally found: *Cissus quadrangularis, C. ternata* and *Ochradenus baccatus.* More species have been recorded in this community e.g. *Convovulus hystrix, Acacia nubica, Indigofera spinosa* etc.

### (4) *Calotropis procera* community

This type is found on sheets of sand drift. We may note, however, that the *Indigofera spinosa* community-type is found in areas that are losing sand by deflation and gradually accumulating coarse (lag) material at the surface whereas the *Calotropis procera* community cover areas are receiving sand. In this community-type *Indigofera spinosa* is the only associate with 60% presence.

*Calotropis* is collected for its stem fibers; the outer tissues are peeled off and worked into ropes of various thicknesses. This community type is consequently subjected to destructive cutting. Th less common species include: *Salsola vermiculata, Convolvulus hystrix, Acacia tortilis*, etc.

### (5) *Acacia nubica* community

This is an open scrubland dominated by *Acacia nubica* shrubs, a type which lies inland to that of *Suaeda fruticosa, S. vermiculata* and *Kochia cana.*

Among the bushes commonly found are: *Acacia tortilis, Cadaba farinosa* and *Lycium shawii. Panicum turgidum* and *Salsola vermiculata* are common in the second layer. A third (ground) layer includes such herbs as *Euphorbia granulata, Heliotropium steudneri* and *Blepharis edulis.* The relatively rich floristic composition of this community (30 associate species, 4 are halophytes) when compared to the other communities, include: *Suaeda monoica, S. Fruticosa, S. vermiculata and Kochia cana (halophytes)* and *Lycium shawii, Cadaba farinosa* etc.

### (6) *Capparis decidua* community

*Capparis decidua* is a species associated with all the community types of the plain. The community type dominated by it has, however, a characteristic habitat, namely: ill-defined delta with soft alluvial deposits. Wind action causes these deposits to heap around *Capparis decidua* producing huge mounds covered and protected by plant growth. *Leptadenia pyrotechnica*, though less abundant, may also build up such hillocks. In between these huge mounds are found: *Acacia tortilis, Cadaba farinosa, Calotropis procera, Indigofera spinosa, Panicum turgidum, Cissus quadrangularis,* etc. The other associate species (16 species) include two halophytes (*Suaeda monoica* and *S. fruticosa*), *Pulicaria crispa, Ruta tuberculata, Salvadora persica* etc.

### Khor (wadi) Arbaat

Khor Arbaat is one of several khors which cut across the Red Sea coastal hills in the Sudan that pour onto the plain. As ecological systems, khors are similar to the wadis of the Egyptian deserts. Kassas and Imam (1954) described the general features of the wadi-bed vegetation and concluded that communities dominated by *Tamarix* spp. or *Acacia tortilis* are among the climax communities. Khor Arbaat is a wadi with an extensive catchment area (and hence considerable water resources) and deep bottom deposits (valley fill). These are conditions suitable for climax vegetation. Kassas (1957) stated that during his first visit to Khor Arbaat in December 1953, he was inspired by the scene of luxuriant growth of *Tamarix mannifera* and *Acacia tortilis*. The former was dominating on sheets of silt, and the latter in areas with coarse sand and gravel.

The khor provides material for studying the gradual development of a *Tamarix mannifera* forest. At a young stage the saplings are crowded into almost pure stands of rich growth. Only rare individuals of *Calotropis procera* and *Aerva javanica* are found. As the community ages, natural thinning reduces the number of individuals per unit area and the associate species increase in number. At a mature stage the *Tamarix* bushes are about 5 m distant but their total coverage is not less than 60%. Associate shrubs are few: *Calotropis procera, Acacia tortilis* and *Lycium shawii*. The ground vegetation includes about 30 species.

Individuals of *Acacia tortilis* may reach 10–12 m high and dominate a multi-layered community with a coverage of 30–50%. The shrub layer includes: *Acacia tortilis, Balanites aegyptiaca, Calotropis procera, Tamarix orientalis* and *Ziziphus spina-christi*. Climbers include: *Cocculus hirsutus, Cissus quadrangularis, Ochradenus baccatus* and *Pentatropis spiralis*. *Loranthus curviflorus* is a common parasite on *Acacia tortilis*. The ground vegetation contains about 10 species.

### II. Erkwit Plateau

In the Erkwit plateau, Kassas (1956) distinguished five main communities dominated and co-dominated by: (1) *Maytenus senegalensis*, (2) *Maytenus senegalensis – Euphorbia abyssinica*, (3) *Euphorbia abyssinica*, (4) *Dracaena ombet – Euphorbia abyssinica* and (5) *Euphorbia thi*. These communities occur in a distinct zonation pattern described below.

#### (1) *Zone I* (*Maytenus senegalensis*)

This zone extends parallel to, and bounded by, the north-east border of the escarpment. By virtue of its position this zone directly faces the water-laden winds and sea mists as they roll inshore. It is represented by five (Gebels): Gebel Yamergermai at the north-western edge and hence less moist, Gebels Sela and Gebel Yoar at the north-eastern boundary and hence wettest. The others lie between these two extremities and include Gebel Essit and Gebel Manaweb.

The vegetation is multilayered. A tree layer represented by *Diospyros mespiliformis* is distantly open. This tree is found everywhere (within the zone) and is particularly abundant on Gebel Manaweb. A shrub layer is well developed though its cover ranges from 30 to 40%. *Maytenus senegalensis* is the dominant shrub. Associate shrubs that are characteristic of the zone are: *Euclea schimperi, Dodonaea viscosa, Rhus abyssinica, R. flexicaulis, Carissa edulis, Phoenix* sp. and *Ximenia americana*. *Acacia etbaica* is a common plant

> all over Erkwit and though present in 70% of zone I, it is not a noticeable member of the community. In other zones it is of greater significance. (Kassas, 1956)

*Olea chrysophylla* (*O. europaea* var. *nubica*) is found in 50% of the stands. This is the highest presence-estimate of this species; in other zones it is rare. In zone I there are a few trees of good size. According to local inhabitants this olive tree never produces fruit. It is particularly subject to cutting for stick-making and may have been at one time more abundant on the Erkwit hills as it is still on the Tokar hills.

*Euphorbia abyssinica* is a rarity within the moist zone I. At the fringes of the zone, where it may be found, it grows into a pole-like growth-form with a long slender stem bearing a few branches at its top. This is apparently due to overcrowding by *Maytenus senegalensis* and its associates. In other zones, where *Euphorbia abyssinica* gains ascendancy, it has a much branched bush.

Among the undergrowth *Coleus barbatus* is the most preponderant; in certain localities it produces a dense growth. *Kalanchoe glaucescens* is found everywhere. It may grow gregariously in small colonies of 20–30 individuals but usually occurs as isolated individuals. It is found all over Erkwit but attains its greatest abundance and best growth of individuals in this zone. Among the characteristic species are: *Cissus cyphopetala, Echinops macrochaetus, Geranium trilophum, Lavandula coronopifolia, Nepeta biloba, Umbilicus botryoides,* and the ferns *Actiniopteris radiata* and *Cheilanthes farinosa*. The following are species with presence-estimates lower than 60% but which are elective to zone I: *Anagallis arvensis, Cyperus bulbosus, Galium* sp, *Grewia ferruginea, Kyllinga pumila, Momordica pterocarpa* and the fern *Onychium melanolepis*.

Mosses are found locally wherever small protected loci are available. Liverworts are here recorded from the summits of Gebel Essit, Gebel Sela-esserir and Gebel Sela. Trunks and twigs of shrubs and trees are densely covered with lichen growth and so are surfaces of rock fragments and boulders.

Gebel Yamergermai (4,020 ft, 1,225 m) and Sela (4,244 ft, 1,294 m) deserve special notes. The western side of Gebel Yamergermai slopes down to Khor Arab which marks the western boundary of the Erkwit oasis. Its north and east sides are covered, from top down to about contour 3,600 ft, 1,097 m, with vegetation dominated by *Maytenus senegalensis*. At lower levels and on the slopes facing south and west the plant cover indicates drier conditions, with less *Maytenus senegalensis* and abundant *Euphorbia abyssinica*.

Gebel Sela combines the virtues of being the highest mountain of the Erkwit oasis and being at the north-eastern edge of the escarpment. The vegetation, though containing all the elements of the zone, shows noticeable differences corresponding with the elevation. Nearest to the top there are: *Diospyros mespiliformis, Euclea schimperi, Launaea schimperi, Maytenus senegalensis, Phoenix* sp., *Rhus abyssinica* and *Ximenia americana* with no obvious dominant. The top 100 ft (c. 30 m) are characterized by the abundance of ferns, mosses and liverworts. The part which lies between 100 and 200 ft (c. 30 and 60 m) from the summit is dominated by *Dodonaea viscosa*. Between 200 and 400 ft (c. 60 and 120 m) from the top, *Euclea schimperi* is mostly dominant though *Dodonaea viscosa* is locally dominant. On the rest of the sloping sides, *Maytenus senegalensis* is dominant as is characteristic of zone I.

### (2) Zone II (*Maytenus senegalensis – Euphorbia abyssinica*)

This zone lies between the previously described moist zone I and the *Euphorbia*-dominated zone III. It differs from the former zone by the abundance of *Euphorbia abyssinica* and from zone III by the abundance of *Maytenus senegalensis*. As a transitional zone it combines certain floristic features distinctive of the two zones on its sides.

*Euclea schimperi* and *Dodonaea viscosa* are among the shrubs commonly found in zone I and zone II. Other species characteristic of zone I and less important in zone II include: *Carissa edulis, Diospyros mespiliformis, Rhus abyssinica,* and *Ximenia americana. Acacia etbaica* is here more abundant. *Acacia tortilis* which is rarely found in zone I is present in 50% of the stands representing zone II.

Among the undershrubs and herbs, *Cissus cyphopetala* and *Kalanchoe glaucescens* are nearly equally common in the two zones. *Coleus barbatus, Nepeta biloba* and the fern *Cheilanthes farinosa* are appreciably less abundant in zone II than in zone I. *Echinops macrochaetus, Elionurus royleanus, Haemanthus multiflorus, Indigofera spinosa, Lavandula coronopifolia, Oxalis anthelmintica* and *Urginea micrantha* are more preponderant in zone II than in zone I.

Mosses are found locally in small patches covering locally protected niches but liverworts are not recorded. Lichens are common on trunks and twigs of trees and shrubs though their growth is less dense than in zone I.

### (3) Zone III (*Euphorbia abyssinica*)

This zone occupies the middle part of the Erkwit Oasis. The outstanding feature is the dominance of *Euphorbia abyssinica*; the plant cover is thinner than in the previous zones, the shrub growth not exceeding 30% and the herbaceous cover being equally thin.

*Maytenus senegalensis*, which is dominant in zone I and co-dominant in zone II, is here of lesser importance. Its individuals are small and less healthy shrubs with pungent spines and reduced leaves. *Euclea schimperi* is equally reduced. Other species that are characteristic of zone I and common in zone II and are here of minor status as members of the community, include: *Carissa edulis, Diospyros mespiliformis, Dodonaea viscosa, Rhus* spp. and *Ximenia americana. Phoenix* sp. is not recorded. *Acacia tortilis*, a species not recorded in zone I and present in 50% of the stands in zone II, is here present in 80% of the stands. *Acacia etbaica* is found everywhere (100% presence). *Dracaena ombet* shows its first appearance, a single tree recorded in stand 3 together with an individual of *Olea chrysophylla*.

Among the undergrowth *Coleus barbatus* is considerably reduced: 50% presence in zone III, 70% in zone II and 100% in zone I. *Echinops macrochaetus, Elionurus royleanus, Lavandula coronopifolia, Kalanchoe glaucescens, Micromeria abyssinica,* etc. are less common in zone III than in zone II. Several herbaceous species common in zone I are not recorded in zone III, e.g. *Cissus* spp., *Haemanthus multiflorus, Leucas nubica, Momordica pterocarpa, Nepeta biloba, Umbilicus botryoides* and *Urginea micrantha*. Among the species found commonly in zone III, rarely in zone II and not recorded in zone I, are *Aloe abyssinica* (60%), *Capparis tomentosa* (40%), *Caralluma penicillata* (40%), *Cucumis prophetarum* (60%), *Echidnopsis nubica* (30%), etc.

Ferns that are characteristic of zone I are not recorded in zone III. Mosses are found locally. Liverworts are not recorded. The lichen growth is considerably thinner and mostly of the crustose type.

Gebel Nafeib (3,787 ft, c. 1,154 m) deserves a special note. It lies on the boundary between zones II and III and rises for about 200 ft (c.60 m) above the level of the plateau. The east-facing slope of the gebel is covered by scrub dominated by *Euphorbia abyssinica* with abundant *Diospyros mespiliformis, Maytenus senegalensis* and *Rhus abyssinica*. The west-facing slope is also dominated by *Euphorbia abyssinica* with rare individuals of the above-mentioned bushes. Mosses are found locally on the east slope and very rare on the west slope.

### (4) Zone IV (*Dracaena ombet – Euphorbia abyssinica*)

This zone lies on the south-west boundary of the area and hence receives the sea mists and water-laden winds only after they have lost the greater part of their moisture. It is, however, less dry than the desert plateau that extends to the west of Erkwit. The zone includes a few high Gebels: Gebel Lagaribab (4,030 ft, c. 1,228 m), Gebel Tatasi (3,967 ft, c. 1,209 m), Gebel Auliai (3,970 ft, c. 1,210 m) and Gebel Dudia (3,915 ft, c. 1,193 m).

The salient feature of the vegetation is the preponderance of *Dracaena ombet* associated with *Euphorbia abyssinica. Acacia etbaica* and *A. tortilis* are common shrubs. *Lycium arabicum* (= *L. shawii*), which is not recorded in the previous zones, is here a common plant. Most of the trees and shrubs characteristic of the wetter zones are not recorded here except for rare individuals that may be found where the local topography allows for water accumulation and protection against insolation.

The ground vegetation shows the disappearance of many species that are common in the previous zones and the presence of a number of species not recorded in the previous zones. The characteristic species are: *Aloe abyssinica, Blepharis edulis, Capitanya otostegiodes, Caralluma penicillata, Euphorbia nubica, E. thi, Fagonia myriacantha, Indigofera spinosa, Otostegia repanda, Salsola baryosma, Seddera virgata* and *Solanum incanum*. Six of these species are not recorded in the previously described zones. There are other species, that are peculiar to zone IV as compared to zones I, II and III. Noticeable among these are species of *Euphorbia* e.g. *E. Consorbina* and *E. mobacanthe*. Mosses and lichens are extremely scarce, e.g. *Mollugo nudicaulis, Striga orobanchoides* etc.

### (5) Zone V (*Euphorbia thi*)

To the west of zone IV, and separating it from the desert plain that extends west of Erkwit, is a fringing zone where *Euphorbia thi* is most common. The zone lies "outside" the Erkwit oasis as it is cut off from the maritime effect by the Erkwit Gebels. It is an erosion pavement of the *hamada* type. The ground is undulated with lowly hillocks covered with rock fragments and boulders. The plant cover of these hillocks represents zone V.

In this zone the plant cover is very sparse (5–10%). *Euphorbia thi* is the most common species. Among the characteristic species are *Acacia etbaica, Barleria acanthoides, Euphorbia cuneata, Fagonia myriacantha, Indigofera spinosa, Lycium shawii, Salsola baryosma* and *Seddera virgata.*

#### *Khors*

> A khor is a dried stream which contains the run-off water. The khor vegetation varies from one zone to another and in accordance with the size of the catchment area, the depth and texture of the bottom deposits etc. Within zone I, khors are not well marked except as valleys between Gebels. The vegetation is similar to that of the zone. (Kassas, 1956)

In zones II and III there is nearly always a line of spaced trees along the sides of the khors. *Acacia raddiana* and *Balanites aegyptiaca* are very common. A few huge trees of *Acacia albida, Ficus sycomorus* and *Ficus* sp. occur locally. *Euphorbia abyssinica* and *Maytenus senegalensis* are occasionally found. *Withania somnifera* and *Argemone mexicana* are the most common undergrowth plants. Among the plants that are occasionally found are: *Boerhavia elegans, Boscia angustifolia, Calotropis procera, Commicarpus africanus, Cyperus rotundus, Datura metel, D. stramonium, Lachnophylis oppositifolius, Launaea* sp., *Panicum turgidum,* etc.

In khors of zone IV, *Acacia tortilis* is very common, associated with species peculiar to the zone. In zone V, *Euphorbia abyssinica* is very common in khors.

## Asian Red Sea Coast

*In Saudi Arabia*

The vegetation of the Saudi Arabian Red Sea coastal desert comprises 13 main communities dominated by: *Salsola baryosma, Hammada elegans, Rhazya stricta, Leptadenia pyrotechnica, Salvadora persica, Dipterygium glaucum, Indigofera spinosa, Calotropis procera Panicum turgidum, Cassia italica, Ziziphus spina-christi, Acacia raddiana* and *Juniperus procera* (treated in the order below).

### (i) *Salsola baryosma* community

*Salsola baryosma* (Chenopodiaceae) is a yellowish-green succulent shrub with a disagreeable odour of rotten fish. It is a plant species of wide ecological amplitude. Its community is of ecological interest as it represents a transitional zone between the littoral salt marsh and the coastal desert ecosystems of the Asian Red Sea coast.

The salt contents of its soil (electrical conductivity= 1.65–5.95 ms $cm^{-1}$) are lower than those of the salt-marsh communities [for example, in the *Halopeplis perfoliata* community the electrical conductivity of the soil is 32–85 ms $cm^{-1}$ (Younes et al., 1983)] but higher than those of the desert communities (for example in the soil of the *Panicum turgidum* community: electrical conductivity= 0.75–1.65 ms $cm^{-1}$).

Being a transitional community, between salt marsh and desert, its floristic composition includes xerophytes such as *Abutilon pannosum*, *Capparis spinosa, Leptadenia pyrotechnica*, and halophytes such as *Suaeda vermiculata, Zygophyllum album, Z. coccineum.* Many annuals and ephemerals, including: *Asphodelus fistulosus* var. *tenuifolius* and *Zygophyllum simplex*, are also recorded here. *Salsola baryosma* is the usual host of the parasite *Cistanche phelypaea.*

#### (ii) *Hammada elegans* community

*Hammada elegans* (Chenopodiaceae) is a stout shrub with glaucous green succulent branches and rudimentary scale-like leaves. It is a common xerophyte in the area. *Hammada elegans* is a sand binder which builds moderate-sized sand hummocks. Common associates include:

*Acacia tortilis, Asphodelus tenuifolius, Capparis decidua, Cometes abyssinica, Dipterygium glaucum, Malva parviflora, Panicum turgidum, Poa* sp., *Rhazya stricta* and *Steinheilia radicans.*

#### (iii) *Rhazya stricta* community

*Rhazya stricta* (Apocynaceae) is a sand-binding undershrub widely distributed in the desert and in the wadis of the Asian Red Sea coast. Its community covers vast areas, but its stands vary with regard to the growth-form of the dominant (*R. stricta*), floristic composition, density, etc. In the stands where the habitat is formed of coarse materials and the amount of rainfall received is small, the individuals of *R. stricta* are stunted and yellowish-green in colour, with plant cover less than 5%. However, in the stands where there is accumulation of fine sediments and the amount of available water is high, the individuals are bigger, dark green in colour and are capable of building sand hummocks. The plant cover usually exceeds 20%. The associate species are mainly xerophytes and include: *Acacia ehrenbergiana, A. raddiana, A. tortilis, Aerva persica, Calotropis procera, Cassia italica, C. senna, Dipterygium glaucum, Farsetia longisiliqua, Hammada elegans* and *Panicum turgidum.*

#### (iv) *Leptadenia pyrotechnica* community

*Leptadenia pyrotechnica* (Asclepiadaceae) is a leafless spinescent xerophytic shrub recorded in many habitats. Being rich in fibre and highly drought-tolerant, *L. pyrotechnica* may be considered as a plant of economic potentiality in the fibre industry of the arid lands (Zahran et al., 2009).

The community dominated by *L. pyrotechnica* is well developed in the midstream parts of the big wadis. Its floristic composition includes mainly xerophytes such as: *Acacia tortilis, Calotropis procera, Capparis decidua, Dipterygium glaucum,*

*Elionorus hirsutus, Paronychia desertorum, Pennisetum divisum, Rhazya stricta* and *Steinheilia radicans.*

### (v) *Salvadora persica* community

*Salvadora persica* (Salvadoraceae) is a glabrous shrub (or tree) with white branches. It is occasionally present in some of the wadis of the northern part, but is very abundant in the southern part of the Asian Red Sea coast. In the Jizan area, at the border between Saudi Arabia and Yemen, *S. persica* is one of the most useful plants to the Bedouins. Its branches have been of medicinal importance ever since the period of the Prophet Muhammed of Islam and to the present day. The branches are used in Saudi Arabia, Yemen and other countries of the Arabian Peninsula as tooth-brushes. Ayensu (1979) stated that "The ethanol-water extract of the stem of *S. persica* exhibits *in vitro* antispasmodic activity and it is used to remedy gonorrhoea, spleen, ache, etc". Al-Arak is the Arabic name of *S. persica*, and accordingly one of the big wadis in the Jizan area is called Wadi Arak, as its plant cover is formed mainly of *S. persica* trees and shrubs building huge sand dunes.

The floristic composition of the *Salvadora persica* community includes: *Abutilon pannosum, Acacia raddiana, A. tortilis, Dipterygium glaucum, Heliotropium* sp., *Hyphaene thebaica, Panicum turgidum* and *Schouwia thebaica.*

### (vi) *Dipterygium glaucum* community

*Dipterygium glaucum* is a deciduous undershrub (Capparaceae) of wide ecological amplitude. It can be seen everywhere in the Asian Red Sea coastal desert, either dominating a community or as an associate of other communities. The high-density stands of the *D. glaucum* community usually occurs in low-lying catchment areas having fine sediments containing a higher amount of available water than those of the nearby higher areas. In such localities the plant cover is up to 60–70% and the floristic composition is rich. In other stands with coarser sediments and little available water, the individual plants of *D. glaucum* are stunted and widely spaced (total plant cover = 5%), and the associate species are few. Xerophytes which are commonly present in the *D. glaucum* community include: *Abutilon fruticosum, Acacia ehrenhergiana, A. tortilis, Aerva persica, Cassia italica, Convolvulus lanatus, Fagonia bruguiera, Leptadenia pyrotechnica, Maerua crassifolia, Malva parviflora, Panicum turgidum, Rhazya stricta* and *Salvadora persica.*

### (vii) *Indigofera spinosa* community

*Indigofera spinosa* is a silvery spiny leguminous undershrub commonly present in the Asian Red Sea coast. Its dominance has been recorded in two different habitats, namely:

(1) lower zones of the rocky slopes of the coastal mountains; and
(2) areas of the coastal desert having a compact soil with high calcium carbonate content.

In the latter habitat, the associates are mixed, including halophytes like *Cressa cretica*, and xerophytes such as *Lycium barbarum, Panicum turgidum*, and *Zygophyllum simplex*. This indicates that, like the *Salsola baryosma* community, the *I. spinosa* community type may be considered as transitional between the littoral salt marsh and the coastal desert. This needs further study.

#### (viii) *Calotropis procera* community

*Calotropis procera* (Asclepiadaceae) is a shrubby desert plant with large broad and fleshy leaves. It is characterized by a milky juice in the leaves, branches, stem, flowers and fruits. This milky sap has uterotonic and cardiotonic properties (Ayensu, 1979), but it also causes blindness.

*Calotropis procera* shrubs are commonly present in the coastal desert and wadis of the Asian Red Sea. Locations where it is dominant are usually in the midstream parts of big wadis having deep sandy sediments.

The floristic composition of the *C. procera* community type is rich, and includes trees such as *Acacia raddiana, Hyphaene thebaica* and *Phoenix dactylifera*; undershrubs and bushes, such as: *Abutilon pannosum, Acacia asak, A. ehrenbergiana, A. tortilis, Aerva persica, Cassia italica, C. senna, Dipterygium glaucum, Farsetia longisiliqua, Hammada elegans, Indigofera spinosa, Leptadenia pyrotechnica, Rhazya stricta, Seidlitzia rosmarinus* and *Tephrosia purpurea.*

The ground layer includes a few perennials such as *Blepharis edulis* and *Citrullus colocynthis*, and ephemerals and annuals including *Aristida* spp., *Astragalus* spp., *Lotus* spp., *Malva* spp., *Schouwia thebaica,* and *Zygophyllum simplex.*

#### (ix) *Panicum turgidum* community

*Panicum turgidum* is a desert bunch grass up to more than one metre high with clustered branches. It is a highly palatable xerophyte with wide ecological amplitude. In the Saudi Arabian Red Sea coastal desert *P. turgidum* predominates in three different habitats:

(1) water runnels dissecting the gravelly desert, lined with fine aeolian material with a very limited moisture content;
(2) sandy plains of the coastal desert storing moderate amounts of water; and
(3) coastal sand dunes where the amount of water stored is relatively high.

The growth-form of *P. turgidum* bunches, and the density and floristic composition of the stands in the three habitats vary. In the water runnels habitat, the individual plants are stunted and widely spaced with a very sparse plant cover (less than 5%) and the associates are mostly short-lived plants appearing only after rainfall, such as *Aristida* sp., *Asphodelus tenuifolius, Astragalus asterias, A. vogelii, Medicago orbicularis, Poa* sp. and *Zygophyllum simplex.* Rarely a few perennial xerophytes, such as *Dipterygium glaucum* and *Farsetia longisiliqua*, may be present.

The second habitat provides the best possibilities for the growth of *Panicum turgidum* and its individuals attain their best growth-form. The plant cover of this habitat is relatively high (40–60%) and the floristic composition of the stands is rich. The associate species include trees (*Acacia raddiana* and *Hyphaene thebaica*), shrubs and undershrubs and bushes, e.g. *Acacia ehrenbergiana, A. tortilis, Aerva javanica, Arnebia hispidissima, Calotropis procera, Cassia senna, Dipterygium glaucum, Erodium glaucophyllum, Indigofera spinosa, Leptadenia pyrotechnica, Maerua crassifolia*. Prostrate perennials include *Citrullus colocynthis* and *Corchorus depressus* and present also are short-lived plants (ephemeral, annual and biennials) e.g. *Anthemis melanopodia, Astragalus eremophilus, Brachiaria leersioides, Cenchrus ciliaris, Corchorus tridens, Lotononis* sp., *Medicago lupulina, M. orbicularis, Paronychia arabica, Setaria glauca, Trifolium procumbens, Trigonella stellata* and *Zygophyllum simplex*.

The third habitat has been recorded in a very limited area near Jizan where *Panicum turgidum* grow from the white coastal sand dunes to the sand embankments covered by *Halopyrum mucronatum*. In this habitat, the plant cover of the *P. turgidum* community is less than 10% and there are no associate species.

### (x) *Cassia italica* community

Four species of *Cassia* (Leguminosae) have been recorded in the Asian Red Sea coast (Migahid and Hammouda, 1978), namely: *C. holosericea, C. italica, C. occidentalis* and *C. senna. C. holosericea* and *C. italica* are very common in the northern part of the coast, while the other two species are rare in the northern part but very common south of Jiddah.

*Cassia italica* (= *Senna italica*, Boulos, 1995) is a blue-green undershrub with soft branches from a woody base. It is a plant of medicinal importance commonly used by the Bedouins as a laxative, and this effect has been proved by Ayensu (1979). The other species of *Cassia* have the same effect.

*Cassia italica* is usually dominant in areas where the soil is compact, formed of fine materials having a high water-holding capacity. The associates are mainly xerophytes such as: *Acacia ehrenbergiana, Calotropis procera, Chrozophora oblongifolia, Cleome arabica, Corchorus depressus, Dipterygium glaucum, Leptadenia pyrotechnica, Rhazya stricta, Ziziphus spina-christi* and *Zygophyllum simplex*.

### (xi) *Ziziphus spina-christi* community

*Ziziphus spina-christi* (Rhamnaceae) is a spiny tree (or shrub) having a historical importance in Christianity. It is believed that the spiny crown and spiny cross of Jesus were made from the spiny branches of *Ziziphus*.

On the Asian Red Sea coast, *Z. spina-christi* grows both as trees and as shrubs. Big trees and a relatively dense plant cover of *Z. spina-christi* can be seen in the big wadis of the southern mountains where the amount of rainfall is relatively high. In the northern wadis where rainfall is much less, stunted and widely spaced individual bushes of this plant were observed. The associate species of the community type dominated by *Z. spina-christi* include: *Abutilon fruticosum, A. pannosum, Acacia tortilis, Aerva javanica, Bassia muricata, Calotropis procera, Cenchrus ciliaris, Commicarpus africanus, Commiphora opobalsamum, Convolvulus hystrix, Eleusine indica, Geranium favosum, Hyphaene thebaica, Leptadenia pyrotechnica, Lycium shawii, Panicum turgidum, Pennisetum divisum, Retama raetam, Rhamnus staddo, Salvadora persica, Senna occidentalis* and *Tamarix aphylla.*

### (xii) *Acacia raddiana* community

*Acacia raddiana* (= *A. tortilis* subsp. *raddiana*) is a spiny tree (or shrub) belonging to the Leguminosae. This community represents the climax stage of the xerophytic vegetation in the Asian Red Sea coast, where 12 species of Acacia have been recorded (Migahid and Hammouda, 1978). *Acacia ehrenbergiana, A. raddiana* and *A. tortilis* are the commonest species. They grow in the main streams of the wadis and on the slopes of the low hills of the whole coast.

In the scrublands dominated by *A. raddiana*, stratification of the xerophytic vegetation is obvious. The tree layer includes the dominant (*A. raddiana*), *A. tortilis, Hyphaene thebaica* and *Ziziphus spina-christi*. The second (suffrutescent) layer includes the greatest number of associate shrubs, undershrubs and bushes, such as: *Abutilon pannosum, Acacia ehrenbergiana, A. tortilis, Calotropis procera, Capparis decidua, Dipterygium glaucum, Farsetia longisiliqua, Indigofera spinosa, Leptadenia pyrotechnica, Lycium shawii, Maerua crassifolia, Panicum turgidum, Pergularia tomentosa, Periploca aphylla, Salvadora persica* and *Zilla spinosa*.

The ground layer includes a variety of plants, as follows:

(1) Creeping and stunted perennials, for instance, *Achillea fragrantissima, Citrullus colocynthis, Convolvulus hystrix, C. prostratus, Cucumis prophetarum, Cynodon dactylon, Fagonia bruguieri, Launaea capitata, L. procumbens, Paronychia desertorum, Steinheilia radicans.*
(2) Thistles, such as *Atractylis flava, Carduus pycnocephalus* and *Echinops spinosissimus.*
(3) Short-lived (annual and ephemeral) plants such as *Asphodelus tenuifolius, Bromus* spp., *Coelachyrum brevifolium, Ifloga spicata, Senecio flavus* and *Zygophyllum simplex.*

### (xiii) *Juniperus procera* community

The climate of Saudi Arabia, in general, is too arid to produce natural timber trees of commercial importance. However, the range of mountains of the Red Sea Coast receive relatively the highest amount of rain (up to 387 mm/year) and its temperature is relatively low (mean maximum= 24.1–28ºC). Such climatic conditions enable the growth of certain non-xerophytic trees and shrubs (Zahran, 1982b).

Allered (FAO) (1968) reported that there are 25 million acres of woodland in the mountains of the westsouthern (Red Sea Coast) region of Saudi Arabia (Hijaz and Asir mountains) forming savanna with grasslands. The dominant tree is *Juniperus procera*. The highest level (c. 3,000 m a.s.l., e.g. as Sudah Summit, 25 km south of Abha), represent the watershed zone of the mountains. In this zone, *J. procera* trees establish a closed canopy with little understory. *Juniperus* trees up to 50 ft (17 m) high and 14 in. (36 cm) diameter at the base and 6 in. (15 cm) diameter at the top were recorded. The ages of these trees range between 70 and 85 years. In the lower zone of *Juniperus* woodland (< 2,000 m a.s.l.) other trees and shrubs e.g. *Commiphora schimperi, Dodonaea viscosa, Dracaena ombet, Olea chrysophylla, Pistacia palestina*, woody herbs e.g. *Euphorbia retusa, Euryops arabica, Lavandula dentata, Psiadia arabica, Rumex nervosus,* grasses e.g. *Themeda triandra* and succulents e.g. *Euphorbia thi, Caralluma* spp. were recorded.

### (xiv) *Acacia* spp. community

In the relatively low zone of the mountains (less than 1,000 m a.s.l.) where there is noticeable change in the climatic and soil conditions, *Juniperus* trees disappear and *Acacia* spp. (*A. etbaica*, and *A. nubica*) predominate. The associate species include: *Adenium arabicum, Commiphora schimperi, Ficus pseudosycomorus, Grewia tenax, Olea chrysophylla.* In the still lower zone the slopes are dominated by *Acacia asak* and *A. mellifera. Acacia etbaica* is abundant; the other associates include: *Aloe vera, Artemisia judaica, Euphorbia cuneata, E. nubica, Francoeuria crispa, Scorzonera intricate* and *Themeda triandra.*

### A. Representative transect for the Vegetation of Higaz mountains

Abdel Ghani (1996) studied the vegetation along about a 450 km transect between Mekka and Medina (21° 26′ to 24° 31′N, 39° 46′ to 39° 42′E) crossing the Hijaz mountains with elevation ranging between 270 and 930 m. A total of 108 vascular plants belonging to 36 families were identified. Altogether 20 communities were recognized on the basis of floristic and ecological features in the three eco-geomorphological units identifed, namely: (1) lowland sector, (2) gentle-sloped sector and (3) upland plateau. The relatively highest number of species (87) has been found in the gentle-sloped sector followed by those recorded in the lowland sector (75 species) and the least number of species in the upland plateau (44 species).

The vegetation of the lowland sector includes eight perennial communities dominated by: *Acacia tortilis* subsp. *tortilis, Lycium shawii, Senna alexandrina, Dipterygium glaucum, Hammada elegans, Indigofera spinosa, Panicum turgidum* and *Rhazya stricta* and one community dominated by the succulent annual plant *Zygophyllum simplex.* Seven communities have been recognized in the gentle-sloped sector co-dominated and dominated by seven perennials (*Acacia ehrenbergiana, Calotropis procera, Acacia hamulosa, Maerua crassifolia, Pulicaria crispa* (= *Francoeuria crispa*), *Suaeda monoica* and *Tephrosia apollinea*) and one ephemeral (*Asphodelus tenuifolius*) species.

In the upland plateau, four communities have been recognized dominated by perennial shrubs and trees, namely: *Acacia tortilis* subsp. *raddiana – Fagonia indica* (co-domination), *Salsola baryosma, Tamarix aphylla* and *Zilla spinosa.*

The following is an ecological account of these communities.

## I. Plant communities of the lowland sector

### 1. *Acacia tortilis* subsp. *tortilis* community

Vegetation dominated by *Acacia tortilis* subsp. *tortilis* in the surveyed area is very sparse, and cofined to the eastern part of this sector. It is also reported from the arid deserts of Arabia (Batanouny, 1987) and from the Eastern Desert of Egypt (Zahran and Willis, 2008). This community abounds in depressions which have shallow coarse-textured soil. Stones and gravels cover a considerable part of the ground surface. In some parts, this community may occur on terraces, in shallow and narrow valleys as well as on gentle slopes.

The total plant cover is low, ranging from 10 to 15% and the number of recorded species, including ephemerals, was 25. *Acacia tortilis* subsp. *tortilis* dominates the tree and shrub layer where its growth is stunted by continuous nibbling by goats. The most common associates are: *Acacia ehrenbergiana, Lycium shawii, Leptadenia pyrotechnica* and *Acacia tortilis* subsp. *raddiana*. The lower shrub layer comprises *Dipterygium glaucum, Fagonia indica, Indigofera spinosa, Rhazya stricta,* together with the perennial grass *Panicum turgidum*, whereas the herb layer is relatively sparse and may be enriched by ephemeral growth in the rainy season.

### 2. *Lycium shawii* community

This community is commonly found in depressions which have shallow coarse-textured soils, with the ground surface covered by angular stones and pebbles. It is divited into two subtypes on floristic and ecological grounds: one growing on rocky slopes and mature runnels, and the second on sandy terraces and lowland along the roadsides. The latter subtype is recognized in the present study. The plant cover shows wide variation in the 10 stands studied, ranging from 20 to 35%, mainly of the dominant species (Zahran et al., 2009).

The tree and shrub layer is dominated by *Lycium shawii* in association with *Acacia tortilis* subsp. *tortilis, Leptadenia pyrotechnica* and *Ochradenus baccatus*. Common associates are *Fagonia bruguieri, Forsskaolea tenacissima, Heliotropium arbainense, Rhazya stricta, Senna italica* and *Tribulus terrestris*.

### 3. *Senna alexandrina* community

The dominant species has a very limited distribution in the study area. It is represented by only five stands in scattered localities. Within this sector, this community is recorded in the foot ridges and sandy plains with fine-textured soils that are occupied by the *Rhazya stricta* community. Moreover, man-made hollows in the terraces occupied by the *Rhazya* community support a relatively dense growth of *Senna alexandrina*. The plant cover ranges from 10 to 15%. Consistent associates are *Dipterygium glaucum, Leptadenia pyrotechnica, Panicum turgidum* and *Rhazya stricta*. Among the less frequent species are *Aerva javanica, Blepharis ciliaris,*

*Chrozophora obliqua, Corchorus depressus, Euphorbia granulata* and *Tribulus terrestris*. In addition, a few individuals of *Andrachne aspera* and *Citrullus colocynthis* were recorded.

**4. *Dipterygium glaucum* community**

This community has a limited distribution and does not contribute much to the vegetation in the area, being confined mainly to sandy plains where the soils are coarse-textured with low water-retaining capacity. This community was also recorded by Abdel Ghani (1993) in the Holy places area of Makkah. The plant cover is low, being only 10% on average, without recognizable stratification of vegetation. Very sparse individuals of *Acacia ehrenbergiana, Maerua crassifolia* and *Ziziphus spina-christi* are recorded. The main recorded associates include: *Aerva javanica, Boerhavia coccinea, Farsetia longisiliqua, Hammada elegans, Heliotropium bacciferum, Pennisetum divisum, Polycarpaea repens, Rhazya stricta* and *Senna italica*.

**5. *Hammada elegans* community**

This community type belongs to the sand and gravel ecosystem (Kassas and Imam, 1959). The dominant *Hammada elegans* is a sand-binding plant that builds moderate sized sand hummocks. It is palatable mainly to camels. The soils supporting this community are deep and coarse-textured. The community, in some places of this sector, occurs in habitats subject to severe erosion mainly by wind. The plant cover ranged between 10 and 15%. The tree layer included also widely spaced individuals of *Calotropis procera, Acacia ehrenbergiana* and *Ziziphus spina-christi*. Associates include: *Abutilon pannosum, Cocculus pendulus, Fagonia indica, Heliotropium bacciferum, Indigofera spinosa, Panicum turgidum, Pennisetum divisum* and *Tribulus terrestris*.

**6. *Indigofera spinosa* community**

This community is not widespread in the study area. It is represented by sparse plant cover at distantly spaced localities in the sandy plains of the foot-hills and in habitats where the surface deposits are admixed gravel and sand receiving little runoff water. The total plant cover is low (about 5%), probably due to grazing, since the plant is highly palatable especially to goats. The floristic composition of this community varies. The tree layer includes very sparse individuals of *Lycium shawii* and *Acacia tortilis* subsp. tortilis. The low shrub layer is dominated by *Indigofera spinosa* in association with *Rhazya stricta, Blepharis ciliaris, Citrullus colocynthis, Dipterygium glaucum, Panicum turgidum, Pennisetum divisum, Pergularia daemia* and *Senna italica*.

**7. *Panicum turgidum* community**

This community has a very limited range of distribution in the study area. The vegetation is dominated by the sand-binding *Panicum turgidum* and is restricted to the drainge lines crossing the sand plain between Abu Urwah and Usfan. This dominant perennial grass forms hillocks of different sizes, and may reach 50 cm high or more. The surface soils of the hillocks are coarse-textured, while deeper layers are fine-textured. *Panicum turgidum* is a highly palatable plant and hence subjected to destruction by grazing. Accordingly, its growth is severely retarded. Batanouny (1987) noted that overgrazing and consequently degradation of the habitat supporting the *Panicum turgidum* community results in the appearance of *Rhazya stricta*. Plant cover is low and not exceeding 10%. Layering of vegetation is clear with a tree and shrub layer formed by *Acacia tortilis* subsp. *tortilis*, *Acacia tortilis* subsp. *raddiana* and *Leptadenia pyrotechnica*. The undershrub layer includes: *Aerva javanica, Aristida mutabilis, Chrozophora obliqua, Dipterygium glaucum, Indigofera spinosa* and *Rhazya stricta*.

**8. *Rhazya stricta* community**

This assemblage is confined to the western part of this sector, and occurs on the sandy terraces and in depressions with coarse-textured deposits where the ground surface is covered with stones and boulders. Both wind and water erosion is fairly pronounced. Due to its physiography, this habitat receives a considerable amount of runoff water and water-borne materials. In the 10 stands of this community, plant cover ranges from 5 to 10%. The tree layer is poor and represented mainly by widely-spaced individuals of *Acacia tortilis* subsp. *raddiana* and *Calotropis procera*. Associated species include *Andrachne aspera, Dipterygium glaucum, Forsskaolea tenacissima, Indigofera spinosa, Panicum turgidum* and *Stipagrostis plumosa*.

**9. *Zygophyllum simplex* community**

Ephemeral growth dominated by *Zygophyllum simplex* is well represented in the study area, dominating the terraces along the main wadi courses and is characterized by relatively high water resources. The ground surface is covered by gravel and pebbles with fine sediments beneath. The low shrub layer provides the bulk of the plant cover with common growth of *Aerva javanica, Dipterygium glaucum, Indigofera spinosa* and *Senna alexandrina*. In the rainy years (as in 1992 and 1993) the ground was covered by *Arnebia hispidissima, Astragalus vogelii, Boerhavia repens, Cenchrus ciliaris, Ifloga spicata* and *Trigonella stellata*.

This community shows two facies: lowland and gentle-sloped sectors, with slight floristic variation. The plant cover in the latter is higher (30–35%) than in the former (15–20%). The number of species (15) is similar in both locations. *Fagonia indica, Farsetia longisiliqua, Panicum turgidum, Senna alexandrina* and *Tribulus pentandrus* are recorded in the lowland sector. On the other hand *Acacia tortilis* subsp. *raddiana, Anastatica hierochuntica, Asphodelus tenuifolius, Diplotaxis acris* and *Suaeda monoica* are growing on the gentle-sloped sector.

## II. Plant communities of the gentle-sloped sector

### 1. *Acacia ehrenbergiana-Calotropis procera* community

This community has a limited range of distribution and usually occurs on deep alluvial plains with sandy soils or in wadis and wide runnels with deep deposits, usually alluvial. It is recorded between Hajar and Wadi Al-Fara's in this sector at altitudes between 400 and 550 m. Cutting of *Acacia* trees is usually followed by soil erosion and invasion by *Calotropis*. The plants cover ranges from 25 to 40%, mainly of *Acacia* trees. The plant growth comprises three structural layers with *Acacia* and *Calotropis* providing the principal frame of the open shrub layer. The low shrub layer comprises seven species with *Rhazya stricta*, and *Senna alexandrina* as common associates. After rain the ground layer is dominated by *Aizoon canariense, Aristida mutabilis, Euphorbia granulata, Malva parviflora* and *Zygophyllum simplex*.

### 2. *Acacia hamulosa* community

The dominant species is widespread on gentle rocky slopes and in the shallow runnels dissecting them, mainly in the area extending from Al-Biyar to Hajar. The soil supporting the growth of this community is covered with barren sharp-angled rock fragments, with accumulated fine sediments between. Generally, this habitat is subjected to erosion, which may be attributed to the combined effects of low water resources and overgrazing. The total plant cover is apparently poor, ranging from 5 to 10%. Despite such a low cover, species numbers are high, reaching 40. Layering of vegetation is a clear feature of this community. Three layers are recognized: (A) tree and shrub layer, represented by *Acacia hamulosa* trees, with occasional individuals of *Acacia tortilis* subsp. *tortilis, Delonix elata* and *Lycium shawii*, (B) low shrub layer, which is of especial importance in the permanent framework of this community and is represented by *Indigofera spinosa, Pergularia daemia, Senna italica* and *Tephrosia apollinea* and (C) herb layer, the richest layer in terms of the number of species. It includes, amongst others, *Astragalus vogelii, Cynodon dactylon, Euphorbia granulata, Forsskaolea tenacissima, Launaea capitata, Schismus barbatus* and *Sisymbrium irio*.

### 3. *Maerua crassifolia* community

This community abounds on the gentle slopes and sandy plains, where the ground surface is covered with gravel, pebbles and stones. Stretches of this community are observed in the western part of this sector, at altitudes from 600 to 700 m a.s.l. The plant cover ranges between 15 and 20% and consists mainly of *Maerua crassifolia. Acacia hamulosa* and *Acacia tortilis* subsp. *raddiana* form the tree layer of this community. Associated species include *Andrachne aspera, Calotropis procera, Cissus quadrangularis, Cucumis prophetarum, Dipterygium glaucum, Senna italica* and *Tribulus terrestris* (which grows climbing on *Maerua*).

**4. *Pulicaria crispa* community**

This assemblage is confined to the main water-course of Wadi Al-Qaha and occurs on deep, alluvial, fine-textured soils. The plant cover ranged between 40 and 50%. This relatively high cover may be attributed to the adequate water supply of this habitat. The floristic composition is diverse, with the tree layer lacking. Numerous associates are recorded, these include *Aizoon canariense, Asphodelus tenuifolius, Diplotaxis acris, Forsskaolea tenacissima, Malva parviflora* and *Zygophyllum simplex*. In addition scattered individuals of *Lycium shawii* and *Zilla spinosa* are recorded in few stands of this community.

**5. *Suaeda monoica* community**

The plant growth dominated by *Suaeda monoica* is not widespread in the study area, favouring sites with deep alluvial, fine-textured, compact and slightly saline deposits overlain by aeolian sand, and is restricted to a stretch of about 15 km extending from Al-Biyar to Al-Dhar'a. It is to be noted that this community type is not recorded further north, though occasional individuals of *Suaeda monoica* may occur within other communities. According to Kassas (1957) this community type follows, in sequence of zonation, that of *Suaeda fruticosa* on the alluvial deposits. Plant cover is distinctly high (60–70%) compared with other community types encountered in the present study. Consistent associates include: *Aristida mutabilis, Farsetia longisiliqua, Hammada elegans, Tribulus pentandrus* and *Zygophyllum simplex*. Other less common associates include *Asphodelus tenuifolius, Calotropis procera, Citrullus colocynthis, Fagonia indica* and *Tephrosia apollinea.*

**6. *Tephrosia apollinea* community**

This community occupies gentle slopes and terraces, where the soils are relatively deep, as well as narrow runnels and affluent wadis. This habitat is subjected to water erosion. The ground surface is covered by variously-sized granite boulders. The plant cover is low (10–15%), mainly of the dominant *Tephrosia* species. The floristic composition, in the 13 stands studied, varied but stratification is not clear. The tree layer includes occasional individuals of *Acacia ehrenbergiana, Acacia hamulosa* and *Maerua crassifolia*. The main associates are *Abutilon pannosum, Indigofera spinosa, Ochradenus baccatus, Pergularia daemia* and *Stipagrostis plumosa*. Ephemerals, such as *Aizoon canariense, Asphodelus tenuifolius, Glinus lotoides* and *Zygophyllum simplex* are recorded where accumulated sediments are found.

**7. *Asphodelus tenuifolius* community**

The ephemeral vegetation dominated by *Asphodelus tenuifolius* is widespread in this sector. It abounds on deep alluvial soils and fine alluvial sediments among stones. This community also occurs on gentle slopes where shallow soils have accumulated. Plant cover is high, at about 40%. This relatively high coverage is attributed to the dense growth of ephemerals including: *Euphorbia granulata, Ifloga spicata, Reichardia tingitana, Rumex vesicarius, Trigonella stellata* and *Zygophyllum simplex*.

### III. Plant communities of the upland plateau sector

#### 1. *Acacia tortilis* subsp. *raddiana-Fagonia indica* community

This community has a wide geographical amplitude within this sector, where it is recorded between Wadi Al-Fara'a and Abyar Al-Mashi at altitudes ranging between 680 and 830 m. It occurs in open beds with more or less plain landscape with deep, fine-textured soils. The plant cover ranged from 20 to 30%. Palatable species were stunted due to grazing. In addition to *Acacia tortilis* subsp. *raddiana*, scattered inviduals of *Lycium shawii* and *Maerua crassifolia* form the tree and shrub layer. *Fagonia indica* is the main component of the low shrub layer. Common associates include *Citrullus colocynthis, Pulicaria crispa, Rhazya stricta, Salsola baryosma* and *Zilla spinosa*. The herb layer comprises the ephemeral growth that appears only in rainy years (as in 1992 and 1993) and its abundance and floristic composition are controlled by local showers as well as irregularity in topography of the area.

#### 2. *Salsola baryosma* community

The community dominated by *Salsola baryosma* has a local distribution in the upper reaches of Wadi Al-Qaha and around Medina. The floristic diversity is low – the total plant cover does not exceed 10%. Widely-spaced *Acacia tortilis* subsp. *raddiana* trees form a distantly open layer. The low shrub layer is the most obvious stratum of plant growth as it includes the dominant *Salsola* and a number of common associates: *Aerva javanica* and *Pulicaria crispa*. Early in this survey, during the rainy year 1992, a rich growth of ephemerals was observed which included *Anastatica hierochuntica, Asphodelus tenuifolius, Morettia philaenana, Sisymbrium irio, Steinheilia radicans* and *Zygophyllum simplex*.

#### 3. *Tamarix aphylla* community

The vegetation dominated by the scrub *Tamarix aphylla* has a very limited range of distribution in the study area. It is confined to Al-Abyar area (700–800 m). representing, one of the natural climax community types of the desert wadis. The mean plant cover may reach 40% with the major part dominated by *Tamarix*. Few associates are recorded; these include *Aerva javanica, Cucumis prophetarum, Dipterygium glaucum, Forsskaolea tenacissima,* as well as some ephemerals such as *Zygophyllum simplex* and *Aizoon canariense*.

**4. *Zilla spinosa* community**

This assemblage is widespread in the eastern part of this sector. It occupies part of wadi beds where the alluvial deposits include considerable amounts of soft material. These deposits are usually finer and deeper than those associated with *Salsola baryosma* communities. The community of *Zilla spinosa* takes the form of pure stands with a cover ranging from 60 to 70%. Very sparse individuals of distantly spaced *Acacia tortilis* subsp. *raddiana* are noted. Two layers form the main framework of this community. *Zilla spinosa* is the dominant species and forms the chief component of the low shrub layer. Common associates are *Pulicaria crispa, Rhazya stricta, Salsola baryosma* and *Senna italica*. The herb layer includes prostrate and dwarf species such as *Aizoon canariense, Aristida mutabilis, Asphodelus tenuifolius, Citrullus colocynthis* and *Forsskaolea tenacissima*.

It is of note here that Batanouny and Baeshin (1982) recorded 247 species (173 genera and 53 families) in the southern – third section (150 km) of the same area studied by Abdel Ghani (1996). Out of these species, one genus, 14 species, one subspecies and 13 varieties were considered new records to the flora of Saudi Arabia. The new records include the following species: *Asteroptrus leyseroides, Boerhavia erecta, Centaurea araneosa, Chenopodium glaucum, Cleome kotschyana, Cuscuta campestris, Fagonia paulayana, F. tristis* var. *tristis, Galium sinaicum, Heliotropium europaeum, H. strigosum, Launaea taraxacifolia, Ochradenus baccatus, Salvia aegyptiaca, Stipagrostis raddiana, Tephrosia purpurea, Tetrapogon cenchriformis, Tribulus bimucronatus, T. pentandrus* var. *micropterus, T. pentandrus* var. *pentandrus, T. terrestris* var. *intermis, T. terrestris* var. *orientalis* and *T. terrestris* var. *robustus*.

**In Yemen**

The dry non-saline coastal desert of the Yemeni Red Sea coast is categorized under five main habitats namely:

A. Coastal lowlands
B. Tihama foothills
C. Lower escarpment
D. Higher escarpment
E. High mountain

The following is an ecological account of these five habitats: references are Newton (1980), Deil and Muller-Hohenstein (1983), Al-Hubaishi and Müller-Hohenstein (1984), Wood (1997) and Zahran et al. (2009).

**A. The coastal lowlands**

The coastal lowlands of Yemen form a 30–50 km wide belt between the Red Sea coast and the Tihama foothills. The altitude ranges from sea level to c. 300 m at the foothills. The surface is level or undulating, sometimes intersected by wide, shallow wadis, draining from the escarpment area to the sea. Near the coast, marine formations dominate, often covered by recent aeolian deposits. Towards the interior, fluviatile silts and sands are deposited along the courses of the wadis and their slightly raised embankments. Water may reach the sea during exceptionally wet periods, but is usually absorbed by the sediments of the plain and nowadays by the irrigation systems. These wadi regions can be separated by larger dunes, higher levels being covered with a layer of gravels and cultivated fields.

The Tihama coastal lowlands have a tropical arid to semiarid climate; the mean annual temperature is very high (c. 30ºC) with fairly small seasonal fluctuations; absolute minima and maxima, however, range between below 20ºC (rarely) in winter and over 40ºC in summer. Rainfall is very inconsistent in time and space. The entire region is under the influence of the monsoon, which brings summer rain with cloudbursts, mainly in July and August. Rainfall during the winter season with perhaps a Mediterranean affinity is very rare. The total amount of precipitation is usually less than 200 mm/year. The average annual air humidity of c. 70% is high, but can be as much as 90% and more at night.

Dew is very common. Winds generally blow from south-west or north-west, sometimes at high speed velocity, causing sand movement and deflation, especially on cultivated fields. The Tihama coastal lowlands is mainly a region of deposition, starting with gravels and stones near the foothills and ending with fine sands and silts near the coast. Only in few places are true soils developed. On marine deposits, saline soils with a substantial content of chlorides and sulphates of sodium dominate (Solonchak). They bear only a sparse vegetation cover, when they are topped by thin sand layers. On fluviatile and aeolian deposits we find a predominance of sandy and silty soils with a very low humus content (0.5–1.0%) and almost no differentiation into strata. Closer to the foothills stony fragments are present in the soil. Dunes and mobile sands are important for the vegetation though they are not true soil.

So we may distinguish four major types of natural environments (ecosystems) in the Tihama coastal lowlands distributed in a fairly regular pattern: (I) the marine salt fans near the sea, (II) the alluvial plains with fluviatile and aeolian deposits, (III) the plains near the foothills with coarse, textured sediments and (IV) the wadi systems.

In the Tihama coastal lowlands, vertical zonation of vegetation is a characteristic feature. It is not so much connected with climatic differences as with such important ecological factors as elevation above the water-table, salinity, texture and mobility of soils. The protected tidal range of the Red Sea with only shallow water is the habitat for a mangrove forest at Al Luhayyah and near Al Khawbah. This forest is composed of only one evergreen sclerophyllous tree *Avicennia marina,* with pneumatophores, sticking out of the saline mud during low tide, assuring the oxygen supply for the roots.

A large zone sometimes up to 5 km from the shoreline is usually sterile, due to its high salinity. Only periodically better drained places may be covered by a few halophytic plants such as *Suaeda fruticosa* and – when topped with thin wind-borne sand-layers – by the grass *Aeluropus massauensis*. So this is a very poor herbaceous and half-woody salt swamp formation.

The next plant formation further inland may be called a subdesert shrubland with halophytic succulents. Here the most important community is a saltbush community with *Suaeda monoica* on slightly elevated, still very compact, marine deposits with high salinity above the level of recent floods. These deposits are often overlain by aeolian sand and the *Suaeda* bushes form small hillocks reaching one metre high and leaving spaces where annuals such as *Zygophylum simplex* appear after heavy rainfall. Vast stretches of land can be covered by this community. The plant cover may reach a surprising density of more than 50%.

The vegetation of the adjacent subdesert plain shows a mosaic pattern, depending on such factors as topography, texture, depth and mobility of the surface deposits and different forms of human interference. Many different plant communities here belong to three plant formations. Among these the first and most important is a drought-deciduous mixed dwarf-shrubland with deciduous and evergreen dwarf-shrubs and only a few microphyllous or leafless shrubs with green stems, with tussock-forming grasses, succulents and other life-forms.

On a few mobile dunes, where the wind is the main agent for deposition and deflation, good sand binders including the grasses *Panicum turgidum* and *Odyssea mucronata* as well as the dwarf-shrub *Indigofera spinosa* and *Salsola spinescens* dominate. Rare but also characteristic of this habitat are the leafless shrubs *Leptadenia pyrotechnica* and *Capparis decidua*. Less mobile sands are locally dominated by *Jatropha villosa* and *Cassia italica* (= *Senna italica*).

On the valley bottoms between the higher and larger dunes phytogenic mounds and hillocks are very common. The evergreen *Cadaba rotundifolia* and *C. glandulosa* collect the moving sand with their leaves and branches, forming hillocks up to two metres high. They are often associated with some succulents like the climbing *Cissus quadrangularis* and *Euphorbia triaculeata*. After rainy periods, on silty and fine sandy deposits around the hillocks, dense ephemeral growth of *Corchorus depressus, Dipterygium glaucum, Euphorbia indica, Fagonia indica, Schouwia purpurea, Tephrosia purpurea* and *Tribulus terrestris* is fairly common.

The same species compose the second important plant formation, the mainly ephemeral weed formation on cultivated land, together with some perennial shrubs such as *Aerva javanica, Cassia senna* and *Indigofera oblongifolia*.

The third zonal plant formation of the subdesert plain is present only near the foothills of the Tihama coastal lowlands and shows some transitions to this adjacent natural region. It consists of drought-deciduous woodlands with only a few evergreen trees such as *Dobera glabra* and *Balanites aegyptiaca* near the mountains and a very sparse ground vegetation on level or only slightly undulated foot plains with thick layers of sands and gravels. The most common community is a thorn woodland with the umbrella-shaped *Acacia ehrenbergiana* and *Acacia tortilis*, sometimes associated with the doum-palm *Hyphaene thebaica* and – especially when disturbed by human activities-*Calotropis procera*. *Cenchrus biflorus, Dactyloctenium scindicum* and *Lasiurus hirsutus* are some of the important grasses in this community.

The riverine plant formations and wadi communities of the coastal lowlands of the Tihama are crossed by a number of wadis. Most of these wadis do not flow as far as the sea, their water being largely absorbed by their deposits. Along these wadis with deep bottom deposits and a much better water supply for plants – as in such wide wadis as the Wadi Mawr with an extensive catchment area in the escarpment and even on the highland – plant cover may be very dense and composed of some very characteristic species. The dominant formation is an evergreen alluvial forest and thicket with fast-growing, mainly evergreen trees and shrubs and a definite stratification in distinct layers. In the tree layer the microphyllous *Tamarix nilotica* dominates, sometimes accompanied by the evergreen *Salvadora persica* and often covered with several climbers such as *Cissus quadrangularis, Momordica balsamina* and *Pentatropis nivalis*. The shrub layer consists of *Cassia senna, Calotropis procera, Leptadenia pyrotechnica* and *Jatropha curcas*.

On the low banks on both sides of the wadi, tall grasses such as *Desmostachya bipinnata* form pure and dense stands; sometimes *Typha angustata* is also found. Finally, on mixed deposits on the lowest terraces or on islands in the wadi beds, often strongly flooded, *Tamarix nilotica* again occurs as the most important pioneer.

### B. The Tihama foothills

The Tihama Foothills are situated between the coastal lowlands and the escarpment areas within the elevation range c. 300–1,000 m above sea level. The landscape comprises hills and sometimes steep mountain slopes composed of ancient rocks of great variety, including granites, gneisses, metamorphic sediments and volcanic rocks. There are also almost flat and stony piedmont plains and small intermountain basins with sandy and silty alluvial deposits. Hills, pediments and basins are dissected by narrow ravines and deeply incised bigger wadis which are often accompanied by different terrace-systems, covered with sands and gravels, or even there may be bare rock. Most of these wadis drain into the Tihama.

The climate of the Tihama Foothills is very hot and dry. The mean annual temerature is about 30ºC and seasonal changes are not very distinct; absolute minima and maxima range from 15ºC in winter to around 40ºC in summer. Rainfall in the foothill is considerably higher than in the Tihama and lies probably between 200 and 400 mm/year, concentrated around July, August and September, though showers in all other months may also occur. However the distribution of rainfall varies throughout the foothills, because sea-facing slopes usually receive a greater share of rain, as do the higher hills further inland. On the other hand, run-off on their slopes is normally very high and water storage is much better in the basin deposits and the valley-fill sediments. So there is no doubt that distribution of rainfall in time and space as well as the possibility of water storage have a most noticeable effect on the plant cover. The air humidity is still very high, especially during night time and in the morning. Winds generally come up from the Tihama during the day. Throughout the Tihama Foot-hills, immature soils without stratification and a low humus content are common and reflect the different parent materials. Though the mineral components of the soils differ largely from place to place, it is the physical nature of the substratum which is of greater importance, because it influences the absorption and retention of moisture. On the steep slopes, soils are stony and very shallow owing to extensive erosion. Only on flatter pediments gravels and sands may form a continuous layer. The basins are regions of deposition with gravels and sands near the hills, followed by loams and silts in the basin center. In all these deposits water from local rain is supplemented by run-off water from the slopes. In the wadi-systems alluvial soils are laid down by stream deposition. They consist of irregularly mixed sands and silts. Most of this land on the wadi soils and in the basins is under cultivation.

In the Tihama Foothills four major ecosystems are also to be found in the natural environment: (I) the steep slopes of the hills and mountains, (II) the level or gently undulated pediments, (III) the flat intermountain basins and (IV) the wadi systems.

Zonation of vegetation in Tihama Foothills is mainly affected by – as almost everywhere else – water. The dominant plant formation on hills, slopes and pediments is a drought-deciduous lowland and submontane woodland with only a few ever-green plants, some succulents, bottle trees, a shrub-layer, composed of many different species and only a sparse ground vegetation. According to soil conditions and the amount of rainfall, different communities usually mingled in a mosaic pattern are found. On the slopes a raingreen woodland is dominated by the scattered umbrella-shaped *Acacia tortilis* and some other *Acacia* species on higher elevations between 500 and 1,000 m above sea level, such as *Acacia mellifera, A. asak* and *A. abyssinica*. Other important trees are *Commiphora myrrha, C. opobalsamum, C. kataf* and *Berchemia discolor*.

Short-stemmed and multi-branched shrubs are usually widely spaced but sometimes form small thickets. Among these on higher elevations (above 500 m) are: *Barleria bispinosa, Dodonaea viscosa, Grewia populifolia, G. velutina, G. villosa, Hibiscus micranthus, Lawsonia inermis* and *Maytenus senegalensis*. On lower hills and on the pediments, where the vegetation cover is only very sparse, and *Acacia tortilis* is the only notable tree, *Abrus bottae, Aerva javanica, Anisotes trisulcus, Cadaba farinose, C. glandulosa, Grewia tenax, Maerua crassifolia, Ormocarpum yemenense* and *Premna resionsa* are the dominant shrubs.

On very stony and rocky places there is sometimes a rich variety of cactus-shaped, thick-leaved or thick-stemmed succulents such as *Adenia venenata, Aloe sabaea, Caralluma quadrangula, Ceropegia variegata, Euphorbia inarticulata, Kleinia odora, Sansevieria ehrenbergii* and the marvellous flowering bottle-tree *Adenium obesum*. Here another very characteristic tree is *Delonix elata*.

The ground cover of annual and perennial herbs and grasses and xerophytic dwarf shrubs is usually scanty and contains: *Blepharis ciliaris, Dactyloctenium scindicum, Ecbolium linnaeanum, Fagonia indica, Indigofera spinosa, Ruellia patula* and *Seddera arabica*. In the silty and sandy basins, which benefit by run-off from the higher slopes, another plant formation, a mainly evergreen woodland with one dominant sclerophyllous tree, *Dobera glabra,* is typical. Among the co-dominant species of this mostly cultivated environment are *Acacia ehrenbergiana, Balanites aegyptiaca, Cadaba rotundifolia, Salvadora persica* and *Zizphus spina-christi*.

The landscape in the Tihama Foothills is characterized by the contrast between the dry, brown hills and the green fertile valleys. Again the wadi systems and their thick deposits offer a special environment for plant growth because of this much better and regular water supply. An evergreen seasonal lowland and submontane forest with microphyllous, sclerophyllous and broad-leaved trees is the most important plant formation. Two main forest communities may be differentiated.

The lowland communities alongside the big perennial wadis are dominated by dense stands of *Tamarix nilotica* which are influenced by the near groundwater and frequent overflow as well as the tall grass *Desmostachya bipinnata*, and the shrubs *Jatropha curcas* and *Leptadenia pyrotechnica*. On low terraces with sandy deposits the prevailing *Salvadora persica, Acacia ehrenbergiana* and *Cassia senna* are the most important species. Both trees and shrubs are frequently covered with creeping *Cissus rotundifolia, C. quadrangularis* and *Commicarpus helenae*. On small sandy terraces near water the perennial herb *Bacopa monnieri* is abundant.

Another type of wadi vegetation is to be found in the narrow and deep incised smaller valleys on the higher slopes between 500 and 1,000 m above sea level. Though there is only rarely a constant flow of water, the high humidity of air, more shadow during daytime and a supplementary amount of run-off water provide good conditions for many evergreen and deciduous, often very tall trees such as *Breonadia salicina, Combretum molle, Ficus salicifolia, Mimusops schimperi, Tamarindus indica* and *Trichilia emetica*.

Common weeds in the Tihama Foothills are the herbs *Aristolochia bracteata, Euphorbia indica, Heliotropium longiflorum, Pulicaria jaubertii* and the shrubs *Abutilon hirtum, Calotropis procera* and *Solanum incanum*.

### C. The lower escarpment

The whole escarpment area ranges from 1,000 to c. 2,200 m altitude. Although this usually very steep access to the highlands sometimes has an extent on only few kilometers from the west to the east, the climatic changes are very distinct and so, too, are the changes in the vegetation cover. Therefore we separate this region and start with the lower parts of the escarpment area, rising to the east to about 1,600 m above sea level.

The landscape is characterized by very steep slopes with rock outcrops and cliffs especially where sedimentary layers of Jurassic sandstone overlie the different basement rocks. The slopes and mountains are dissected by wadis and deep gorges with a strong headward erosion. The most important climatic feature is a considerable increase of rainfall, above all on the western slopes, which are exposed to the rain-bearing clouds. Here, the average annual precipitation may reach 600 mm or more, while in rainshadow areas 400 mm or even less is probable. Further up to flanks, springs occur and in some smaller valleys water can be found in some parts of the valley bottom and in rock pools throughout the year. Not only the higher amount of moisture but also the lower mean temperature of about 25°C give better conditions for plant life than in the Tihama Foothills. The humidity of air is only moderate, except for the early morning hours.

Soils are usually very shallow, rocky and stony lithosols. Because of the steepness of slopes, soil erosion is extensive and bare rocks are abundant. Only on cultivated terraces medium-textured anthropogenic soils are developed. These sometimes show a stratification although the content of organic matter in the upper layer is only about 3%. In the V-shaped valleys, wadi-soils and alluvial deposits are usually rare. Only three ecosystems are to be discussed in the lower escarpment area: (I) the more humid, west-exposed slopes and mountain sides, (II) the drier east-exposed slopes and (III) the wadi systems.

Zonation of vegetation is obvious in this mountainous habitat. Although different plant communities are to be found in the lower parts of the escarpment area, they almost all belong to one main plant formation: a drought-deciduous submontane woodland, mixed with few evergreen trees and shrubs, some succulents, lianas and bottle-trees; there is only a sparse ground vegetation of different herbs and grasses. This deciduous woodland differs in its floristic composition according to the exposure of the slopes. The main species of the tree layer are *Acacia asak* and *A. mellifera* and some of the *Commiphora* species already present in the Tihama Foothills such as *C. kataf* and *C. myrrha*. Very characteristic here is *Commiphora abyssinica*, one of the myrrh resin-producing trees.

On the western slopes with higher rainfall, scattered individuals of *Berchemia discolor, Ficus salicifolia, Phoenix reclinata* and *Trichilia emetica* can be found among these trees, although there are only few relic woodlands on some of the more inaccessible and therefore not cultivated slopes. On terraces and fields, trees such as *Cordia abyssinica, Terminalia brownii* and others provide shade for the crops. *Breonadia salicina* trees are found growing naturally or have been planted and cared for by farmers for their timber. Sometimes these trees are lopped like the also highly valued *Ziziphus spina-christi*, especially during the dry seasons of the year.

Dominant shrubs in these woodlands are *Cadia purpurea, Carissa edulis, Grewia tenax, G. velutina, G. villosa, Hibiscus micranthus, Ochna inermis, Oncoba spinosa* and *Pterolobium stellatum*. In the ground vegetation throughout the year the grass *Hyparrhenia hirta* is very widespread.

In rain-shadow areas, the woodlands have a more park-like aspect and are much poorer in their floristic composition. *Acacia mellifera* and *Commiphora abyssinica* dominate in a tree layer not higher than 5 m, but most of the other tree species mentioned above are absent. Here the most common shrubs are *Abrus bottae, Acalypha fruticosa, Anisotes trisulcus, Barleria bispinosa* and *Jatropha spinosa*. The ground layer consists mainly of herbs including *Blepharis ciliaris, Commicarpus plumbagineus, Forsskaolea tenacissima* and such different grasses as *Cenchrus ciliaris, Cynodon dactylon, Dactyloctenium scindicum* and *Themeda triandra*.

On rocky outcrops a large number of succulents and climbers can be found such as the bottle-tree *Adenium obesum, Aloe vera, Kalanchoe lanceolata, Adenia venenata* and *Cissus rotundifolia*. However the most characteristic succulent for this environment is *Euphorbia cactus* which forms large and sometimes pure stands near settlements.

In the wadi systems of the lower escarpment area a tropical evergreen seasonal forest with broad-leaved evergreen trees prevails, with closed, sometimes very dense, canopy in the most important plant formation.

Again the larger valley bottoms are an exceptional environment for plants because of frequent floods in the rainy seasons and a high water table in the deposits throughout the year. The smaller V-shaped valleys are largely uncultivated because of the steepness of the slopes and the absence of higher amounts of alluvial deposits. Here forest "islands" may be still present with sometimes huge individuals of *Acacia abyssinica, Combretum molle, Ficus populifolia, Mimusops schimperi, Terminalia brownii* and *Trichilia emetica.*

The larger valley bottoms are heavily cultivated. Here the trees grow around the edges of the fields and the flood basin. Most frequent are *Ceiba pentandra, Ficus salicifolia, F. sycomorus* and *Tamarindus indica* especially in the Wadi Sharaz between Hajjah and Kuhlan. Coffee trees are usually shaded by the huge *Ficus vasta* and *Cordia abyssinica.* Because of the dense stands of the trees, shrubs are rare and the ground cover very sparse.

### D. The higher escarpment

The higher part of the escarpment area ranges between 1,600 and c. 2,200 m although the peaks and ridges of the watershed zone may reach several hundred metres higher. The main reason for this boundary however lies not in relief properties or geological structures but in a very simple but important climatic factor. In these altitudes between 2,000 and 2,200 m above sea level, frost occurs more or less regularly during the winter months, limiting the life conditions for all tropical lowland and most of the tropical submontane plant species. The temperatures are generally lower, the annual average c. 20°C or less; rainfall varies here, too, but can locally reach 800 mm/year or more, again with two annual rainy seasons, a shorter one in March/April and a longer one in July until September, with usually high rain intensities. Cloud and mist in the early morning hours, and in the evenings, are very frequent throughout the year, humidity again being high. Only the eastern slopes have a lower average rainfall. The conditions for plant life and production however are certainly the best of all Yemen.

The terrain is still very steep, rock outcrops are numerous, and in the upper layers of sand and limestones of the Kuhlan- and Amran-series, almost vertical walls are frequently found. The slopes are dissected by a large number of ravines and gullies, V-shaped valleys with the bigger wadis deeply incised. Gently undulating slopes and small intermountain plains are very rare. Except where reclaimed by terraces, light brown stony soils on the slopes are either eroded or very shallow. The terrace soils are loamy, medium-textured, often calcareous, containing organic matter of c. 3% and more. On these terraces, owing to good soil conditions and the best rain supply, crop production can be very high. The same three ecosystems as in the lower escarpment – only at higher altitudes – are to be taken into account: (I) the well-watered western slopes with sometimes vertical cliffs of bare rock, (II) the more arid eastern slopes and again (III) the different wadi systems.

After thousands of years of clearing and forest exploitation, it is very difficult to name the most important plant formations of the higher escarpment area. However, a few woodland relics remain in usually inaccessible places. They belong to either an evergreen broad-leaved woodland or thicket with many sclerophyllous trees and shrubs, seldom exceeding an average height of 5–6 m, or – on the driest sites on the eastern slopes – to a drought-deciduous mountain woodland. In only very few places remnants of a real tropical broad-leaved cloud forest with some epiphytes and parasitic plants and a ground cover of hygromorphic herbs can be found.

Although trees and shrubs on terrace walls and rocky places show a very scattered mosaic of different communities it is possible to differentiate between two main woodland communities. The first one, a xerophyllous community on the drier slopes, again is dominated by an *Acacia* sp., *Acacia negrii,* which sometimes is associated with *A. abyssinica* and *A. gerrardii*. In the shrub layer *Barleria prionites, Cadia purpurea, Carissa edulis, Dodonaea viscosa, Hibiscus micranthus, Myrsine africana, Plectranthus barbatus* and *Phoenix reclinata* are rather common. In open stands along dry roadsides *Withania somnifera* and *Lycium shawii* may occur. The ground cover is characterized by spiny dwarf shrubs and inedible herbs such as *Euphorbia schimperiana, Felicia abyssinica, Indigofera spinosa, Polygala tinctoria, Striga hermonthica* and such grasses as *Andropogon distachyus, Hyparrhenia hirta* and *Pennisetum setaceum.*

The second one, a hygrophilous community on the wetter slopes, consists of a large number of mainly evergreen trees and shrubs such as *Buddleja polystachya, Dichrostachys glomerata, Ehretia abyssinica, Olea chrysophylla, Pterolobium stellatum, Rhus abyssinica, Rosa abyssinica* and *Sageretia thea.* Some of the species names indicate that these plants also occur in the highlands of Ethiopia. On rocky places, *Ficus palmata, Centaurothamnus maximus* and *Primula verticillata* are very common. The ground cover contains herbs of Mediterranean affinity such as *Campanula edulis, Celsia bottae, Crassula alba, Crinum yemense* (geophyte) and *Scadoxus multiflorus*. Between the stones of the terrace walls particularly in places where mist occurs almost every day, *Dorstenia foetida* and *Selaginella yemensis* are as characteristic as a number of ferns such as *Asplenium trichomanes, Adiantum capillus-veneris, A. incisum* and *Actinopteris semiflabellata.* A very dinstinctive plant of this “cloud” – zone is the tallest of the succulent Euphorbiaceae, *Euphorbia ammak.* There are many transitional subtypes of these two communities, due not only to changes in the ecological conditions but primarily to different direct or indirect human influences.

The most important plant formation in the wadi systems of the higher parts of the escarpment differs only slightly from those formations discussed previously for the lower escarpment. It is still a tropical evergreen forest with evergreen trees and shrubs dominating; the tropical lowland species are absent. The stands are floristically poorer. The most characteristic high trees left now are *Ficus* species such as *Ficus vasta*, the smaller *F. palmata* and – more rarely – *F. sycomorus* and *F. salicifolia.* The dominant tree species however is *Cordia abyssinica.* Common shrubs and herbs are *Achyranthes aspera, Kanahia laniflora* and *Mirabilis jalapa.* Whenever possible, the wadi bottoms and the adjacent slopes are cultivated.

Among the dominant species of the weed flora and the ruderal communities along roadsides and around settlements are *Acanthus arboreus, Argemone mexicana, Datura stramonium, Euphorbia schimperiana, Gomphocarpus fruticosus, Mirabilis jalapa, Rumex limoniastrum, Ruta chalepensis* and others.

### E. The high mountains

The highlands and the highest mountain chains are the result of block faulting along a north-south axis parallet to the Red Sea. The Precambrian basement, mainly granites and gneiss, has been lifted so high that sometimes these rocks lie at the surface, especially in the southern part of the Yemen Arab Republic. Most of the upland massifs, however, have a capping of Jurassic and Cretaceous sand- and limestones. Very often these sediments are still overlain by extensive outflows of volcanoes of Tertiary and Quaternary age. This high level plateau has an average altitude of c. 2,300 m above sea level, but some massifs, bordering this plateau in the west and in the east, frequently exceed 3,000 m. The highest peak, Jabal An Nabi Shu'ayb (the mountain of the prophet Shu'ayb) also the highest peak in the entire Arabian Peninsula – rises to 3,766 m. On the Red Sea side the highland falls steeply to the coastal plain; on the desert side in the east the slopes are less abrupt.

As a result of these different bedrocks, the volcanic activities and the system of the only slightly incised wadis, the topography of the highlands is very complex: plateaux, intermountain basins, isolated hills and mountains, alluvial plains, colluvial slopes and steep ones with bare rocks are the most common relief forms. In spite of this, the climate of the highland varies little. Rainfall usually comes in heavy intermittent showers in March and from July to September, amounting to only 300–500 mm/year with a high variability from year to year. The average air humidity is below 50%. So again water is the limiting factor for plant life. However, a very important factor is the thermal one. The mean monthly temperature changes from between over 20ºC in summer and about 10ºC in winter, but frost is frequent during winter time, excluding all really tropical plant species in the vegetation cover. On the higher ridges and peaks not only is the average annual temperature still lower (about 10ºC) but frost, snowfall and hail are of regular occurrence in the cold season.

The principal soils of the highland are those of the intermountain plains and wadi systems. They consist of usually fine to medium-textured, calcareous strata with a low humus content and a high pH over 8. On lava fields, hills and slopes and in the higher mountains only extremely stony soil layers of shallow depth are to be found. Many sites are even bare of soil.

The following main ecosystems and their vegetation cover are apparent in the highlands and the high mountains: (I) the more or less level plains and basins with rather deep loamy soils which receive run-off water, including the wadi systems, (II) the gently undulated pediments and lower isolated hills with shallow stony soils, (III) the plains of subrecent volcanic deposits and still no soil development, and (IV) the stony and rocky high mountain slopes and peaks with a water supply below average.

During a long period the highland has been even more heavily cultivated than the escarpment area. Therefore it is very difficult to reconstruct the natural plant cover. It can be postulated only as a "savannah" – like open woodland with a large number of spiny and thorny species. Leaf-reduction and succulent life forms are again very common. The main plant formation is a drought-deciduous mountain woodland. But only a few woodland relics are still to be found and these very often are not natural. Only in the north of Yemen, between Huth and Sa'dah, owing to the smaller population, some nearly natural woodland communities have been preserved. The dominant tree species are again *Acacia* spp. such as *A. negrii* and *A. gerrardii*. Occasionally may be found *Buddleja polystachya, Cordia abyssinica, Ficus palmate, Olea chrysophylla* and *Juniperus excelsa*[4] which forms extensive woodlands in the Asir mountains in Saudi Arabia but is very rare here owing to tree felling and goat overgrazing over many centuries. Among the shrubs *Carissa edulis, Ehretia abyssinica, Grewia mollis, Myrsine africana* and *Rosa abyssinica* are the most common.

Usually the mountain plains are cleared and cultivated, and the bordering rocky slopes and lava fields are over-grazed. In the generally sparse vegetation cover, nevertheless, a large number of different small shrubs, herbs and grasses show that the highland communities were once floristically very rich. The most important species are: *Aristida adscensionis, Commicarpus sinuatus, Echium longifolium, Euphorbia fruticosa, E. schimperi, Euphorbia schimperiana, Fagonia indica, Helichrysum fruticosum, Hyparrhenia hirta, Kleinia semperviva, Lavandula coronopiflia, L. pubescens, Lycium shawii, Reichardia tingitana, Salvia merjamae* and *S. schimperi*. Near well-watered places the plant cover can be very dense with *Flaveria trinerva, Mentha longifolia* and *Xanthium spinosum* as the most characteristic species. At higher altitudes but still below the timberline, which probably lies somewhere between 2,800 and 3,000 m, the plant cover is usually denser and there are many endemics, such as *Crinum yemense, Cichorium bottae, Delosperma harazianum, Macowania ericifolia*,[5] and *Teucrium yemense* (*C. album*, Wood, 1997).

As well there are species which can also be found in the highlands of Ethiopia such as *Campanula edulis, Crassula alba, Felicia abyssinica* and *Pterocephalus frutescens*. Some ferns such as *Ceterach officinarum, Cheilanthes pteridioides* and *Adiantum capillus-veneris* and the attractive *Primula verticillata* grow on wet places below shady rocks; *Centaurothamnus maximus* is rather common on the otherwise bare cliffs. In the highest mountains above the timberline, the dominant plant formations are alpine pastures and meadows, rich in forbs and grasses. On several occasions during the winter season they may be covered with snow for a few hours or days. In these natural grasslands *Eleusine floccifolia* and *Pennisetum setaceum*

---

[4] Wood (1997) stated that *J. excelsa* (= *J. procera*) is the Yemen's only native conifer found between 1,800 and 3,300 m mostly on mountains with medium rainfall.

[5] Wood (1997) stated that *Macowania ericifolia* (Compositae) occurs also in Ethiopian (Eritrean) mountains in the African Red Sea coast.

are important grasses and *Dianthus uniflorus, Micromeria biflora* and *Craterostigma pumilum* frequently form small patches.

There is not much left of the azonal plant communities in the highland because even the slightly incised large wadis of the plains are heavily cultivated. Nevertheless there are some small remnants of an evergreen tropical highland forest with only one tree species forming dense thickets: *Tamarix nilotica* which is accompanied by *Typha angustata* (= *T. domingensis*) when water is regularly flowing. On higher altitudes another riverine (mesophyte) tree has probably been introduced: *Salix alba*. The herbs in this community are the same as those mentioned above for places with an exceptionally good water supply. The small gorges and valleys of the dissected hills and slopes have a visibly denser plant cover. The species composition however differs from those of the communities on the hills and slopes themselves.

## 2.6 Coastal Deserts of the Gulfs of Suez and Aqaba, Sinai, Egypt

### *2.6.1 General Remarks*

The Sinai Peninsula is a triangular plateau in the northeast of Egypt with its apex, in the south, at Ras Mohammed, where the eastern coast of the Gulf of Suez meets the western coast of the Gulf of Aqaba (latitude 27° 45′N). Its base, in the north, is along the Mediterranean Sea – the eastern section of the Egyptian Mediterranean coast that extends for about 240 km between Port Said and Rafah (latitude 31° 12′N). The area of the Sinai Peninsula (61,000 km$^2$) is about 6% of that of Egypt. More than half the peninsula is between the Gulfs of Aqaba and Suez. Sinai Peninsula is the Asian part of Egypt (Zahran and Willis, 1992) (Fig. 2.1).

The eastern and western edges of the Sinai horst are different from one another. The western coastal plain, known as El-Qaa, is wide. It borders the Gulf of Suez, which has a depth of 100 m and a length (from El-Shatt southwards to Ras Mohammed) of about 400 km. The eastern coastal plain extends from Aqaba southwards to Ras Mohammed for about 235 km. bordering the Gulf of Aqaba.

The mountains which form the igneous core of the Sinai Peninsula rise to considerably greater heights than any of those of the African part of Egypt. The highest peak, Gebel St Katherine, is 2,641 m above sea level. Many other peaks and crests rise above the 2,000 m contour, conspicuous among which are Gebel Umm Shomer (2,586 m), Gebel Musa (2,285 m), Gebel Al-Thabt (2,439 m) and Gebel Sebal Pile (2,070 m). "The core of the peninsula is highly dissected; its gaunt mountains and deep rocky gorges form one of the most rugged tracts on the earth's surface" (Said, 1962).

Because of its high altitude, the southern section of Sinai receives ample rainfall which has produced wadis. Most of the wadis run in long hollows and appear as hanging valleys. Some wadis flow to the Gulf of Aqaba, e.g. Wadis Ghayib, Nasb and Watir, all of which are steep valleys. Running to the El-Qaa plain, in the west, are for example Wadis Feiran, Sidri, Sudr and Gharandal, all of which are wide and have relatively rich vegetation.

The higher part of the limestone plateau which flanks the igneous core to the north is called El-Tih. At the southern end of Gebel El-Tih is the Ugma Plateau which is 1,620 m above sea level. The central portion of the plateau forms fairly open country draining to the Gulfs of Aqaba and Suez.

The western coastal plain of Sinai is relatively broad, extending in western Sinai from the Gulf of Suez to the great western El-Tih escarpment and its continuation southwards in the granitic ridges of southern Sinai. The plain is broad in its northern part but narrows south of Gebel Hammam Faraon (about 80 km south of El-Shatt). The northern division has an average width of 30 km. It is very gently undulating and locally dotted with low hills of limestone. The surface of the plain is largely covered with drift sand which forms parallel crescentic dunes. In more southern parts, the coastal plain, which is traversed by several wadis, is of sandy marls and gypsum, covered in some parts with gravels.

The foreshore areas of the western coast of Sinai are subject to flooding with water of the Gulf of Suez. In the dry season these areas become covered with a thin mantle of white salts. The southern half of the coastal plain is crossed by well-defined ridges such as Gebel Hammam Faraon and Gebel Araba, the summits of which are over 500 m above sea level. Opposite these ridges the coastal plain is very much narrower.

The hills forming the eastern boundary of the foreshore plain become progressively higher southwards. The plateau of the southern subregion of Sinai drops by a number of steps to that plain. Along the Gulf of Aqaba the coastal plain is relatively narrow.

The central subregion of the Sinai Peninsula is called the El-Tih Plateau, but this is a misnomer because this plateau is only the southern part of central Sinai. The middle part of this central section is known as the Al-Ugma Plateau and the northern part is the area of domes. These domes echo the alpine movements which occurred in the Eastern Mediterranean Region (Abu Al-Izz, 1971). The important peaks of central Sinai are:

1. Gebel El-Halal (890 m), an anticline with its axis running from northeast to southwest, parallel to the axis of the domes. For 7 km, Wadi El-Arish crosses this mountain with a narrow course.
2. Gebel El-Maghara (500–700 m), which includes several secondary domes. It is an asymmetrical fold covering an area of about 300 $km^2$. It is the largest Jurassic exposure in Egypt.
3. Gebel Yalaq (Yellag), about 1,100 m, is a great asymmetrical anticline with its southern side steeper and with abundant faults.

El-Tih Plateau slopes from the above-mentioned heights down to the north. It is dissected by drainage channels, most of which flow to the Mediterranean Sea. These channels are generally much shallower and more open than the wadis of the southern mountainous subregion.

## 2.6.2 Climate

Being part of Egypt at the extreme northeast of Africa, the Sinai Peninsula belongs, climatically, to the dry province. According to Ayyad and Ghabbour (1986), the Sinai Peninsula can be divided into two main climatic zones: arid and hyperarid. The arid zone includes the northern subregion: summer is hot, winter is mild and rainfall usually occurs in winter. It is distinguished into two provinces (UNESCO/FAO Map, 1963): (1) the coastal belt province under the maritime influence of the Mediterranean Sea with a relatively shorter dry period (attenuated) and annual rainfall ranging between 100 and 200 mm, and (2) the inland province with a relatively longer dry period (accentuated) and an annual rainfall of 20–100 mm. The hyperarid zone, however, covers the central and southern subregions of the peninsula. It is also distinguished into two provinces: (1) hyperarid province with hot summer, mild winter and winter rainfall which is considered an extension of the Eastern Desert and includes central Sinai or El-Tih Plateau together with the western and eastern coasts of the Gulfs of Aqaba and Suez; (2) hyperarid province with cool winter and hot summer located around the summits of the Sinai mountains.

During the winter months some areas of Sinai experience short periods of brief but heavy rainfall that may cause the wadis to overflow.

Air temperature in Sinai is subject to large variations, both seasonally and spatially. Minimum winter temperature ranges from 19°C at Sharm El-Sheikh to 15°C at El-Tor, 14°C at El-Arish, 9°C at Nakhl to 0°C at St Katherine (–4°C during January 1987). Maximum summer temperature also shows a large variation, and ranges from near 20°C at St Katherine, with its high elevation the coolest in the peninsula, to more than 50°C at El-Kuntilla. During summer the Gulf of Suez region is much warmer (35°C) than the northern Mediterranean region (30°C).

The amount of rainfall in Sinai decreases from the northeast to the southwest. The relatively highest amount of rain is in Rafah (304 mm/year) followed by that of El-Arish (99.7 mm/year), but only about 10.4 mm/year occurs in the southwest. Rainfall decreases in the plateau region to about 23.3 mm/year, but then increases in the southern mountainous region to about 62 mm/year in St Katherine where precipitation may occur as snow that may last for 4 weeks (Migahid et al., 1959). In some years more than one snowfall may occur whereas in others snow may be absent. Precipitation may occur as hail on the high peaks. Water derived from melting snow or hail is usually insufficient to infiltrate the desert soil at the foot of the mountains appreciably.

Rain decreases in southerly (towards Ras Mohammed), easterly (towards the Gulf of Aqaba) and westerly (towards the Gulf of Suez) directions and averages only 12 mm/year in the Gulf of Suez (23.6 mm/year in Suez, 22 mm/year in Abu Redis and 9.3 mm/year in El-Tor).

Rainfall occurs in Sinai mainly during the winter season (November – March) and during spring or autumn. It decreases markedly or is completely lacking from May to October. However, summer rain resulting from the influence of the Red Sea depressions causes floods. Tropical plants of Sudanian origin may germinate in the moist wadi beds following summer rain. Also, summer fogs and dew are frequent in

Sinai north of El-Tih but absent in southern Sinai. In some years dew provides more moisture than does rainfall. Lichens are particularly effective in capturing moisture from fog and dew. Overall the annual average rainfall for the entire Sinai Peninsula is 40 mm, of which 27 mm is estimated to come from individual storms of 10 mm or more.

The mean annual maxima and minima of relative humidity at El-Arish, El-Tor and Suez are: 79 and 56%, 70 and 50%, 73 and 30% respectively, and the mean values of annual evaporation of these three areas are: 4.3, 10.2 and 9.4 mm/day (all Piche) (*Climatic Normals of Egypt*, Anonymous, 1960).

## 2.6.3 Vegetation Types

### 2.6.3.1 Gulf of Suez Coast

The western coast of the Sinai Peninsula (eastern coast of Suez Gulf) is bounded by the Gulf of Suez in the west and the limits of the coastal desert and wadis that drain into it in the east. It extends from El-Shatt (latitude 30ºN) in the north to Ras Mohammed (latitude 27º 40′N) in the south for about 340 km. The width of this coastal area depends mainly upon the geomorphology of the area and reaches its maximum (1–2 km) south of El-Tor while at Ras Bakr (about 90 km south of El-Shatt) the hills extend to the shore-line.

Four types of habitat have been recognized in the western coast of Sinai, namely: mangrove swamps, littoral salt marshes, oases and coastal desert (Zahran and Willis, 2008).

#### (a) Mangrove swamps

Mangrove swamps are absent from the whole stretch of the eastern coast of the Gulf of Suez (as in the western coast). However, at the cap of the Sinai Peninsula where the Suez Gulf meets the Aqaba Gulf (Ras Mohammed), there is a shallow and narrow lagoon extending from the Gulf of Suez landward. This lagoon provides a suitable site for the growth of mangal vegetation. Thickets of a pure community dominated by *Avicennia marina* are present in this lagoon which has a muddy substratum (Ferrar, 1914; Zahran, 1965, 1967, 1977). Shrubs of this mangrove grow near to the banks of the lagoon as well as in its shallow water channel. *A. marina* predominates also in some stands in the southern section of the Gulf of Aqaba, the northernmost stand is Nabq mangrove swamp: 28º 12′N.

#### (b) Littoral salt marshes

This is the salt-affected land that runs parallel to the east side of the Gulf of Suez, being strongly influenced by the saline water of the Gulf. Its width depends on the maximum reach of the Gulf's water as well as on the topography of the coast.

In the halophytic vegetation of these littoral salt marshes ten communities with different dominants have been recognized, namely: *Halocnemum strobilaceum, Arthrocnemum macrostachyum, Aeluropus massauensis, Zygophyllum album, Nitraria retusa, Tamarix nilotica* (common communities), *Halopeplis perfoliata, Limonium pruinosum, Cressa cretica* and *Juncus rigidus* (less common communities).

### 1. *Halocnemum strobilaceum* community and 2. *Arthrocnemum macrostachyum* community

*H. strobilaceum* and *A. macrostachyum* are usually closely associated, *A. macrostachyum* being an abundant associate in stands of *H. strobilaceum* which is abundant in stands dominated by *A. macrostachyum*. Many stands are co-dominated by these two succulent halophytes.

The vegetation of the stands of these two communities (cover ranges between 40 and 90%) includes *Aeluropus massauensis, Atriplex leucoclada, Cressa cretica, Schanginia hortensis* and *Suaeda vermiculata* as less common species and *Nitraria retusa* and *Zygophyllum album* as common associates.

### 3. *Aeluropus massauensis* community

This species and the community which it dominates are largely restricted to a stretch of about 70 km (between 11 and 81 km south of El-Tor). It grows in two forms, the normal mat form covering sheets of moist salty soil and a peculiar cone-like form, noted only in one stand at 81 km south of El-Tor. The two forms have also been observed in the Red Sea coastal land. The cover of this community is high (60–90%) contributed by the dominant grass; associates (*Cressa cretica, Nitraria retusa* and *Zygophyllum album*) have negligible cover.

### 4. *Zygophyllum album* community

Along the Sinai coast of the Suez Gulf, *Z. album* is abundant and recorded as an associate species in almost all of the communities in addition to the one which it dominates. It is dominant in two types of habitat: dry salt-affected areas in the zone inland to that dominated by *A. massauensis* and also the areas having phytogenic sand mounds along the coast. The cover of the stands is thin (5–15%); however, the number of the associates is relatively high (15) and includes a mixture of xerophytes and halophytes. These are: *Nitraria retusa* (abundant), *Hammada elegans* (common), *Suaeda vermiculata* and *Tamarix nilotica* (occasional) and species rarely present which include *Alhagi maurorum, Anabasis articulata, Atriplex leucoclada, Halocnemum strobilaceum, Hyoscyamus muticus, Neurada procumbens, Pergularia tomentosa, Phoenix dactylifera* (semi-wild), *Salsola tetrandra, Zilla spinosa* and *Zygophyllum coccineum*.

### 5. *Nitraria retusa* community

*N. retusa* is a widely distributed species along the whole stretch of the western coast of Sinai. It builds sand mounds and hillocks of considerable size. The community dominated by this shrub occupies the most landward zone of the salt marsh ecosystem separating it from the non-saline desert ecosystem.

The vegetation of the *N. retusa* community contains a considerable number of associates (18) – xerophytes and halophytes. These are: *Zygophyllum album* (abundant), *Atriplex leucoclada* (common), *Aeluropus massauensis, Alhagi maurorum, Salsola tetrandra, Schanginia hortensis, Suaeda vermiculata* and *Tamarix nilotica* (occasional), *Arthrocnemum macrostachyum, Cressa cretica, Halocnemum strobilaceum, Hammada elegans, Juncus rigidus, Limonium pruinosum, Lygos raetam, Phoenix dactylifera, Zygophyllum coccineum* and *Z. simplex* (rare).

*N. retusa* is present along the whole eastern coast of the Suez Gulf but its dominance starts at 100 km south of El-Shatt and continues to 276 km. The cover of this community ranges between 5 and 50%. In one part of this coastal area, extending between 257 and 263 km from El-Shatt mainly dominated by *N. retusa*, the plants build huge hillocks that cover more than 50% of the area.

### 6. *Tamarix nilotica* community

*T. nilotica* is a widely distributed bush (or tree) in the western coastal area of Sinai. Its community is the most elaborately organized of the salt marsh ecosystem, growing in a variety of habitat conditions and showing varied physiognomy. *T. nilotica* bushes are sand binders, capable of building hillocks.

*T. nilotica* is usually dominant in the deltas of the large wadis, e.g. Wadis Sudr, Gharandal, Sidri and El-Tor. This community may also occupy the most landward zone of the salt marsh ecosystem in areas where *Nitraria retusa* is absent.

The flora of the *T. nilotica* community includes a large number of associates (22); 16 xerophytes and 6 halophytes. *Launaea spinosa* is interesting as it dominates a community in the northern section of the western coast of the Suez Gulf but is rare on the eastern coast. In the delta of Wadi Sidri there is an open salt marsh scrub dominated by *T. nilotica*. The water channel of the wadi is blocked by the hillocks built by *T. nilotica*. The silty terraces that bound the wadi channel are also dominated by its growth but hillocks are not built here.

*Nitraria retusa* and *Zygophyllum album* are the abundant associates, while *Z. coccineum, Hammada elegans* and *Lygos raetam* are common. Four species, namely, *Hyoscyamus muticus, Phoenix dactylifera, Reaumuria hirtella* and *Salsola tetrandra* are occasional. Rare species are *Achillea fragrantissima. Alhagi maurorum, Atriplex leucoclada, Diplotaxis acris, Fagonia glutinosa, Gymnocarpos decander, Halogeton alopecuroides, Mesembryanthemum forsskaolii, Polycarpaea repens, Schanginia hortensis* and *Zilla spinosa*.

### 7. *Halopeplis perfoliata* community

This community is recorded in only one locality – the coast of Abu Zenima (124 km south of El-Shatt); elsewhere, *H. perfoliata* is not even an associate species. Associates of this community are: *Arthrocnemum macrostachyum, Halocnemum strobilaceum* and *Zygophyllum album* (common) and *Nitraria retusa* (rare). Pure stands of *H. perfoliata* are also present.

### 8. *Limonium pruinosum* community

Like the preceding community, the presence of the *L. pruinosum* community is confined to a limited stretch (about 0.5 km) in the Ras Bakr area (91 km south of El-Shatt) which is a high beach lacking a definite salt marsh, the hills being close to the rocky shore, and covered with rock detritus. The cover of this community is thin (2–5%); associates are *Hammada elegans, Nitraria retusa, Salsola tetrandra, Suaeda vermiculata* and *Tamarix nilotica.*

### 9. *Cressa cretica* community

*C. cretica* is an associate species in several localities of the eastern coast of the Suez Gulf. Dominance of this mat-forming halophyte is restricted to two sites in the south part of the coast – at 1 and 17 km south of El-Tor. *C. cretica* usually grows in pure stands on the moist – saline sandy shore-line.

### 10. *Juncus rigidus* community

*J. rigidus* is not a common halophyte in the eastern coast of the Suez Gulf, being recorded as an associate species in the *Nitraria retusa* community only. It is dominant in four localities: 25 km south of El-Shatt where it is associated with *Nitraria retusa* and *Zygophyllum album*, and south of El-Tor where there are three localities in which it occurs either in pure stands or associated with *Aeluropus massauensis, Cressa cretica* and *Nitraria retusa.* The cover of the stands of this community is up to 100%.

### (c) The oases

The coastal eastern stretch of the Suez Gulf is characterized by oasis-like depressions, e.g. Ayon Musa (Musa Springs) and Hammam Musa (Musa Bath). These two oases are at 20 and 240 km south of El-Shatt respectively. Scrubland of *Tamarix nilotica* typifies these oases, with abundant growth of *Alhagi maurorum, Cressa cretica, Desmostachyta bipinnata, Juncus rigidus, Nitraria retusa* and *Zygophyllum album.* In areas where water is exposed, species such as *Phragmites australis* are present. These oases are bounded by the desert country supporting xerophytes which include *Asteriscus graveolens, Chenopodium murale, Diplotaxis acris, Fagonia glutinosa, Gymnocarpos decander, Halogeton alopecuroides, Hyoscyamus muticus, Peganum harmala, Plantago amplexicaulis, Pulicaria undulata, Reaumuria hirtella, Zilla spinosa* and *Zygophyllum coccineum*. Date palm (*Phoenix dactylifera*) occurs in groves as well as individual trees. The presence of these trees is an indicator of a fresh-water zone among the ground water layers of the oases (Abdel Rahman et al., 1965).

### (d) Coastal desert

The eastern coastal desert of the Gulf of Suez is characterized by several wadis that run from the mountains of southern and central Sinai and flow into the Gulf. Among these are: Wadis Sudr, Gharandal, Tayiba, Matulah, Nukhul, Baba, Sidri and Feiran. An account of the vegetation of four of these wadis and of the El-Qaa plain in the south of this coastal area is given.

#### (i) Wadi Sudr

This is one of the most developed wadis of the northern section of the western coast of Sinai. It is bounded by Gebels Raha (c. 600 m) in the north and Sinn Bishr (c. 618 m) in the south. The main trunk of the wadi extends roughly in a NE-SW direction for about 55 km and flows into the Suez Gulf at Raṣ Sudr (c. 55 km south of El-Shatt).[6]

The vegetation of Wadi Sudr includes a variety of communities and species with wide tracts covered by plants. In the main channel of its downstream part there is an open scrub of *Tamarix aphylla* with frequent *T. nilotica*. The course of the wadi here reaches its greatest width and receives maximum water revenue and surface deposits. This vegetation is present in the downstream 17 km, but further upstream *T. aphylla* is replaced by *T. nilotica*, *Lygos raetam* and *Hammada elegans* being important elements of the vegetation.

The courses of the upstream tributaries of Wadi Sudr are more defined, narrow and with the central part (water course) devoid of vegetation and fine sediments, being occasionally swept by torrents. The vegetation in these tributaries is confined to side terraces and dominated by *Tamarix nilotica* and/or *Lygos raetam*. In some of these tributaries where the mountains of the El-Tih Plateau bounding the course are high, the channel is devoid of side terraces and the sediments are compact. In these tributaries there is an open growth of *Acacia raddiana; Capparis cartilaginea* is occasional in the crevices and fractures of the cliffs and the rocky sides of the wadi.

The vegetation of the large affluents and runnels draining the gravel formations is a *Panicum turgidum* grassland. In the finer affluents and runnels the vegetation is dominated by *Artemisia judaica* and *Jasonia montana* or *Zygophyllum decumbens*.

In Wadi Sudr there is a fresh-water well – Bir Sudr – in the flood channel of a main tributary of the wadi. Its fresh water is utilized for domestic purposes. The vegetation of Bir Sudr is, however, halophytic, dominated by *Juncus rigidus* with abundant *Tamarix nilotica, Nitraria retusa* and groves of date plam.

---

[6]El-Shatt is a coastal city facing Suez on the eastern coast of the Gulf of Suez.

### (ii) Wadi Gharandal

Wadi Gharandal extends east-west for about 80 km. its origin is to the north of Gebel Pharaon and it drains into the Gulf of Suez at Hammam Pharaon 80 km south of El-Shatt. The inland portion of the wadi is in the El-Tih Plateau (Eocene) whereas the outlet section runs through the Pleistocene and Recent coastal belt of the Gulf of Suez (Said, 1962).

The main channel in the downstream portion of the wadi is covered with thickets of *Tamarix nilotica* where the course is well defined and bounded by low hills. Where the water course is broad, sand deposits are frequent and the underground water is deep. *Zygophyllum album* is dominant in these parts and *Hammada elegans* is abundant. This downstream portion of Wadi Gharandal, which extends eastwards for about 14 km, is especially rich in water springs that form localized swamps. Water runs in the lower parts of the main channel creating swamps and salt marsh habitats dominated by *Phragmites australis* and *Typha domingensis*. These swamps are surrounded by lawns of *Cyperus laevigatus* and of *Juncus rigidus, Alhagi maurorum* and *Desmostachya bipinnata* which form localized patches in the higher parts. *Tamarix nilotica* thickets occur where the sand deposits are deep. *Nitraria retusa*, however, forms discontinuous narrow patches on the terraces bordering the main channel. This downstream portion of Wadi Gharandal may be considered as one of the most outstanding agricultural settlements of Sinai where horticultural crops are prosperously cultivated. Palm groves are dense in this area.

The vegetation of the upstream affluents of Wadi Gharandal is dominated by *Hammada elegans* and *Zilla spinosa*. In the finer runnels *Achillea fragrantissima, Artemisia judaica, Jasonia montana* and *Zygophyllum decumbens* are dominants. *Capparis cartilaginea* makes dense growth on the side hills bordering Wadi Umm Lasseifa (an affluent of Wadi Gharandal). This dense growth is unique to that wadi as it contrasts with the usual sporadic distribution of *C. cartilaginea*. In the delta of Wadi Lasseifa is a well of good water quality, used for domestic purposes by the bedouins.

Ain Hegiya is a spring in Wadi Hegiya (another affluent of Wadi Gharandal), associated with swamp and salt marsh vegetation: *Typha domingensis, Phragmites australis, Cyperus laevigatus, Juncus rigidus* and *Tamarix nilotica*. In the water creeks *Veronica beccabunga* is common.

### (iii) Wadi Sidri

The main channel of Wadi Sidri runs NE-SW for about 80 km and receives a number of tributaries and feeders. During its course Wadi Sidri cuts across rocks of different origin: Eocene-Cretaceous limestone in the downstream part and Nubian sandstone and igneous and metamorphic rocks for most of its length. The wadi flows into the Gulf of Suez at about 150 km south of El-Shatt.

In the downstream part the surface deposits are deep sand and the bed is covered with boulders. The vegetation is dominated by *Hammada elegans*. Limited patches of *Tamarix aphylla* scrub occur on side terraces of the main channel and at the confluence of side tributaries.

The midstream part of the wadi cuts into Eocene-Cretaceous limestone in the west for about 20 km and then makes its course in igneous and metamorphic rocks with scattered patches of Cretaceous Nubian sandstone for another 20 km. The vegetation of this part of the main channel is dominated by *Hammada elegans* with abundant *Acacia raddiana* in the eastern part and *Artemisia judaica* on the west. *Capparis cartilaginea* grows regularly at different heights on the flanking slopes and cliffs of the wadi whereas *C. aegyptia* is sporadic.

The difference in the origin of the substrata of the tributaries and runnels of the eastern and western parts of the wadi affects the type of vegetation. The tributaries and runnels of the west, draining limestone formations, have very coarse deposits of whitish gravels and boulders. The large tributaries are dominated by *Hammada elegans* whereas the smaller affluents are dominated by *Artemisia judaica* or co-dominated by *A. judaica* and *H. elegans. Cleome droserifolia* and other calcicolous species such as *Fagonia mollis, Iphiona mucronata* and *Reaumuria hirtella* occur in these affluents. On the other hand, the tributaries of the eastern parts cut mainly across basement complex formations with occasional patches of Nubian sandstone. These tributaries are relatively narrow, bordered by high mountains. The surface deposits are sandy with dark rock fragments on the surface. The vegetation here is an open scrub of *Acacia raddiana* with *Hammada elegans* dominating the undergrowth.

The upstream tributaries of Wadi Sidri drain the northeastern fringes of the southern mountainous area formed of Nubian sandstone. The surface deposits are of deep loose reddish sand, mostly with no gravels or boulders. The dominant is *Hammada elegans. Lygos raetam* and *Panicum turgidum* are abundant and *Lycium shawii* is occasional. On the sides of the course is sporadic growth of *Acacia raddiana*, which is largely replaced by *Lygos raetam* in the Nubian sandstone tributaries.

### (iv) Wadi Feiran

Wadi Feiran is the longest and broadest wadi of southern Sinai. It rises from the high mountains surrounding the Monastery of St. Katherine at 2,500 m or so above sea level. It descends steeply to the north, then turns to the west until it terminates in the Suez Gulf about 165 km south of El-Shatt.

The downstream part of Wadi Feiran extends for about 20 km, covered by sediments of rock boulders and fragments in a sandy-clay matrix. *Hammada elegans* dominates in this habitat, growing in distantly spaced patches forming huge hummocks. In addition trees of *Acacia raddiana* are widely spaced on gullies and rocky slopes. Common associates include *Anthemis pseudocotula, Artemisia judaica, Cleome arabica, Diplotaxis acris, Fagonia arabica, Farsetia aegyptia, Lygos raetam, Mentha longifolia* ssp. *typhoides, Pituranthos tortuosus, Zilla spinosa* and *Zygophyllum simplex. Moricandia sinaica* is rare in this xeric habitat. About 22 km east of the wadi mouth, fine sandy-clay soil constituents increase. Here, *Aerva javanica* v. *bovei* and also v. *forsskalii* appear in addition to the above-mentioned common associates.

Feiran Oasis is about 43 km east of the mouth of Wadi Feiran and appears as a deep, fertile extension of the wadi surrounded by high red mountains crowded with trees (*Acacia raddiana, Phoenix dactylifera* and *Tamarix aphylla*). The oasis extends over a distance of 10 km. Abundant ground water and deep sandy-clay deposits (wadi terraces), as well as the natural protection of the locality against wind, favour the utilization of the oasis as a productive area, e.g. to cultivate fruit trees.

### (v) El-Qaa Plain

The El-Qaa Plain is a depression of about 1,125 $km^2$ on the southern section of the eastern coast of the Gulf of Suez desert. It lies between sea level and 200 m, sloping gently towards the southeast and stretches in a NW-SE direction for some 120 km, reaching an average width of 20 km. This plain is bounded on the east by the western outskirts of the rugged montane area of south Sinai and on the northwest by several isolated blocks (Ahmed, 1983). The surface of the plain is slightly undulating and covered with outwash deposits originating from the neighbouring highlands. It is dissected by shallow drainage lines (wadis) originating mostly from the eastern and western high lands, which flow towards the central channels, running more or less parallel to the coastal highways crossing the plain. These wadis are: Thegheda, Ratama, Hebran and El-Tor. Two of these are described below.

(1) Wadi Thegheda is a secondary channel in the northeast of the El-Qaa Plain. The flood channel of the wadi has an average width of about 60 m. The bed forms gradually elevated terraces running along fault planes striking in a NE–SW direction. The upstream part of the wadi does not extend much into the montane area. In the uppermost site, relicts of the old wadi-terraces are extensively eroded to considerable depths by the action of torrential floods, giving a network of pathways for the small volume of seeping water. Seepage and eventual evaporation of this brackish water create swampy and wet salt marsh habitats. The swampy habitat is dominated by *Phragmites australis* with abundant *Typha domingensis* whereas *Juncus subulatus* dominates the wet salt marsh. *Phoenix dactyifera* groves are also present.

The slightly elevated rocky parts of the wadi covered with boulders and stones are dominated by pure stands of *Zygophyllum coccineum*, with scattered plants growing well in cracks and concavities filled with transported sand and silt. In the shallow ill-defined water course, where rock fragments are covered with coarse sand and fine sediments, *Alhagi maurorum* dominates, associated with *Zygophyllum coccineum; A. maurorum* is an indicator of underground water. In stands where soil is saline, *Cressa cretica* replaces *Z. coccineum.*

*Hammada elegans* dominates in areas of Wadi Thegheda where the substratum is rocky, with surface deposits containing a substantial fraction of sandy calcareous materials intermixed with rocks of various sizes. Associate species include:

*Fagonia arabica, F. schimperi, Iphiona mucronata, Zilla spinosa* and *Zygophyllum coccineum.*

(2) Wadi El-Tor is one of the prominent features of the western coast of Sinai. It is marked particularly by a huge, steeply curved, meander with its apex facing east in its central region some 15 km northwest of El-Tor city. This part of Wadi El-Tor cuts across deep alluvial deposits that form wadi terraces, 2–5 m high, dissected by affluents draining the surrounding water-collecting areas. At some 4 km from the entrance to the central part, the flood course bends gradually southward and then to the southwest. The downstream part extends southwest to the sand beach, which represents the coastal fan of the wadi on the Suez Gulf.

In the flood channel (wadi bed) of the upstream part of Wadi El-Tor there is a community co-dominated by *Artemisia judaica* and *Zilla spinosa.* Associates are *Anthemis pseudocotula, Fagonia kahirina, F. mollis, Hammada elegans, Lygos raetam, Matthiola elliptica* and *Panicum turgidum.* In this part the wadi terraces, 40–60 cm high, are dominated by *Ephedra alata* associated with *Aerva javanica* v. *bovei* and v. *forsskalii, Iphiona mucronata, Ochradenus baccatus, Pituranthos tortuosus, Zygophyllum coccineum* and *Z. simplex* (Anonymous, 1982). In the rocky run-off habitat, where boulders and gravels cover the substratum, *Ephedra alata, Gymnocarpos decander* and *Zygophyllum simplex* are abundant.

In the downstream part of Wadi El-Tor communities dominated by *Zygophyllum coccineum* and by *Hammada elegans* occur. The common associates of the *Z. coccineum* community include *Fagonia glutinosa, Francoeuria crispa, Iphiona mucronata* and *Ochradenus baccatus* together with *Zilla spinosa* (abundant). The stands of *H. elegans* are either pure or *Z. coccineum* is the only associate. In the saline areas of this part of the wadi *Tamarix nilotica* dominates; associates include *Cressa cretica, Cynodon dactylon, Cyperus laevigatus, Juncus acutus* and *Zygophyllum album.* Groves of *Phoenix dactylifera* are also present.

### 2.6.3.2 Gulf of Aqaba Coast

The western coast of the Aqaba Gulf (the eastern coast of the Sinai Peninsula) extends for about 235 km from Aqaba southwards to Ras Mohammed where it meets the southern part of the eastern coast of the Suez Gulf. The eastern foothills of Sinai descend sharply towards the Gulf of Aqaba and the width of the coastal plain is greatly reduced. However, the southern part of the coastal plain is broader than the northern. Alluvial fans derived from magmatic and metamorphic rock cover most of this plain. In the southern section a large area near the beach is of fossil coral reef.

The vegetation of the western coast of the Gulf of Aqaba may be divided into: mangal, littoral salt marsh and coastal desert types. The mangal vegetation is represented by limited shore-line swamps dominated by *Avicennia marina* in the coastal area of Nabq (about 50 km north of Ras Mohammed). The coral reefs of the shoreline in the southern section of Aqaba Gulf as well as the warm temperature (mean annual temperature 26°C) enable the mangrove plants to dominate. North of Nabq no mangroves have been recorded. The absence of these plants from the northern section of the Aqaba Gulf western coast may be attributed to the following factors.

1. The mean temperature of the coldest month is below the necessary requirement for their successful growth. Mangroves, in general, are a tropical formation and a high temperature in the coastal areas is a prerequisite for their presence (Chapman, 1975). These plants require a mean temperature of not less than 15ºC in the coldest month of the year. In the southern section of the Aqaba Gulf the mean temperature of the coolest month is 18.2ºC, within the range of the mangrove tolerance. In the northern section, however, temperature appears to be lower.
2. The steep cliffs of the coastal hills in the northern section prevent the development of a suitable shore-line for the growth of mangrove.

The littoral salt marsh vegetation of the Gulf of Aqaba western coast is in zones roughly parallel to the coast. The first zone, close to the shore-line, dominated by *Limonium axillare*, is subject to periodic flooding with sea water. In the Nabq area a few shrubs of *Avicennia* may grow as associates in the *L. axillare* community. The second landward zone is co-dominated by *Nitraria retusa* and *Zygophyllum album*.The soil is highly saline and the water-table shallow. The third zone is dominated by the xerophytic shrub *Salvadora persica* which grows in a prostrate form and builds sand mounds.

The coastal desert and wadis of the western coast of the Aqaba Gulf bear xerophytic vegetation of the following species: *Abutilon fruticosum, Aerva javanica, Artemisia judaica, Blepharis edulis, Capparis cartilaginea, Cleome chrysantha, Crotalaria aegyptiaca, Cymbopogon schoenanthus, Cyperus jeminicus, Eremopogon foveolatus, Gymnocarpos decander, Hammada elegans, Heliotropium arbainense, Hibiscus micranthus, Lasiurus hirsutus, Launaea spinosa, Lindenbergia sinaica, Otostegia fruticosa* ssp. *schimperi, Panicum turgidum, Pterogaillonia calycoptera, Salsola cyclophylla, S. schweinfurthii, Seidlitzia rosmarinus, Solenostemma oleifolium, Taverniera aegyptiaca* and *Zilla spinosa*. In the large wadis there are xerophytic trees and shrubs, e.g. *Acacia raddiana, A. tortilis, Calotropis procera, Capparis decidua, Leptadenia pyrotechnica, Moringa peregrina, Salvadora persica, Tamarix aphylla* and *T. nilotica*. Groves of *Hyphaene thebaica* are also present (Danin, 1983).

*Acacia raddiana* usually dominates in the alluvial fans at the foot of the coastal hills. In these alluvial fans and after rain storms a dense vegetation of *Pulicaria desertorum* appears and persists as long as there is enough water. In other alluvial fans much *Schouwia thebaica* develops also after strong storms and lives even longer than *P. desertorum*.

The fossil coral reefs of the Aqaba Gulf western coast support a vegetation dominated by *Pulicaria desertorum* and *Schouwia thebaica*. This habitat receives sufficient run-off to support long-living shrubs and trees. In the coral reef wadis *Capparis decidua* and *Leptadenia pyrotechnica* are dominants (Danin, 1983).

#### 2.6.3.3 The Montane Country

##### (a) Physiography

The montane country of Sinai is of triangular shape, the apex of the triangle being near the cape of the peninsula (Ras Mohammed). It is encompassed by the Gulf of Suez on the west and the Gulf of Aqaba in the east. In the north this montane country is bounded by the calcareous plateau of El-Tih which slopes down northward into a wide coastal plain with sand dunes.

The ranges of mountains that form the montane country of Sinai are in the southern and central subregions of the peninsula and comprise Gebel St. Katherine (the highest peak in Egypt, 2,641 m) as well as Gebels Musa, El-Tih, Halal, El-Ugma, Yalaq, El-Maghara and others. On the eastern side of the peninsula, the mountains are so close to the Gulf of Aqaba that there is almost no coastal land. On the other hand, the mountains on the western side of Sinai are relatively far from the Suez Gulf and there is a wide coastal belt along the whole stretch.

The montane country of Sinai is dissected by narrow wadis with deep slopes and is characterized by the presence of springs around which there are oases and human settlements.

Rain water falling on the mountains runs over the slopes and into the narrow deep wadis where it forms perpetual streams or pools. Some of this water percolates into the substratum and is stored in rock crevices. It can be obtained by digging wells or it may appear at the surface as springs or streams of fresh water. Snow is another source of water in the Sinai montane country as it covers the summits of mountains higher than 1,000 m during winter. When it melts with the advance of warm weather, water runs down the mountain slopes and into the wadis. Because of its altitude the total amount of rainfall of this montane country is about 60 mm/year, mostly orographic rain. If snowfall is added to rainfall, the water supply in this montane area is enormous in comparison with that of other desert regions of Egypt.

The climate of the Sinai montane country is determined primarily by the altitude, the effect of which masks that of latitude (Zahran and Willis, 2008). There is a wide difference in temperature between summer and winter. August is the hottest month (mean temperature 24.5°C) and January the coldest (mean temperature 8.7°C). Throughout the winter, the mean monthly temperature is below 10°C. Absolute minima of less than 0°C (–4°C during the winter of 1987 and –6°C during the winter of 1966) are of frequent occurrence between November and March at the highest altitudes. Similarly summer temperature is relatively lower than in any of the inner and coastal deserts of Egypt. The diurnal range of temperature is very wide, varying from 16°C in winter to 20°C in summer. With regard to relative humidity, this montane country is the driest part of Egypt in all seasons, as the relative humidity is less than 40% (Anonymous, 1982).

##### (b) Vegetation

The flora and vegetation of the montane country of Sinai proper have been studied by many workers, e.g. Täckholm (1932, 1956, 1974), Zohary (1935, 1944).

Migahid et al. (1959), Boulos (1960), Ahmed (1983), Danin (1981, 1983), Danin et al. (1985), El-Gazzar et al. (1995), El-Demerdash et al. (1996), El-Bana (2003) and Zahran and Willis (2008).

A large number of plants grow in the different habitats of the montane country of Sinai, most of which are chasmophytes. In areas of high water resources there are oases and cultivated gardens.

The characteristic habitats of this montane country include small, narrow blind rocky wadis, upstream parts of large wadis (that originate within the hills and run eastwards to flow into the Gulf of Aqaba, westwards to flow into the Gulf of Suez and northwards to flow into the Mediterranean Sea), gullies, terraces, rock crevices and slopes of mountains at different levels. A rocky substratum is the general feature of these habitats. Sediments are coarse with or without fine particles.

The rock habitat is unfavorable to the growth of plants because of the high resistance to root penetration, a thin depth of soil and deficient water content. For these reasons only certain plants, chasmophytes, can tolerate the adverse conditions of the habitat. Some of the rock plants of Sinai are firmly attached to the smooth surface of the rock by means of hook-like roots, e.g. *Galium sinaicum* and *Origanum syriacum.* Other plants grow in rock crevices, which, though narrow, are sometimes very deep. Soil and plant litter accumulate in these crevices, retaining water and forming a fertile substratum through which plant roots penetrate. Deep crevices support several species of shrubs and trees, e.g. *Capparis cartilaginea, Cupressus sempervirens, Ephedra alata, Ficus pseudosycomorus* and *Moringa peregrina* (Migahid et al., 1959). Rock plants may also be found in surface notches and depressions in which soil and decaying matter are retained. Another rocky medium is terraces and flat plateaux on the surface of which a small depth of soil is deposited.

The mountainous district of Sinai is the coolest owing to its high elevation. The flora is diverse and includes Irano-Turanian, Mediterranean and Sudanian species that are isolated from their main areas of distribution. The nearest regions for several of the isolated species in these mountains are in Iran or in Mount Hermon (Anti-Lebanon). *Crataegus sinaica* and *Scrophularia libanotica* occur both in Sinai and on Mount Hermon (Danin, 1983). *Primula boveana* is a rare endemic which has been isolated in Sinai since the Tertiary. Its nearest relatives are in eastern Africa, Yemen and the Zagros mountains of Iran. There are several other endemics, most of which are restricted to smooth-faced rock outcrops.

The flora of the Sinai mountains is dominated by Irano-Turanian species and the most common plant is *Artemisia inculta*. It is accompanied by *Gymnocarpos decander* in fissured rocks at lower elevations and by *Zilla spinosa* and *Fagonia mollis* in stony alluvium. *Anabasis setifera, Atraphaxis spinosa* v. *sinaica* and *Halogeton alopecuroides* are associates in soil derived from dark volcanic rocks. *Stachys aegyptiaca* and *Pyrethrum santolinoides* accompany *Artemisia* at the foot of smooth-faced rock outcrops at low elevations and on stony slopes at higher sites.

Rock vegetation of these mountains is rich in semishrubs, shrubs and trees (Danin, 1983). Characteristic species are *Cotoneaster orbicularis, Crataegus sinaica, Ficus pseudosycomorus, Pistacia khinjuk, Rhamnus disperma, Rhus tripartita* and *Sageretia brandrethiana.* The common annuals include *Boissiera squarrosa, Eremopoa persica, Gypsophila viscosa, Lappula sinaica* and *Paracaryum intermedium.*

Wadi El-Raha is a short, broad wadi ending blindly in a granitic mass. It is very close to St Katherine Monastery.

The rocky slopes of the mouth of Wadi El-Raha support two communities, one dominated by *Alkanna orientalis* and one by *Varthemia montana.* Associate species of the first community are *Achillea fragrantissima, Stachys aegyptiaca, Varthemia montana* and *Zilla spinosa.* In the stands of the *V. montana* community, associates are *Achillea santolina, Alkanna orientalis, Cynodon dactylon, Lavandula stricta, Stachys aegyptiaca* and *Stipa capensis.* The species grow at heights of 4–10 m from the bed level of the wadi. They are rooted in crevices of the granite rock and are widely spaced, *Zilla spinosa* tends to decrease progressively with height up the slopes whereas *Varthemia* and *Stachys* increase. In the wadi bed *Z. spinosa* dominates, with cover of 20–30%. Associates include *Achillea fragrantissima, Artemisia inculta, Diplotaxis harra, Fagonia mollis, Francoeuria crispa, Gomphocarpus sinaicus, Peganum harmala* and *Reseda pruinosa.*

A gully at a higher level on the northern side of Wadi El-Raha (about 4–5 m broad) is typical of other gullies of the mountains. The surface of the gully bed is covered with boulders to which *Galium sinaicum* sticks firmly. On the flat soil of the gully bed, between the boulders, are *Ephedra alata, Fagonia mollis, Gomphocarpus sinaicus, Lavandula stricta, Teucrium polium* and *Zilla spinosa.* The following grow in fissures of the boulders: *Alkanna orientalis, Artemisia inculta, A. judaica, Ballota undulata, Capparis spinosa, Fagonia mollis, Parietaria alsinifolia, Scirpus holoschoenus, Stachys aegyptiaca, Teucrium polium* and *Varthemia montana.*

The rocky run-off slopes of the El-Raha plain support vegetation, at different levels relative to ground water availability, comprising the following: *Ajuga iva, Centaurea aegyptiaca, Delphinium* sp., *Echinops spinosissimus, Farsetia aegyptia, Gymnocarpos decander, Hyoscyamus muticus, Lotus* sp. and *Stachys aegytiaca* (Ahmed, 1983).

At the head of Wadi El-Raha is an oasis supporting a number of species of wild and cultivated fruit trees. In the middle of this oasis is a well containing fresh water 3 m below the soil surface. During rain periods, the water surface rises to within only 0.5 m of the surface. The cultivated fruit trees and shrubs include, besides the date palm, carob, pomegranate, peach, almond and apricot.

Wadi El-Arbaeen is another narrow steep wadi, the mouth of which lies opposite that of Wadi El-Raha. On its bed are scattered boulders and large stones. There are successive broad terraces reducing to a deep narrow channel flooded by spring water. Round these springs are hygrophytic shade plants as well as aquatic and salt marsh species, e.g. *Adiantum capillus-veneris, Equisetum ramosissimum, Mentha longifolia* ssp. *typhoides* and *Origanum syriacum. Ficus pseudosycomorus* is rooted in crevices near this vegetation. Where the water supply is abundant it allows the development of oases with cultivated gardens where palm trees, pomegranats, almonds, plums, grapes, apples, pears, peaches and *Cupressus sempervirens* are cultivated.

The herbs *Solanum nigrum* and *Verbascum schimperianum*[7] grow abundantly in these oases.

The bed of Wadi El-Arbaeen has a rich flora which includes *Ammi majus, Anchusa aegyptiaca, Brachypodium distachyum, Carduus arabicus, Euphorbia peplus, Lactuca orientalis, Onopordum ambiguum, Plantago ciliata, Pulicaria arabica, Sisymbrium irio* and *Sonchus oleraceus*. Also *Asperugo procumbens, Hypericum sinaicum* and *Verbascum schimperianum* are recorded in the southern mountains.

In the mouth of Wadi El-Arbaeen the vegetation is thin (cover about 10%), dominated by *Peganum harmala* associated with *Zilla spinosa* (abundant), *Achillea fragrantissima* and *Stachys aegyptiaca* (common). *Alkanna orientalis, Artemisia judaica, Ballota undulata, Origanum syriacum* v. *aegyptiacum, Phlomis aurea* and *Teucrium polium* are rare.

St. Katherine Monastery lies at the bottom of a narrow flat depression surrounded by steep high mountains on all sides except the west leading to the prophet Aaron's tomb at the meeting point of Wadi El-Raha and Wadi El-Arbaeen. In the mountains surrounding the monastery the flora includes e.g. *Achillea fragrantissima, Alkanna orientalis, Andrachne aspera, Echinops glaberrimus, Fagonia arabica, Ficus carica* v. *rupestris* (semi-wild), *F. pseudosycomorus, Gomphocarpus sinaicus, Heliotropium arbainense, Hyparrhenia hirta, Iphiona mucronata, Launaea spinosa, Orobanche muteli* v. *sinaica, Peganum harmala, Pituranthos tortuosus, Scrophularia libanotica, Solanum nigrum* and *Varthemia montana*. These plants occur at different levels on the mountain slopes, the more xerophytic species tending to be more abundant in the lower zones.

Gebel Musa is located southeast of St. Katherine mountain. The northern, windward, slope of this mountain is richer in vegetation than the southern slope of the opposite mountain on the other side of the monastery. On the northern slope there are many shrubs on the rocks, e.g. *Cupressus sempervirens, Ephedra alata, Ficus carica* v. *rupestris* and *F. pseudosycomorus*. The following species are also common: *Artemisia inculta, Astragalus fresenii, Atraphaxis spinosa* v. *sinaica, Bromus tectorum, Callipeltis aperta, Crataegus sinaica, Echinops glaberrimus, Isatis microcarpa, Lactuca orientalis, Nepeta septemctenata, Origanum syriacum, Oryzopsis miliacea, Phagnalon sinaicum, Phlomis aurea, Pituranthos triradiatus, Plantago ciliata, Pyrethrum santolinoides (Tanacetum sinaicum) Scandix stellata* and *Silene leucophylla* (Danin et al., 1985).

About midway to the summit of Gebel Musa is a broad flat area named Farsh El-Gebel, in which springs and fresh-water streams are present. At one side is a runnel sloping down steeply towards the Farsh El-Gebel and having a thick layer of silt at the surface superimposed on the rocky substratum. On the alluvial soil is a community of *Thymus decussatus* (endemic) and *Artemisia inculta*. The plant cover is high (70%) in the densest upper part, but decreasing to 50% towards the foot of the slope and to 40% in the lowermost flat part at Farsh El-Gebel. The flora also

[7] *V. schimperianum* is endemic to the montane country of Sinai (Täckholm, 1974).

includes: *Phlomis aurea* (abundant), *Pyrethrum santolinoides* (common), *Scirpus holoschoenus, Stipa parviflora, Teucrium polium* and *Varthemia montana* (rare).

In the flat part of Farsh El-Gebel the vegetation is thinner than in the runnels, cover not exceeding 40%. This may be related to the more deficient water supply, rain water being evenly distributed over the whole area but accumulating by run-off in the runnels. Here *Aristida coerulescens* v. *arabica* dominates, *Artemisia inculta* is abundant whereas *Phlomis aurea* and *Pyrethrum santolinoides* are common.

Around the fresh-water spring of Farsh El-Gebel is a dense vegetation dominated by *Scirpus holoschoenus*. Associates include *Stipa capensis* (abundant), *Anagallis arvensis, Bromus rubens, Galium sinaicum, Juncus bufonius, Phlomis aurea, Polypogon monspeliensis* and *Veronica anagallis-aquatica* (common). Algae form a green scum alongside the spring.

The rocky slopes of the mountains on the two sides of the upstream section of Wadi Feiran are characterized by a number of trees and shrubs, e.g. *Acacia* spp., *Ficus pseudosycomorus, Moringa peregrina, Tamarix nilotica, Capparis cartilaginea* and *Ephedra alata*. Except for *Ephedra* and *Ficus*, these species are absent in the cooler, less arid, granite mountains of southern Sinai. They grow abundantly at lower levels of the slopes but their cover decreases on higher ground.

Gebel Ugma is one of the mountains of the central subregion of Sinai. Its vegetation on the slopes changes with elevation. In the wadis at the foot of the mountain, *Artemisia inculta* is dominant on the gravelly terraces where the chalky material is leached. *Hammada elegans* is abundant on the sandy terraces. In the chalky substratum of the small wadis, species such as *Salsola delileana, S. tetrandra, Halogeton alopecuroides, Krascheninnikovia ceratoides* and *Reaumuria hirtella* occur. The large wadis are vegetated by *Achillea fragrantissima, Atriplex halimus, Lygos raetam* and *Zilla spinosa*.

Slopes up to 600 m are bare except for *Salsola tetrandra*, most plants being dead. At 600–1,500 m the number of shrubs of *S. tetrandra* increases, perhaps because microhabitats at higher elevations have improved moisture regimes (Danin, 1983). In wet years *Atriplex leucoclada* is abundant, particularly on the lower slopes. *Artemisia inculta* and *Halogeton alopecuroides* are common at high elevations of north- and south-facing slopes respectively. The highest belt of vegetation on Gebel Ygma includes the *Chenolea arabica-Atriplex glauca* (both species are not recorded by Täckholm, 1974) community on hard chalk (Danin, 1983; Danin et al., 1985). In wet years this belt is covered with annuals – *Anthemis melampodina* and *Leontice leontopetalum*. Scanty shrubs of *Tamarix* sp. grow in this high belt.

Gebel El-Tih is in the central part of the El-Tih plateau. It has slightly inclined strata. Small outcrops of smooth-faced limestone occur in the flanks at high elevations. Many springs, including Ain Sudr, Moyet El-Gulat, Ain Shallal and Ain Abu Ntegina, occur in canyons draining the mountain. Annual rainfall is 50–100 mm and mean annual temperature 16–20ºC.

The smaller wadis in the area of the El-Tih plateau mountain are dominated by *Anabasis articulata, Artemisia inculta, Hammada scoparia, Gymnocarpos decander, Salsola tetrandra* and *Zygophyllum dumosum*. However, the larger wadis are co-dominated by *Lygos raetam* and *Achillea ftragrantissima*.

The inclination of the strata influences weathering patterns, water regime and vegetation. At higher altitudes, the *Z. dumosum* community is replaced by *A. inculta, H. scoparia* is dominant on marl outcrops having a salt regime.

The steep escarpment of Gebel El-Tih supports a pioneer community dominated by *Anabasis setifera* and *Halogeton alopecuroides.*

The vegetaton of Gebel El-Tih varies with the type of rock. *Artemisia inculta* or *Zygophyllum dumosum* dominates those slopes consisting of hard rock containing little marl. Small horizontally bedded hard strata are dominated by *Hammada scoparia.* Softer rocks support *Z. dumosum* along with *Salsola cyclophylla* and other xerohalophytes (Danin, 1983; Danin et al., 1985). This habitat also supports a *Halogeton alopecuroides-Salsola schweinfurthii* community on the slopes of the mountain and *Hammada scoparia* in runnels at the summit of the plateau. Outcrops of smooth-faced limestone are restricted to the dip-slopes of inclined rock strata. *Halogeton poore*[8] grows in this habitat.

The other species of Gebel El-Tih area include *Pistacia atlantica*[9] (on which grows the woody *Loranthus acaciae*, a very common parasite of *Acacia* in the Egyptian Eastern Desert), *Anabasis articulata, A. setifera, Gymnocarpos decander, Noaea mucronata, Reaumuria hirtella, R. negevensis* and *Varthemia iphionoides.*[10]

The district of Gebel Halal (892 m) and Gebel El-Maghara (750 m) includes several folds of Cenomanian-Turanian age with limestone, chalk, dolomite and marl outcrops. Extensive erosion in Gebel El-Maghara has exposed a sequence of 2,000 m of Jurassic limestone, shales and sandstone. Large outcrops of smooth-faced limestone and dolomite occur at Gebel Halal. The wadis are filled with sand-covered alluvium.

Rainfall in this district is 50–100 mm/year distributed during January-February; the rest of the year is almost rainless. Mean annual temperature is in the range 16–20ºC; the highest in June–July and the lowest in January–February. Average temperature rarely exceeds 30ºC and rarely goes below 10ºC. Extremes of up to 40ºC, however, are recorded (Boulos, 1960; Danin, 1983).

The large wadis of the northern limestone district of the El-Tih plateau where Gebels Halal and El-Maghara are situated are dominated by *Acacia raddiana, A. gerrardii* ssp. *negevensis, Tamarix aphylla* and *T. nilotica.*

Gebel Halal is characterized by an erosion crater. The old strata at the bottom of the crater are covered with alluvium derived from the weathering of adjacent ridges. Areas of the crater floor, used for Bedouin encampments, are dominated by *Anabasis syriaca,*[11] a xerophyte that can tolerate high concentrations of nitrogen (Danin, 1983). The slopes inside the crater are covered with stony alluvium which, with the favourable water regime, supports sparse, semi-shrub vegetation. The southern slopes of some alluvial hills bear branches of *Caralluma sinaica.*

[8]Not recorded by Täckholm (1974) nor by Boulos (1995).

[9]Not recorded by Täckholm (1974) but recorded by Boulos (1995).

[10]Not recorded by Täckholm (1974) but recorded by Boulos (1995).

[11] Not recorded by Täckholm (1974) but recorded by Boulos (1995).

Most of the semi-shrubs on the slope are xerophytes, e.g. *Anabasis setifera, Atriplex leucoclada, Halogeton alopecuroides, Reaumuria hirtella* and *Suaeda palaestina.*[12]

In the northwestern flanks of Gebel Halal bedded limestone, chalk, marl and limestone and hard dolomite are exposed. Wadis cutting through these flanks produce slopes facing in various directions. Limestone on south-facing slopes and marl chalk on all slopes support mixed or monospecific communities dominated by *Atriplex glauca, Halogeton alopecuroides, Reaumuria hirtella, R. nevegensis, Salsola schweinfurthii, S. tetrandra* and mostly supports communities dominated by *Artemisia inculta* and *Noaea mucronata.* In the spring of rainy years the vegetation is accompanied by the geophytes *Anemone coronaria, Ranunculus asiaticus* and *Tulipa polychroma* and many annuals.

The most interesting species of Gebel Halal is *Juniperus phoenicea*, which grows in the crevices of the smooth-faced outcrops of the hard limestone and dolomites of the northwest slopes as well as in the wadis. Some *Juniperus* trees may reach 10–12 m and individuals of 8–4 m are common. According to Täckholm (1956, 1974), Boulos (1960, 1995), El-Hadidi (1969) and Danin (1983), *J. phoenicea* is absent from all regions of Egypt except this area of Sinai. It occurs throughout the Mediterranean coastal region except for Libya and Egypt. Hundreds of trees of *J. phoenicea* are present in Gebel El-Maghara, dozens are in Gebel Yalaq and thousands are in Gebel Halal. "Vines" of *Ephedra aphylla* cover many trees of *J. phoenicea.* Another species accompanying *J. phoenicea* in rock habitats is *Origanum isthmicum* which is, according to Danin (1969) and Boulos (1995), endemic to Gebel Halal of Sinai. Danin (1969) reported that the entire world population of 1,000–2,000 individuals of *O. isthmicum* occurs within an area approximately 5×2 km on the northwest flanks of Gebel Halal. Other notable associates are *Astoma seselifolium, Ephedra campylopoda, Rubia tenuifolia* and *Sternbergia clusiana* (not recorded by Täckholm, 1974). All of these plants are absent from other regions of Egypt.

The lowest parts of the northwest flanks of Gebel Halal bear a community dominated by *Zygophyllum dumosum.* The wadis at the foot-hills of Gebel Halal support the growth of *Acacia gerrardii* and *A. raddiana.*

Gebel El-Maghara (750 m), some 110 km southwest of El-Arish, consists of Jurassic rocks surrounded by marine Lower Cretaceous exposures which form a conspicuous topographic low separating it from the outer slopes which are occupied by on Upper Cretaceous formation (Ball, 1916). The vegetation of Gebel El-Maghara has been studied by Zohary (1935, 1944), Boulos (1960), and others, the following are the common species of the rocky and sandy habitats.

### (i) Common plants of rocky habitat

*Achillea fragrantissima, Allium artemisietorum, Anabasis setifera, Anastatica hierochuntica, Anthemis melampodina, Asparagus stipularis, Ballota undulata,*

[12]Not recorded by Täckholm (1974) but recorded by Boulos (1995).

*Callipeltis cucullaria, Caralluma sinaica, Centaurea eryngioides, Cocculus pendulus, Colutea haleppica, Cornulaca monacantha, Ephedra alata, Eryngium glomeratum, Euphorbia erinacea, Globularia arabica, Gomphocarpus sinaicus, Gymnocarpos decander, Helianthemum lippii, H. ventosum, Juniperus phoenicea* (at high altitudes), *Linaria floribunda, Lycium europaeum, Matthiola livida, Micromeria sinaica, Muscari racemosum, Noaea mucronata, Notholaena vellea, Ochradenus baccatus, Oryzopsis miliacea, Parietaria alsinifolia, Paronychia sinaica, Pennisetum ellatum, Rorippa integrifolia* (endemic, Täckholm, 1974), *Salsola tetrandra, Salvia aegyptiaca, Scrophularia xanthoglossa, Silene setacea, Stachys aegyptiaca, Stipagrostis ciliata, Tetrapogon villosus, Tricholaena teneriffae, Urginea maritima, Varthemia montana* and *Zosima absinthifolia.*

### (ii) Common plants on sandy habitat

*Adonis cupaniana, Aizoon canariense, Anabasis articulata, Anchusa milleri, Andrachne telepioides, Arnebia decumbens, Astragalus sinaicus, Atriplex leucoclada, Avena alba, Bromus fasciculatus, Calendula micrantha, Carduus getulus, Carthamus glaucus, Citrullus colocynthis, Cleome arabica, Convolvulus elarishensis, C. oleifolius, Cucumis prophetarum, Diplotaxis acris, Echiochilon fruticosum, Erodium laciniatum, Erucaria uncata, Fagonia mollis, Farsetia aegyptia, Francoeuria crispa, Frankenia pulverulenta* (on moist sandy soil), *Hammada elegans, Heliotropium undulatum, Herniaria hemistemon, Hippocrepis unisiliquosa, Ifloga spicata, Iris sisyrinchium, Koniga arabica, Lappula spinocarpos, Launaea angustifolia, Linaria tenuis, Lotus glinoides, Lygos raetam, Malva parvifora, Moltkiopsis ciliata, Neurada procumbens, Ononis reclinata, Panicum turgidum, Papaver hybridum, Pergularia tomentosa, Phagnalon barbeyanum, Polycarpon succulentum, Pteranthus dichotomus, Pterocephalus papposus, Pulicaria undulata, Reseda decursiva, Savignya parviflora, Schismus barbatus, Senecio desfontainei, Silene villosa, Sonchus oleraceus, Telephium sphaerospermum, Teucrium polium, Thymelaea hirsuta* and *Urospermum picroides. Cuscuta brevistyla* is a common parasite in Wadi El-Maghara on a variety of species, e.g. *Artemisia inculta, Centaurea sinaica, Gymnocarpos decander, Helianthemum kahiricum, Lycium shawii, Salvia aegyptiaca* and *Stipa capensis. Sedum viguieri*[13] grows in dense tufts at high altitudes and on slopes, in protected moistened areas. *Juncus rigidus* and *Lamarckia aurea* are common in salt ground near the wells of Wadi El-Maghara. *Orobanche cernua* and *O. ramosa* are common root parasites on a variety of plants.

[13]Täckholm (1974) and Boulos (1995) recorded *Sedum sempervivum* (= *Rosularia lineata*).

# Chapter 3
# Climate–Vegetation Relationships: Perspectives

## 3.1 Introduction

Climatic changes not only produce latitudinal shifts in different ecological belts, equally significant is the diverse pattern of minute but extremely important changes observed in the mountains of the arid land regions of the world. These areas of greater elevation enjoy a higher local rainfall even today and during periods of wetter climate, rainfall values increase appreciably creating large islands of, relatively, rich vegetation and attractive landscapes (Butzer and Twidale, 1966). The preceding pages of Chapters 1 and 2 of this book confirm this statements. The variation in climatic conditions prevailing on the Afro-Asian Mediterranean-Red Sea coastal lands recognizably affect their plant life, the species of which exhibit special adaptation in structure, ecology, and associations (Kassas, 1996). Here are some concluding remarks on the relationships between climate and plant life of these coastal lands, discussing it under the two themes of climatic features and vegetation forms.

## 3.2 Climatic Features

Climate is the master of all environmental factors, it controls not only the growth and development of the plants but also their geographical distribution. This is obvious in that the major vegetation forms of the earth are related to the different climatic types. However, other environmental factors (land-forms, habitat types, soil characteristics, etc.) are also locally effective. The climatic features associated with the Afro-Asian Mediterranean and Red Sea coastal lands are described below.

### *3.2.1 Afro-Asian Mediterranean Coastal Lands*

Branigan and Jarrett (1975) state that the lands around the Mediterranean Sea, including the North African and SW Asian coastal lands, represent most clearly defined climatic units in the world. In summer, following the northward march of the sun, the Mediterranean region lies in the belt of the north-east trade winds and

M.A. Zahran, *Climate–Vegetation*, Plant and Vegetation 4,
DOI 10.1007/978-90-481-8595-5_3, © Springer Science+Business Media B.V. 2010

of high pressure and it is, therefore, dry. In winter, with the retreat of the sun south of the equator, the region falls under the general influence of the westerlies and the depression associated with them, and is wet. Many of the depressions travel the whole length of the Mediterranean from west to east, and most of the winter rain typical of the region owes its origin to them. It may be said, in general, that the Mediterranean region has a temperate[1] and changeable climate for the winter half of the year and a more uniform climate of the "hot dry" type during the summer.

The main characteristics of the Mediterranean climate may be summarized under four headings: winter rain, summer drought, mild winter temperature (mean $\geq$ 6.1°C) and hot summer (mean $\geq$ 21.1°C). It is also noteworthy that long periods of sun-shine and cloudless skies are experienced at all seasons.

Ayyad and Ghabbour (1993), following Meig's climatic system, stated that the climate of the North African Med. coastal lands lies in the area of "warm coastal desert": summer's warmest month has a mean temperature less than 30°C and winter coldest month has a mean temperature above 10°C. Though occasional short rain-storms occur in winter, most of the days are sunny and mild. Winds are generally gentle, but violent sand-storms and dust pillars are not rare. Dry hot dust laden winds from the south known as *khamasin*[2] in Egypt, *sirocco* in Libya and Tunisia blow occasionally for about 50 days during spring and early summer. These winds develop as a result of the passing of lower pressure over the Mediterranean.

The mean annual rainfall in the North African Med. coastal lands varies from about 140 mm in the arid climate of Egypt and Libya to almost 600 mm at the transition between upper semiarid and subhumid climates in northern Tunisia and the highlands of Cyrenaica. The seasonal distribution of rainfall can be distinguished into two patterns: one in which most of the rain (60% or more) occurs during winter (November–February) and the summer is virtually dry and one in which rainfall is more or less evenly distributed between autumn, winter and spring with some rainy days during summer. The first pattern of rainfall distribution characterizes the arid and semiarid coastal ecosystems of Egypt and Libya while the second pattern occurs in Tunisia.

In the SW Asian Med coastal lands, represented by Palestine-Israel-Syria section, the Mediterranean climate is marked by a mild, rainy winter and prolonged dry and hot summer. Geographical latitude and altitude, the blocking effect of the mountain ranges and distance from the sea are among the factors which modify the climate (Zohary, 1962). The effect of latitude manifests itself in the abrupt north to south decrease in annual rainfall so that within an average of 4 latitudinal degrees (33–38°N) the rainfall drops from about 1,000 to 25 mm/year. Temperature, increases from north to south, with the mean annual temperature rising from below 16°C in the north to approximately 23°C in the extreme south of the Palestine-Israel coast. In a west to east direction, from the sea shore landwards annual rainfall and mean

[1]Temperate–having climate intermediate between tropical and polar; moderate or mild in temperature.

[2]Khamasin is the Arabic for fifty.

temperature undergo similar but less regular changes, this is because of the interference of the Israeli and Jordanian mountain ranges. As a result of their interception of rains, part of the Jordan valley is turned into a rain-shadow desert. The tempering influence of the Mediterranean Sea is greatly limited by the mountain ranges, leaving most of the Palestine-Israel area open to a widen range of seasonal and diurnal temperatures. Climatically Zohary (1962) classified the SW Asian Med. coastal land, represented by the Palestine-Israel area, into three zones:

1. an arid zone with 200–25 mm/year rainfall,
2. a semi-arid zone where annual rainfall ranges, between 200 and 400 mm,
3. a sub-humid zone with an annual rainfall of 400–1,000 mm.

### *3.2.2 Afro-Asian Red Sea Coastal Lands*

The Red Sea Basin occupies an area of about 810,000 $km^2$ including essentially the Red Sea coastal plain (210,000 km) and the highlands (600,000 $km^2$) bordering them on both sides in Africa and Asia. The entire area of the Red Sea and its coastal lands are generally hot deserts with high temperature and low precipitation. Two climatic types are recognized: the monsoon[3] type in the south with summer rainfall and the Mediterranean type in the north with winter rainfall. Rainfall regions are almost comparable on both African and Asian sides having a bimodal distribution pattern with spring and autumn peaks that mimics the Mediterranean pattern. The annual precipitation is variable, from virtually less than 5 mm in the tropical coastal lowlands to more than 200 mm in some locations where the shores of the Red Sea are near the mountain ranges. On the highlands, the annual precipitation varies from 300 mm or slightly less to over 1,000 mm. Such variation of rainfall has its effects on the diversity, density and distribution pattern of the natural vegetation (Le Houerou, 2001). Temperatures in the low lands are among the highest on the planet reaching an annual mean of about 30°C, with maxima exceeding 60°C. The potential evapotranspiration is also among the highest on the earth, between 1,800 and 3,000 mm/year. The decrease in mean annual temperature and potential evapotranspiration with altitude is about 0.55°C and 55 mm/100 m, respectively. Rainfall increases by 10% for each increase of 100 m, i.e. doubling for each 1,000 m up to 2,000–3,000 m, but then decreasing. In this context, Dallman (1998) stated that the elevation is important in relation to temperature and rainfall. In the montane countries, temperature decreases by roughly 5°C/1,000 m. Rainfall is typically

[3]Monsoon = A seasonal wind of south Asia that blows from the SW in summer bringing heavy rains and from the north east in winter. The rainy season where the SW monsoon blows is April–October.

greater with increase in elevation. The flora and vegetation are likewise similar and occasionally identical for comparable degrees of elevation, temperature and aridity. The two areas, north-east and south-west of the Red Sea as a whole including their highlands, share 60% of the total number of the genera and some 30% of their approximately 5,000 species (Le Houerou, 2001).

Over the whole Red Sea Basin, winter may extends from mid-October to mid-April with summer occupying the rest of the year (Anonymous, 1993). The January mean daily temperature varies from about 20°C in the far north to about 29°C in the far south. The corresponding July figures are 35 and 40°C, respectively. Rainfall over the Red Sea coastal lands as a whole is sparse, and sporadic and often very localized (Zahran and Al-Kaf, 1996). A particular location may receive no rain for years, and then can experiences a brief heavy rainfall which may not be repeated for a similar length of time. For example in Hodayda in Yemen (Asian Red Sea coast), the total annual rainfall in 1989 was 207.3 mm occurring in January (31.4 mm), February (30.5 mm), March (56.4 mm) and April (89.0 mm). In 1988, the total annual rainfall was 127 mm occurring in April (39 mm), August (61.4 mm) and September (27.5 mm). These 2 years were exceptionally wet. The mean annual rainfall on the Hodayda coast usually ranges between 35 and 60 mm, mainly between October and April.

The African Red Sea coastal lands in the Sudan receives the main part of rainfall during the winter season: October–January, i.e. Mediterranean affinity (Kassas, 1957). As the prevalent continental northerlies pass over the warm water of the Red Sea, they absorb a considerable amount of moisture, causing convectional rainfall. This is especially affected by winter cyclonic activity over the Red Sea (Hefny, 1953). Though the coast is shut off from summer monsoon influence by mountain barriers, local penetration due to low topography can bring summer cloudbursts in July–August. The convectional rain of winter decreases landward (westward). For example, Sallom (19° 22′N, 37° 06′E, Alt. 170 m), which lies on the coastal plain at a distance of about 22 km from the shore, receives less winter rain (56 mm/year) than Suakin (19° 07′N, 37° 20′E, Alt. 5 m) (122 mm/year) to its south east and Port Sudan (19° 37′N, 37° 13′, Alt. 5 m) (92 mm/year) to its north-east. In contrast, Gebeit (18° 57′N, 35° 5′E, Alt. 795 m), within the Red Sea hills 60 km inland, receives a negligible amount of winter rainfall (12 mm, in October–February) because it lies at about the northern boundary of the tropical continental climate region where the southerlies bring summer (May–September) rainfall (111 mm/year).

Variability of the annual rainfall, a characteristic feature of the arid lands, is obvious on the Sudanese Red Sea coast (Kassas, 1957). The Suakin rainfall, for example, ranges from 617 mm/year in 1896 to 33 m/year in 1953 and in Port Sudan, rainfall ranges from 422 mm/year in 1925 to 19 mm/year in 1910 (a variability of about 56%). The moisture-laden northerlies cause convectional rain at the coast which decreases inland. But as they cross the coastal plain, the northerlies encounter the Red Sea hills causing orographic rain and other forms of condensation as dew. The seaward (east) slopes of the hills receives a greater share of this rain. Again the higher hills farther inland receiver, by virture of their altitude, more of this moisture than the low hills and buttes.[4] This will have a noticeable effect on the plant cover and its altitudinal zonation pattern.

## 3.3 Vegetation Forms

Eleven major forms of vegetation are recognized along the Afro-Asian Mediterranean and Red Sea coastal land's (Table 3.1): mangrove, reed swamp, salt marsh, sand dune, rocky ridge, desert, Mediterranean steppe grassland, broad-leaved evergreen forest, stunted woodland (matorral), coriferous forest and scrubland vegetation. The fact that the distribution of these vegetation forms along the coastal lands is affected mainly by the prevailing climatic condition is of ecological interest. Four vegetation forms: (reed swamp, salt marsh, sand dune and desert) occur in all sections of the Afro-Asian Mediterranean and Red Sea coastal lands, while the others have different distribution patterns. Mangrove vegetation is a tropical formation, and hence, its presence is restricted to the coastal belts of the Afro-Asian Red Sea but absent from those of the Mediterranean Sea. Rocky ridge vegetation occurs in the Egyptian and Libyan sections of the North African Mediterranean coast whereas Mediterranean steppe grasslands are localized in the coastal lands

**Table 3.1** Distribution of the major vegetation forms along the Afro-Asian Mediterranean and Red Sea coastal lands

| | Mediterranean coastal land | | | | | | | Red sea coastal land | | | | | |
|---|---|---|---|---|---|---|---|---|---|---|---|---|---|
| | North Africa | | | | | SW Asia | | SS | African | | | Asian | |
| Coastal vegetation forms | M | A | T | L | Eg | NS | PI | Eg | Eg | Su | Er | SA | Y |
| 1. Mangrove vegetation | – | – | – | – | – | – | – | + | + | + | + | + | + |
| 2. Reed Swamp vegetation | + | + | + | + | + | + | + | + | + | + | + | + | + |
| 3. Salt marsh vegetation | + | + | + | + | + | + | + | + | + | + | + | + | + |
| 4. Sand Dune vegetation | + | + | + | + | + | + | + | + | + | + | + | + | + |
| 5. Rocky Ridge vegetation | – | – | – | + | + | – | – | – | – | – | – | – | – |
| 6. Desert vegetation | + | + | + | + | + | + | + | + | + | + | + | + | + |
| 7. Mediterranean Steppe Grasslands | + | + | + | – | – | – | – | – | – | – | – | – | – |
| 8. Forests | | | | | | | | | | | | | |
| 8.1. Broad-leaved evergreen forests | + | + | + | – | – | – | + | – | – | – | – | – | – |
| 8.2. Stunted woodlands (matorrals) | + | + | + | + | – | – | + | – | – | – | – | – | – |
| 8.3. Coniferous forests | + | + | + | – | – | – | + | + | – | – | – | + | + |
| 8.4. Scrubland vegetation | – | – | – | – | – | – | + | + | + | + | + | + | + |

M = Morocco, A = Algeria, T = Tunisia, L = Libya, E = Egypt, PI = Palestine-Israel, NS = north Sinai, SS = South Sinai, SW = South west, Er = Eritrea, SA = Saudi Arabia, Y = Yemen, + = present, – = absent.

[4]Buttes = isolated steep-sided flat topped hills.

of the three western countries of the North African Mediterranean coast (Morocco, Algeria and Tunisia). The same distribution pattern is seen in broad-leaved evergreen forest and stunted woodland (matorral) but both also occur in the escarpment of the Palestine-Israel coast of the SW Asian Mediterranean. Coniferous forests are ecologically related to cold climate territories and, therefore, occur at high altitudes of the North African and, SW Asian Mediterranean, and the southern Sinai and Asian Red Sea coastal lands. Scrubland vegetation is abundant in the Afro-Asian Red Sea coastal lands, including Southern Sinai, poorly represented in the Palestine-Israel section of the SW Asian Med. and almost absent from North African coastal lands. Details of each of these eleven vegetation forms may explain the relationship between them and their floristic composition with climatic conditions particularly rainfall and temperature.

### *3.3.1 Mangrove Vegetation*

Mangrove vegetation is one of the characteristic ecological features of both African and Asian Red Sea coastal belts with almost the same latitudinal distribution, but are totally absent from the coastal belts of the North African and South-East Asian Mediterranean. It is, actually, a climatic phenomenon, mangrove forests are tropical formations that only grow in the swamps of warm and hot coastal belts between the two tropics: the Tropic of Cancer (23° 30′N) and Tropic of Capricorn (23° 30′S). Low temperature is fatal to the young seedlings of these plants, a cogent explanation for the absence of this vegetation form from the coastal belts of the cool and cold areas of the world (Chapman, 1977).

Three mangrove species are recorded along the Afro-Asian Red Sea coastal belts: *Avicennia marina, Rhizophora mucronata* and *Bruguiera gymnorrhiza. A. marina* is well developed along both coasts of the Red Sea starting from latitude 27° 14′N southwards in Egypt, the Sudan, Eritera and Djibouti (African Red Sea coast), and in Saudi Arabia and Yemen (Asian Red Sea coast). North of this latitude, *A. marina* forest is practically absent except for a few isolated patches with stunted bushes in the swamps of Ras Muhammed in Sinai (27° 40′N) and on the Nabq coast at the southern part of the western coast of the Gulf of Aqaba (28° 10′N). A comparable distribution pattern is recognized along the Asian Red Sea coast. The most extreme northern presence of *A. marina* is recorded in the coast at Duba (27° 22′N) (Zahran et al., 1983; Zahran, 2007). These findings confirm Chapman's statement (1977): "Mangal reaches its optimal development in the tropics but it extends to the subtropics described as warm temperate where the air temperature seems cooler, particularly at night, than the range of tolerance of the seedlings of the mangrove plants".

*R. mucronata* and *B. gymnorrhiza* mangroves have a different distribution pattern: both are absent from the northern sections of the Afro-Asian Red Sea coastal belts (north of 23°N). In Egypt *R. mucronata* starts to appear in the mangrove swamps of the Shalatein coast (23°N) and continues southwards to Mersa Halaib (22°N) on the Sudano-Egyptian border. *R. mucronata* has been also recorded further south in Suaken (Sudan) and in Djibouti (19° 15′–19°N), Zahran et al. (2009). Along the Asian Red Sea coastal belt, *R. mucronata*, occurs on Jizan coast, Saudi Arabia (16° 20′N) and southwards at Hodayda in Yemen (15°N). *B. gymnorrhiza* was recorded in Suaken (Sudan), Hodayda (Yemen) and Djbouti coasts. Nowadays, it is absent from these coasts due to the influence of man (Andrews, 1948, 1950–1956; Draz, 1956; Zahran et al., 1983, 2009).

### *3.3.2 Reed Swamp Vegetation*

Reed swamp vegetation inhabits the lakes, lagoons and water creeks of the Afro-Asian Mediterranean and Red Sea coastal lands. The most widespread and characteristic species (helophytes) are: *Phragmites australis* and *Typha domingensis*, are recorded in almost all stands of this vegetation form. Other common swampy species include: *Cyperus articulatus, C. dives, Eleocharis palustris, Lemna* spp, *Samolus valerandi, Scripus mucronatus, Spirodela polyrrhiza* and *Wolffia hyalina*. There are reed-swamps species (helophytes) with restricted geographical distribution; many of them occur only in the swampy areas of the Afro-Asian Mediterranean coastal lands while others occur only in Afro-Asian Red Sea coastal lands. For example, *Typha elephantina* has been recorded in a swampy area in the downstream part of Wadi Gizan along the Asian Red Sea coast of Saudi Arabia (Zahran, 1982b), but has never been recorded in the swamps of the Afro-Asian Med. Coast. Andrews (1950–1956) recorded some helophytes growing in the swampy habitat of the Red Sea coast of the Sudan: (*Cyperus amauropus, C. schimprianus, C. cristatus* var. *nigricans, C. rotundus* var. *nubicus, C. microbolbos* etc.), and *Cyperus bulbosus* was recorded from the Red Sea coast of the Sudan and Egypt but absent from the Med. coast. The are also the helophytes recorded from the Afro-Asian Mediterranean coastal land but absent from those of the Red Sea coast including: *Cyperus papyrus, C. compressus* (= *C. conglomerates*), *C. rotundus* var. *rotundus, C. capitatus, C. difformis, C. alopecuroides, Juncus subulatus, Carex divis, C. distans, Eleocharis palustris, Fimbristylis bisumbellata, Fuirena pubescens* (= *Scirpus pubescens*), *Scirpus maritimus, S. littoralis, S. holoschoenus, Phalaris canariensis, P. aquatica.* etc.

### *3.3.3 Salt Marsh Vegetation*

Littoral salt marsh vegetation occurs in coastal lands subjected to maritime influences: periodic flooding with sea water, sea water spray, sea water seepage etc. They can be found, in general, if any of the following physiographic conditions is fulfilled: the presence of estuary, the shelter of spits, off-shore barrier islands, large or small protected bays with shallow water etc. (Chapman, 1974).

Littoral salt marshes of the arid and semi-arid coastal lands are really fringes of inland deserts and their landward area being defined by their desertic qualities.

Climate and terrain can be used to mark off their inland boundary. The halophytes, the plants of this vegetation, may be considered as high specialized plants most of which have great tolerance to salt. The salt tolerance of the halophytes increases both during their growth and development from generation to the next (Keith, 1958).

Zonation pattern is a universal feature of the littoral salt marsh vegetation (Chapman, 1974; Kassas, 1957) controlled mainly by tidal movement, land relief, soil salinity and underground seepage of sea waer. The species of the lower zones, near to the sea, may be very different from those of the higher zones and "in many instances, only a few inches increase in level results in a profound change" (Chapman, 1974). The zonation sequence of the littoral salt-marsh vegetation is complete only where the shore rises gently and gradually into the land.

Littoral salt marsh vegetation is well developed along the Afro-Asian Med. and Red Sea coastal lands but the geographical distribution of its floristic elements varies. Some of these species have wide ecological amplitude and grow under the different climatic conditions of these coastal lands, whereas others have a narrow range and grow either in the Mediterranean or the Red Sea coastal lands. Representative characteristic species of these three groups of halophytes are:

(a) Halophytes occurring in both the Afro-Asian Med. and Red Sea coastal lands include: *Arthrocnemum macrostachyum, Aeluropus lagopoides, Cyperus laevigatus, Cressa cretica, Halopeplis perfoliata, Halocnemum strobilaceum, Imperata cylindrica, Juncus rigidus, Limonium pruinosum, Nitraria retusa, Salicornia fruticosa* (= *Sarcocornia fruticosa*), *Sporobolus spicatus, Suaeda vermiculata, Tamarix aphylla, T. nilotica, T. tetragyna* etc.
(b) Halophytes occur only in the Afro-Asian Med. coastal lands include: *Atriplex halimus, A. portulacoides, A. leucoclada* var. *inamoena, Frankenia revoluta, Inula crithmoides, Juncus acutus, Limoniastrum monopetalum, Salsola tetrandra, Schoenus nigricans, Sporobolus pungens, Suaeda paleastina* and *S. pruinosa*.
(c) Halophytes restricted to the littoral salt marshes of the Afro-Asian Red Sea coastal lands include: *Aeluropus littoralis, Limonium axillare, Salsola spinescens, Sevada schimperi, Suaeda monoica, Tamarix amplexicaulis* and *T. macrocarpa*.

The geographical distribution of six of these halophytes deserves extra consideration, these are: *Halocnemum strolbilaceum, Nitraria retusa, Suaeda monoica, Halopeplis perfoliata, Tamarix macrocarpa* and *Salicornia fruticosa*.

1. *Halocnemum strobilaceum* dominates a community common in the littoral salt marshes of both areas but its presence in the Afro-Asian Red Sea coasts is restricted to the northernmost 400 km but rare or even absent in the southern sections (south of latitude 27°N).
2. *Nitraria retusa* similarly dominates a community common in the littoral salt marshes of the Afro-Asian Med. and Red Sea coastal land but its presence is

restricted to the northern 700 km stretch, i.e. north of Mersa Alam in Egypt and Umlog in Saudi Arabia, southwards, it is absent along the whole Red Sea coasts.

3. *Suaeda monoica* is a succulent halophyte comparable in habit and habitat to *Nitraria retusa*. The two species have an ecological range that extends beyond the limits of the salt marsh to the fringes of the coastal desert plain. Both may form and protect mounds and hillocks of sand, though *S. monoica* hills can be larger (Zahran and Willis, 2008). However, these two species seem to have different geographical areas. Along the African Red Sea coast, *N. retusa* occurs in the northern 700 km stretch south of Suez but is absent in the south. The reverse in true for *S. monoica* gradually replacing *N. retusa* over the 300–700 km stretch south of Suez. Further south, *S. monoica* is a salient feature of the southern section of the Red Sea coast in Egypt as well as in the Sudan and Eritrea (Kassas, 1957; Hemming, 1961). The same geographical distribution of *N. retusa* and *S. monoica* has been noticed on the Asian Red Sea coast (Zahran, 1982b). *S. monoica* has also been recorded by Zohary (1962) in the inland desert of Palestine-Israel area.
4. *Halopeplis perfoliata* is a succulent halophyte recorded on the littoral salt marshes of both the Afro-Asian Med. and Red Sea coastal lands. Its presence and distribution along the Red Sea coastal belts is of ecological interest. Along the African Red Sea coast within the 950 km stretch from Suez to Mersa Kilies, *H. perfoliate* is recorded in one locality (55 km south of Ras Gharib, 295 km south of Suez) but is otherwise very rare or absent. South of Mersa Kilies, *H. perfoliata* and its community become common features of the littoral salt marsh vegetation. It is also very common further south in the Sudan and Eriterea (Kassas, 1957; Hemming, 1961). Almost, the same geographical distribution of *H. perfoliata* has been recorded by Zahran (1982b) on the Asian Red Sea coast.
5. *Tamarix macrocarpa* (= *T. passerinoides*) is absent from the Afro-Asian Med. coastal lands and is rare along the Egyptian Red Sea coast. Its presence is limited to a narrow area: El-Mallaha[5] 20–40 km south of Ras Gharib (260–280 km south of Suez) where there is an inland depression separated from the shore by an elevated raised beach and fed with seawater through underground passage. *T. macrocarpa* has not been recorded from the Yemeni Red Sea coast (Wood, 1997) but has been further north in Saudi Arabia by Migahid and Hammouda (1978).
6. *Salicornia fruticasa* (= *Sarcocornia fruticosa*) is a widespread succulent halophyte along Afro-Asian Med. coastal land but is absent from the Asian Red Sea coast. On the African Red Sea coast, *S. fruticosa* is recorded from only one site, the El-Mallaha saline depression in Egypt, where it dominates pure stands with 50–100% cover. Andrews (1950–1956) recoded *S. fruiticosa* from the Sudanese Red Sea littoral salt marshes.

[5]El-Mallaha is the Arabic wording for the highly saline land.

### *3.3.4 Sand Dune Vegetation*

Along the Afro-Asian Mediterranean and Red Sea coastal belts lie chains of sand dunes which, due to their proximity to the seas, are more humid and exposed to maritime influences such as sea-water spray. As rainfall is considerably higher along the Afro-Asian Med. coasts than those of the Red Sea, the fresh water of rain is usually stored in the dunes and is frequently obtained by digging carefully to a 3–4 m depth. Such fresh (rain) water, having a lower specific gravity than the saline water below, can form a layer above it. There may be a hard pan of limestone rock underlying the sand which prevents percolation of rain water, resulting in the sand dunes acting as reservoirs of fresh water.

Plants growing in the sand dune habitats (psammophytes) are highly specialized and can have the ability to elongate vertically on burial with sand (Girgis, 1973). They are also subjected to partial exposure of their underground organs often without being seriously affected. Many psammophytes develop extensive superifical roots that make use of dew.

Plant diversity of the sand dune vegetation of the Afro-Asian Mediterranean coastal lands is rather richer than that of the Afro-Asian Red Sea. This is, presumably, due to the relatively high amount of rainfall in the Mediterranean relative to the levels in the Red Sea coastal lands. The geographical distributions of species recorded from these coastal lands are mainly affected by rainfall and temperature. The psammophytes restricted to the North-African and SW Asian Med. coastal lands include: *Ammophila arenaria, Lygeum spartum, Elymus farctus, Lotus polyphyllus, Thymelaea hirsuta, Lycium europaeum, Silene succulanta, Asparagus stripularis, Allium roseum, Convolvulus althaeoides, Pancratium maritimum, Phlomis floccosa, Salvia lanigera, Zygophyllum aegyptium* (an endemic), *Echium angustifolium* subsp. *sericeum, E. glomeratum Artemisia monosperma, Moltkiopsis ciliata, Scrophularia hypericifolia, Salsola kali, Cakile maritima, Ipomoea litteralis, Lotus creticus, Medicago marina, Stipagrostis scoparia* (= *Aristida scoparia*), *Lolium gaudini, Astragalus fruticosum* (= *A. tomentosum*), *A. spinosum, Euphorbia paralias, Prosopis farcta* (= *Lagonychium farctum*), and *Ziziphus lotus*. All of these are absent from the Red Sea coastal lands. Psammophytes inhabiting both the Afro-Asian Med. and Red Sea coastal lands include: *Cyperus conglomeratus, Retama raetam, Cutandia memphitica, Panicum turgidum, Pennisetum divisum, Eremobiun aegyptiacum, Echinops spinosissimus, Halplophyllum tuberculatum, Neurada procumbens, Deverra tortusa* (= *Pituranthos tortuosus*), *Cynodon dactylon, Stipagrostis plumosa, Alhagi graecorum, Tamarix aphylla* and *Zygophyllum album*. Only two species are restricted in their distribution to the coastal sand dunes of the Afro-Asian Red Sea coastal lands, namely: *Atriplex farinosa*, and *Halopyrum mucronatum. A. farinosa* is widespread along the whole length of the African and Asian Red Sea coastal lands inhabiting the low sand dunes close to the sea. However, *H. mucronatum* has been recorded only from very limited areas of both Red Sea coasts: from Mersa Abu Ramad (33 km) north of Mersa Halaib at the Sudano-Egyptian border, from the Eritrean coast (African Red Sea coast) and from the Saudi Arabian coast at Jizan.

### 3.3.5 Rocky Ridge Vegetation

The rocky ridge is a unique geomorphological landscape of the western section of the North African Mediterranean coastal land in Libya and in Egypt. Embabi (2004) stated that, although the surface of the Mediterranean coastal plain in Egypt slopes gently towards the sea, the plain is characterized by a sequence of low carbonate ridges (bars) which are roughly parallel to the present coastline. These ridges are separated from each others by sabkha-lagoonal depressions interrupted in parts by secondary transverse bars. The number of these coastal ridges ranges between 4 and 8 reaching altitude of 10–110 m. All extend for long distances along the coastal plain parallel to the shoreline. The first (nearest the sea), and the second bars are almost continuous and prominent (El-Shazly and Shata, 1969).

Submerged ridges are also recognized in the offshore areas. These are composed of oolitic limestone, marl with shell fragments and calcareous sand, lying 0.71–15.79 m below sea level. Butzer (1960) recognized eleven ridges lying at up to 11 m below sea level and two others at 40 m below sea level some 6 km offshore.

The coastal ridges of the North African Med. Coastal lands are veneered with a hard brown calcareous duricrusts (calceretes) which are either exposed or covered by a thin layer of recent aeolian sand. The tops of these ridges are composed of hard oolitic limestone with horizons of palaeo-soil (Rashed, 1998). This may indicate that they have been deposited during a generally arid period with two relatively short periods of wet climate (El-Asmar and Wood, 2000).

El-Morsy (2008) and Zahran and Willis (2008) found that the rocky ridge vegetation in Libya and Egypt is formed mainly of elements restricted geographically to the Mediterranean region and absent from the Red Sea coastal lands, there are few species recorded from both coastal areas. The characteristic Mediterranean elements include: *Thymus capitatus, Glaucium grandiflorum, Ononis natrix, Panicum repens, Asparagas aphyllus, Bellevalia fruticosum, Echium sericeum, Helianthemum peruviana, Lygeum spartum, Lycium europaeum, Thymelaea hirsuta, Salvia lanegera, Chenolea arabica* (= *Bassia arabica*), *Globularia alypum, Scilla peruviana, Reaumuria hirtella, Iris sisyrinchum, Stipa parviflora, Haloxylon scoparium, Noaea mucronata, Plantago albicans, Ephedra alata, Carduus getulus, Bupleurum nodiflorum, Arisarum vulgare, Aegilops kotschyi, Lotus glaber,* and *L. creticus*. Apart from these, the flora of rocky ridge vegetation comprises species of wide geographical distribution and may grow in the other habitats of the Red Sea costal lands e.g. *Deverra tortusa, Gymnocarpos decander, Aeluropus lagopoides, Echinops spinosus, Helianthemum lippii, Globularia arabica, Plantago albicans, Stipa capensis* etc.

### 3.3.6 Desert Vegetation

The coastal desert vegetation is the most prominent and widespread type in the Afro-Asian Mediterranean and Red Sea Coastal lands. It predominates in desert areas extending between the inland borders of the littoral salt marshes and/or coastal sand dunes and the coastal plateaus of the Med. coasts or the ranges of hills and mountains of the Red Sea coasts on the inland side. Due to its intermediate position, the ecological features of the desert vegetation, especially on its fringes, show transitional characters.

The coastal desert vegetation is far away from maritime influences and has non-saline habitat but climate and soil aridity are the main environmental factors. It presents a complicated pattern owing to the difference in topography, characters of the surface deposits and the relationship with mountain groups. The habitat is essentially a gravel-covered plain traversed by the mid- and downstream extremities of the main wadis. The downstream parts of these wadis, particularly along the Afro-Asian Red Sea coastal lands and south Sinai can form deltaic basins.

Within these coastal desert ecosystem soil transporting agencies, mainly climatic (wind and rain), are actively operating. The alluvial deposits carried by rainfall range in particle sizes from silt to coarse gravels and boulders and often build compact terraces on the sides of the water courses (wadis). The building and destruction of these terraces are mainly physical processes independent of the vegetation. Aeolian deposits carried by wind, on the other hand, are sandy superimposed as sheets, mounds or hills of various heights and extent. These sandy bodies are usually built around plants (phytogenic) and their sizes seems to depend upon the plant speices.

Unlike the zonation of littoral salt marsh vegetation, coastal desert vegetation shows a mosaic pattern and distinct seasonal aspects due to the preponderant growth of therophytes (ephemeral, annuals, and biennials) after rain. These short-lived plants usually become dry and disappear vegetatively during the dry seasons but their seeds are dormant in the soil until the next rains when they germinate to give new seedlings. This aspect of seasonal phenology is essentially a climatic feature, characteristic of desert vegetation but not seen in the salt marsh plants. High soil salinity in the latter habitat prohibits the growth of most therophytes.

Desert vegetation contains the largest number of species, and the floristic composition of its communities is usually much more elaborate than the simple composition of the salt-marsh vegetation. Its framework is formed of perennial xerophytes associated with therophytes during the rainy seasons. Thus, the desert vegetation can be classified into two main types: ephemeral and perennial vegetation types.

**(A) Ephemeral vegetation type**

This type include all short lived plants (therophytes) that appear in the deserts after rain. It indicates soil conditions that do not allow water to be stored over the year: soil westness is maintained during only part of the year. This may be due either to the scantiness of rainfall or also to surface deposits being too shallow. Green patches of ephemeral vegetation are a recurring phenomenon seen in the wadis and plains of the coastal deserts after the rains usually forming a mosaic of patches each of which may be dominated by one or more species. One type of ephemeral growth may recur in the same patch for several years, possibly due to the availability of seeds where parent plants have existed.

Depending on the growth form of the dominant (or most abundant) species, three types of ephemeral vegetation are recognized in Afro-Asian Red Sea and Med. coastal deserts (Zahran and Willis, 2008; Zohary, 1962; El-Morsy, 2008). One is dominated by succulent plants, a second by grasses and a third by herbaceous species. *Zygophyllum simplex, Aizoon canariense* and *Spergula fallax* are widespread succulent ephemerals in the Afro-Asian Med. and Red Sea coastal deserts. *Trianthema crystallina* and *Tribulus pentandrus* represent succulent ephemerals restricted in their distribution to Red Sea coastal deserts, whereas *Mesembryanthemum* spp. (*M. nodiflorum, M. crystallinum*) represent succulent ephemerals restricted to Med. coastal deserts. The tissues of these succulent plants can store water for use later in the growing season. They have shallow roots and survive throughout a season longer than the other ephemeral.

The vegetation of the ephemeral grasses is of special importance for the nomadic herdsmen for whom it is valuable pasture. *Bromus fasciculatus, Eragrostis cilianensis, Schismus barbatus, Setaria verticilata, Stipa capensis* and *Stipagrostis hirtigluma*, are among the widespread ephemeral grasses growing in both Afro-Asian Med. and Red Sea coastal deserts. Ephemeral grasses restricted to the Med. coastal deserts but absent from the Red Sea coasts include: *Bromus diandrus, Dactyloctenium aegyptium, Eleusine indica* subsp. *indica, Eragrostis aegyptiaca, Hordeum spontaneum, Lolium rigidum, and Triticum aestivum* whereas those recorded only from Afro-Asian Red Sea coastal deserts are represented by: *Aristida mutabilis, Brachiaria deflexa, Cenchrus pennisetiformis, Digitaria ciliaris, Eragrostis ciliaris* and *Trichoneura mollis* (= *Diplachne arenaria*).

The herbaceous ephemeral type of vegetation may be dominated by one of a great variety of species or can be mixed species with no obvious dominant. These plants, like the grasses, can provide valuable grazing sites. Within Red Sea coastal deserts, patches of ephemerals dominanted by one of the following species are recorded: *Arnebia hispidissima, Astragalus vogelii* subsp. *vogelii, A. eremophilus, Ifloga spicata, Cleome brachycarpa, Plantago parviflora, Senecio flavus, and Tribulus pentandrus*. The herbaceous ephemerals of the Afro-Asian Med. coastal deserts also contain very many species, such as: *Arnebia decumbens, A. linearifolia, Asphodelus viscidulus, Astraglus boeticus, A. tribuloides* var. *tribuloides, A. mareoticus, A. annularis, Filago contracta, Malva sylvestris, M. aegyptia, Plantago squarrosa, Raptianus raphanistrum, Senecio vulgaris, Linum decumbens, Lythrum hyssopifolia, Erucaria crassifolia, Sinapis alba, S. arvensis* subsp. *arvensis, Brassica rapa, Lobularia arabica, Centaurium spicatum, Notobasis syriaca*. Species recorded in both Afro-Asian Med. and Red Sea coastal deserts include: *Asphodelus tenuifolius, Filago desertorum, Malva parviflora, Neurada procumbens, Shouwia thebaica, Senecio glaucus* subsp. *coronopifolius,* and *Tribulus terrestris*.

**(B) Perennial vegetation type**

Xerophytic perennial vegetation of the Afro-Asian Med. and Red Sea coastal deserts can be classified into two main subtypes: suffrutescent[6] and frutescent.[7] The vegetation of the suffrutescent subtype consists of two layers: an upper layer (30–120 cm) that includes the dominant species and a ground layer (< 30 cm) that includes associated therophytes and cushion-forming perennials such as *Cleome droserifolia, Cucumis prophetarum* subsp. *prophetarum, Fagonia indica, F. mollis, Blepharis edulis* etc. (Red Sea coastal desert), *Astragalus alexandrinm, Fagonia bruguieri,* and *F. cretica, F. glutinosa* (Med. coastal desert), and *Citrullus colocynthis, and Polycarpaea repens* (from both Med. and Red Sea coastal deserts). Frutescent perennial vegetation comprises three layers-the two of the suffrutescent subtype and a higher (< 120 cm) one including the dominant. This vegetation includes some trees more than 5 m such as *Acacia raddiana* and *Balanites aegyptiaca.*

I. **Suffrutescent perennial vegetation**

Three units of different species are recognized in the suffrutescent perennial vegetation: succulent half-shrubs, grasslands and woody plants.

(a) The succulent half-shrubs are represented by the following species: *Anabasis articulata* and *Hammada elegans* (= *Haloxylon salicornicum*) in both Med. and Red Sea costal deserts; *Haloxylon scoparium, H. negevensis* and *H. persicum* are restricted to the Mediterranean, whereas *Zygophyllum coccineum* and *Z. decumbens* only occur on the Red Sea coastal deserts.

(b) Grassland perennial vegetation is represented by:

(i) Perennial grasses recorded in both Med. and Red Sea coastal deserts: *Panicum turgidum, Pennisetum setaceum, Saccharum spontaneum* subsp. *aegyptiaca, Lasiurus scindicus, Centropodia forsskaolii, Cenchrus ciliaris, Hyparrhenea hirta* and *Polypogon viridis.*

(ii) Perennial grasses recorded in the Med. but absent from the Red Sea coastal deserts include: *Panicum repens, Saccharum spontaneum* subsps. *spontaneum, Sorghum halepense, Stipagrostis scoparia, S. lanata, Agropyron cistatum, A. obtusa, Dichanthium annulatum, Lolium perenne, Leersia hexanda, and Oryzopsis miliacea.*

(iii) Perennial grasses recorded from the Red Sea but absent from the Med. coastal deserts include: *Pennisetum divisum, Stipagrostis*

[6]Suffrutescent = undershrubby.

[7]Frutescent = shrubby.

*acutifolia, S. raddiana, Dicanthium foveolatum, Dactyloctenium scindicum, Cenchrus setigerus,* and *Enneopogon desvauxii.*

(c) Woody vegetation is represented by shrublets with various distribution patterns. Some species are widespread and grow in both coastal desert areas such as: *Zilla spinosa, Cornulaca monacantha, Calligonum comosum, Artemisia herba-alba, A. judaica, Achillea fragrantissima, Ephedra alata, Gymnocarpos decander, Iphiona mucronata,* and *Pulicarea crispa.* Those restricted to Med. coastal deserts include: *Achillea santolina, Peganum harmala, Ranunculus asiaticus, R. millefolius, Noaea mucronata, Echium angustifolium, E. glomeratum,* and *Ephedra aphylla,* while those restricted to the Red Sea coastal deserts include: *Launaea spinosa, Abutilon pannosum, A. fruticosum, Hibiscus micranthus, Cocculus pendulus, Aerva javanica, A. lanata, Periploca aphylla, Cadaba rotundifolia, C. farinosa, Solenostemma arghal, Commicarpus helenae, Fagonia indica, F. thebaica,* and *F. boulosii* (an endemic, Boulos, 1995) etc.

II **Frutescent perennial vegetation**

Details of this vegetation type are dealt with in Section 3.3.8.4 (scrubland vegetation) of this book.

### *3.3.7 Mediterranean Steppe Grasslands*

There are extensive areas of purely Mediterranean grasses in Morocco, Algeria and Tunisia, i.e. restricted only to the western section of this coastal land. This grassland is formed mainly of the tough tussocky grasse *Stipa tenacissima,* called: esparto by the Spaniards and alfa by the Arabs (Branigan and Jarrett, 1975). *S. tenacissima* is indigenous to the south of Spain and the western section of North Africa but absent elsewhere. It grows well in dry, sunny situations on the Algerian and Tunisian coasts. The main shoots can grow to a heights of 1 m and the leaves to 15–90 cm in length. When young, *S. tenacissima* serves as a feed for cattle and sheep, but after a few years' growth it acquires great toughness of texture. On account of the tenacity of their fiber and their flexibility, the leaves have for centuries been employed in the making of such useful articles as baskets, sandals, ropes and mats. Today, they are used as raw material for paper making.

The vast alfa steppes of North Africa are associated with an abundant growth of undershrubs such as *Artemisia* spp. and *Rhanterium* spp. in addition to scattered trees of *Pistacia atlantica. Lygeum spartum* which is very common grass in the sand dune and rocky ridge habitats of the eastern section of the North African Med. coastal land is also a common associate of *S. tenacissima.* When aridity is combined with a high water table, creating saline flats, the salt tolerant bushes of *Atriplex, Salsola* and *Suaeda* species constitute an extensive formation.

### *3.3.8 Forests of the Afro-Asian Mediterranean and Red Sea Coastal Lands*

In the Mediterranean basin, including the North African and SW Asian coastal lands, the term "forest" should not necessarily bring to mind an image of high, dense stands of trees with closed canopies. Mediterranean forests are highly diverse in their architecture, appearance and woody plant species composition (Blondel and Aronson, 1999). Mediterranean forests are highly varied in growth form, morphology, physiology and phenology of the dominant trees in each region. Four leaf types occur in varying combination. Firstly leaves may be sclerophyllous and evergreen leathery in texture and often spiny or prickly. A second group has laurel-like leaves that are somewhat softer and shiny but still evergreen, like the foliage in many tropical forest trees. Thirdly, they may be semi-deciduous and remain on their stems over winter, with reduced or terminated growth and photosynthesis. Leaves are not shed until spring when they are replaced by a new crop of leaves. Examples of such species are *Quercus faginea* and *Q. infectoria*. The fourth group has typically deciduous leaves such as predominate in northern temperate forest trees e.g. *Carpinus, Corylus, Ostrya* and *Zekova* species.

Mediterranean forests also differ in the structure or physiognomy they assume under human management. Example include forest where all conifers have been removed but not the oaks. The largest and most diverse evergreen sclerophyllous forest in the Mediterranean area today are the "lauriphyllous" forests, relicts of a now virtually extinct Tertiary flora previously widespread in southern Europe and Northern Africa. These forest include species of the tropical family Lauraceae (e.g. *Apollonia barbujana, Laurus azorica, Persea indica* and *Octea foetens*) as well as several endemic broad-leaved evergreen trees such as *Arbutus canariensis, Myrica faya* and *Visnea mocanera*. All of these trees share their broad-leaved schlerophyllous leaf shape with the laurel (*Laurus nobilis*) which still occurs widely throughout the Mediterranean Basin. At high altitudes, open formations of coniferous forest containing species of *Abies, Cedrus, Cupressus, Juniperus* and *Pinus* are found with an understorey of spiny shrubs such as species of *Astragalus* and *Genista*.

In the Red Sea basin, including the southern mountainous section of Sinai, the term forest may be applied to the open scrubland vegetation that characterizes the montane countries where the amount of the orographic rain is relatively higher than the annual rainfall of the surrounding coastal deserts. This amount of rain associated with suitable temperature enable woody xerophytic trees and shrubs of, for example *Acacia* spp., or *Balanites aegyptiaca* etc. to grow and dominate forming a vegetation form in the wadis and low altitudes of the slopes of the mountains. At still higher altitudes, where temperature is low, coniferous forest of *Juniperus* and *Pinus* may also occur.

The vegetation forms that may represent the forests of the Afro-Asian Mediterranean and Red Sea coastal lands are: broad-leaved evergreen forests, stunted woodland (matorral), coniferous forest and scrubland vegetation (Table 3.1). The geographical distribution and floristic composition of these forms may reflect the relationship between vegetation and climatic factors prevailing in these coastal lands.

### 3.3.8.1 Broad-Leaved Evergreen Forests

This type of forest lies entirely within the limits of olive tree cultivation and may be regarded as the most characteristic type of the Mediterranean vegetation. The chief trees are the evergreen oaks of various species, the most numerous being holm oak (*Quercus ilex*), the cork oak (*Q. suber*) and kermes oak (*Q. coccifera*). The last species (*Q. coccifera*) is the most widespread in the three western North African countries (Morocco, Algeria and Tunisia) and also occur in the Levant on the Palestine-Israel coast. *Q. coccifera* has more stunted appearance than other evergreen oaks and its timber is of little economic value. Unfortunately, most of the regional broad-leaved evergreen forests have been cut down and replaced by cultivated trees or by the land left derelict so that there are only a few representative vegetation areas with too little rain for agriculture; exceptions may occur where oaks are of economic value as in the cork oaks (*Q. suber*).

The broad-leaved evergreen forests of the North African Mediterranean coastal land are formed of widely spaced oak trees with dense mats of xerophytic woody shrubs, undershrubs and stunted trees in between such as brooms (*Sarothamnus scoparius, Cytisus* spp., *Genista* spp., *Spartum* spp., *Leguminosae*), *rosemary* (*Rosmarinus officinalis, Labiatae*), myrtle (*Myrtus communis*, myrtaceae), gorse (*Ulex europeaus*, Leguminosae), privet (*Ligustrum vulgare*, Oleocae), rockrose (species of *Helianthemum, Tuberaria, Citrus* etc.), and Laurel (*Laurus mobilis, Apollonia barbujana, Laurus azorica, Persea indica* and *Octea foetens,* Lauraceae) forming a ground vegetation and above them rise the taller figs (*Ficus carica, F. alba*), wild olive (*Olea europaea*) terebinth (*Pistacia terebinthus*) and the strawberry trees arbutus (*Arbutus unido, A. canariensis, Murica faya and Visnea mocanera*), while bulbous plants such as narcissi (*Narcissus poeticus*), tulips (*Tulipa stylosa*) and lilies (*Lilium* spp.) are scattered throughout the forest.

In certain coastal areas where summer drought is too intense or the soil too thin for major tree growth, oaks show stunted growth or they may disappear completely to be replaced by other woody trees such as wild olive (*Olea europaea*), locus tree (carob, *Ceratonia siliqua*) and mastic tree (*Pistacia lentiscus*).

### 3.3.8.2 Stunted Woodlands (Matorrals)

The stunted woodlands of the Mediterranean Basin is the undergrowth of broad-leaved evergreen forest which remains after the uncontrolled destruction of trees and/or depredations of goats causing degeneration (Branigan and Jarrett, 1975). This vegetation forms an extensive and virtually impenetrable tangle of woody, thorny shrubs and dwarfed twisted trees rising to a height of not more than 3 m. It covers vast areas of Morocco, Algeria, Tunisia and Libya (North African Med. Coast) as well as the Palestine-Israel area of the SW Asian Med. Coast. "Woodland is, generally, found in areas that are suitable for agriculture use, so that much of it has been converted into pasture or cleared for growing grains" (Dallman, 1998).

This stunted woodland has diverse names, with almost every region or country with its own name or names to designate the diverse local names: *garrigue* or *gariga* and *maguis* or *macchia* in France and Italy, *xerovuni* in Greece, *matorral* and *tomillares* in Spain, *choresh* or *maquis* in the Palestine-Israel region. In many countries, the term *maquis* (or *mecchia*) refers to the first major stage in forest degradation, followed by *garrigue, phrygana* or *batha* which are all of still lower stature and complexity than *maquis*. The distinction between *garrigue* and *maquis* is considered by many geographers and phytoecologists to depend on the substrate. *Garrigues* are said to occur primarily on limestone and include the full range of species associated with holm-oak (*Quercus ilex*), while *maquis* is reserved for those formations occurring on acid, siliceous soils. In addition to the cohort of species found on nearby *garrigues, maquis* includes such calciphobe[8] marker (indicator) species such as strawberry tree (*Arbutus unido, A. canariensis*) and heath family trees (species of *Calluna, Erica* etc.), as well as certain rock-roses e.g. *Cistus ladanifer*, lavenders (e.g. *Lavandula vera*), and other shrubs.

The most characteristic feature of the matorrals is that they include a fine-grained mosaic of almost all the growth forms recognized by plant ecologists. Their under storey is rich with the full range of plant life-forms of Raunkiaer (1934): therophytes (ephemerals, annuals and biennials), cryptophytes (perennial plants with buds below the soil), geophytes (plants with perennating organs buried in the ground e.g. bulbs, rhizomes, corms, etc.), chamaephytes (plants with resting buds slightly above the ground but less than 25 cm), hemicryptophytes (with their buds at the level of the ground) and pahnerophytes (trees, shrubs or vines whose perennating buds are more than 25 cm above the level of the soil).

Matorrals are dominated by shrubs with evergreen, broad and small, stiff and thick (sclerophyllous) leaves, an overstory of small trees may present and with or without an understorey of annuals and herbaceous perennials (Di Castri, 1981). Matorral has also been defined by Zohary (1962) as any sclerophyllous evergreen dense vegetation type that may attain 4–6 m in height. The prominent dominant species include various species of oaks, carobs and lentisk in addition to different species of *Arbutus. Daphne, Laurus, Phillyrea, Myrtus, Rhamnus* and *Viburnum*, all of which are sclerophyllous. The understorey of the matorrals of the North-African and SW Asian Med. Coastal lands where human pressure is higher, tends to have fewer shrubs and more hemicryptophytes (Blondel and Aronson, 1999).

Though matorrals appear predominantly evergreen, yet some of their woody plants are in fact winter-deciduous. Examples are the maples (species of *Acer*), and species of *Quercus, Pistacia,* and *Rhus*. The brooms are notably common in Morocco, and have the distinction of bearing evergreen stems that are photosynthetically active all year round. Most of these species in the so-called retamoid[9] group have small deciduous leaves that fall readily during drought.

*Maquis*, is the most widespread form of vegetation in the Mediterranean Basin with a Med. climate. It is highly varied with respect to dominant species and height of vegetation. Dallman (1998) classified maquis into two types: high maquis and low maquis. High maquis includes tall shrubs and small trees of varying height

[8]Calciphobe = Chalk – hating plants, plants shunning chalk or limestone.

[9]*Retama raetam* is a soon leafles desert shrub (Täckholm, 1974).

and favours shady slopes and sheltered locations. Dominant plants such as kermes oak (*Quercus coccifera*) and holm oak (*Q. ilex*) have dense branches and small, dark green, leathery, evergreen leaves. Strawberry tree (*Arbutus unedo*) and heather tree (*Erica arborea*) are also common. Interspersed among these are shrubs such as myrtle (*Myrtus communis*) and Spanish broom (*Genista hispanica*). Low maquis, consists mainly of evergreen shrubs including rock rose and oleander. Rock rose in the genus *Cistus* (Cistaceae) is a dominant aromatic shrub common in Morocco and Iberia. *Cistus* species bear large numbers of five-petalled, rose like flowers over a period of several weeks in the spring. Flower colours range from white to pink through violet and in most species last for no more than 1 day. Sage-leaved cistus (*Cistus salviifolius*) has white flowers and narrow leaves while *C. crispus* has pink flowers. Associated with *Cistus* is the genus *Cytinus*, a parasitic species that grows on the roots of *Cistus*. These low-growing plants have yellow, white or pale pink flowers with conspicuous orange or red-coloured bracts. Oleander (*Nerium oleander*[10]) is a poisonous plant (Apocynaceae), with a milky sap., avoided by herbivorous animals. *N. oleander* is abundant in many wadis often shaded by plane trees (*Platanus orientalis*) in the eastern part of the Mediterranean. *N. oleander* is one of the dominant species growing on stony and gravelly banks or beds of permanent and ephemeral runnels in the mountains and plains of Palestine-Israel area, SW Asian Med. coastal lands (Zohary, 1962). Baumann (1993) found that oleander leaves are stuffed into mouse holes, the mice die after they nibble the leaves.

Zohary (1962) reported that *maquis* vegetation in Palestine-Israel Med. coast is represented by the following species: *Genista shacelata, Olea europaea, Calycotome villosa, Pistacia lentiscus, Ceratonia siliqua, Quercus calliprinos, Q. ithaburensis, Crataegus azarolus, Rhamnus palaestina* and *Amygdalus communis*.

Garrigne consists of low shrubs dotted overhills with intervening bare, stony or shady patches. The term garrigue is a widely used French term for the vegetation type, also called *Batha* in the Palestine-Israel Med. coast where it accounts for more than 40% of all hilly, upland terrain. Much of the garrigue has been heavily grazed for thousands of years (Shmida and Barbour, 1982). This may explain why many of the most prominent species of the present status of this vegetation type are spiny plants not attractive to grazing animals. They often form distinct cushion shapes, leaving rooms from bulbs and orchids in the intervening spaces. Among these cushion plants are: the Greek spiny spurge (*Euphorbia acanthothamnos*) and thorny burnet (*Sarcopoterium spinosum*) both native to the eastern coast of the Med. Basin. In SW Med. coast of Palestine-Israel area, the Mediterranean batha (garrigue) vegetation is represented by (Zohary, 1962): *Asparagus aphyllus, Salvia triloba, Cistus villosus, Thymus capitatus, Poterium spinosum, Satureja thymbra* and *Teucrium polium*.

---

[10] *Nerium oleander* has been introduced to Egypt as an ornamental plants but never seen naturally growing in Egypt's Med. Coastal land.

Garrigue vegetation contains many plants producing two types of foliage referred to as seasonally dimorphic. Jerusalem sage (*Phlomis fruticosa*), for example, has relatively large and soft leaves which emerge in the winter rainy season. Later in spring and summer, a more drought-tolerant foliage is produced. Garrigue vegetation contains many aromatic shrubs that emit oils with a pungent odor particularly in the heat of day (Dallman, 1998). Plants of the mint family are common in the Mediterranean Basin e.g. rosemary (*Rosmarinus officinalis*) and Spanish lavender (*Lavandula stoechas*) both with blue to violet flowers. Jerusalem sage (*Phlomis fruticosa*) has very attractive bright yellow flowers. *Lavandula stoechas* is an aromatic plant growing wild on Palestine-Israel Med. coast but introduced elsewhere long ago as an ornamental plant. Its oil has long appealed to man, but in fact in nature discourages consumption by foraging animals.

Bulbs are most abundant in the garrigue of North Africa particularly in Morocco, including: narcissus (*Narcissus bulbocodium, N. tazetta*), crocus (*Crosus biflorus*), tulip (*Tulipa doerfleri*) and iris (*Iris* species).

The areas covered with maquis and garrigue in the North African and SW Asian Medi. coastal lands have been estimated by Dallman (1998), in million hectares, as follows: Morocco (5.2), Algeria (2.4), Tunisia (0.8), Libya (0.5), Palestine-Israel (0.1), Lebanon (0.1) and Syria (0.4).

#### 3.3.8.3 Coniferous Forests

Coniferous means related to or belonging to the plant group Coniferae. Conifer is any gymnosperm tree or shrub of the group coniferae typically producing cones and evergreen leaves. This group includes the pines (*Pinus* spp.), spruce (*Picea* spp.), firs (*Abies* spp.) larches (*Larix* spp.), yews (*Taxus* spp.), juniper (*Juniperus* spp.) cedars (*Cedrus* spp.), cypressus (*Cupressus* spp.). All conifers belong geographically to the North Hemisphere in areas with cold climate latitudinally and altitudinally.

Conifers make their appearance in the region of the true Mediterranean climate where mountains are high enough and/or air temperature is low enough to enable conifer trees and shrubs to grow and reproduce. The lower limit of conifers varies in altitude with climatic conditions. On the west Mediterranean coast there is usually a zone of mixed deciduous coniferous trees from about 450 m upwards but in cooler areas e.g. the Balkan Peninsula, conifers may be found in lower altitudes of the mountainous slopes. One species of pines, the pinaster (*Pinus pinaster*), frequently occurs near sea level (Branigan and Jarrett, 1975). Coniferous forest are actually adapted to a low-temperature climate, the opposite to mangrove which never appears except in hot coastal swamps of tropical seas and oceans (see Section 3.3.1).

Three coniferous forests dominated by pines are recognized by Dallman (1998) in the Mediterranean Basin: Stone (or Umbrella) pine, Maritime pine and Aleppo pine, all commonly growing close to the sea.

(a) Stone or umbrella pines (*Pinus pinea*) develop into an umbrella shape and most consistently grow up to 25 m high. It has been propagated since ancient times for its edible seeds called pignole, pignon or pine nuts. The seeds are bone on large, shiny, red-brown cones that take 3 years to mature, ripening in early spring to release seeds on very hot summer days or after a fire. Nuts are harvested by pulling the closed cones off the trees with a long hooked pole. One cone may contain 100 nuts. Dallman (1998) referred to *P. pinea* growing in the Iberian Peninsula, Italy, Greece and Turkey and Greco (1966) recorded it among the eight pine species of Algeria in the western section of the North African Med. Coast.

(b) Maritime pine (*Pinus pinaster*) grows primarily in the western Med. coast from Morocco in the south to Spain, France and Italy in the north. It is more frost sensitive than the Stone pine and since the sixteenth century it has been Europe's primary source of turpentine. *P. pinaster* trees have branches extending from the top third of a reddish trunk that reaches a height of 30 m. It has dark foliage and can form dense woods with an understorey of evergreen plants of the maquis such as tree heather (*Erica arborea*), strawberry tree (*Arbutus unedo*) and rock rose (*Cistus monspeliensis*). Plantings of *P. pinasters* have been used to stabilize sand dunes.

(c) Aleppo pine (*Pinus halepensis*) is found along the hotter parts of the Mediterranean coastal lands mainly in North Africa (Morocco, Algeria and Tunisia) and SW Asia (in Palestine-Israel) coastal high lands (mountains). *P. halepensis* is the most drought tolerant, and is the most susceptible to fire of these three pines. It has a round or pyramidal crown, most common and most widely distributed pine near coasts. It is usually found in groves of scattered trees mixed with maquis or garrigue vegetation of lavender (*Lavandula stoechas*), rosmary (*Rosmarinus officinalis*), thyme (*Thymus* spp.) and rockrose (*Cistus monspeliensis*) growing in the intervening open places. Aleppo pine is rich in resin used to flavour wines.

Apart from the trees of *Pinus* species, the coniferous forests of the North African and SW Asian Med. Coastal lands includes trees of cupressus (*Cupressus* species), cedars (*Cedrus* species), juniper (*Juniperus* species) and fir (*Abies* species). For example, Greco (1996) recorded in the flora of the Algeria coastal mountains eight species of *Pinus*: *P. burta, P. canariensis, P. coulteri, P. halepensis, P. insignis, P. pinea* and *P. radiate,* six species of *Cupressus*: *C. arizonica, C. atlantica, C. demakhar, C. dupreziana, C. glabra*, and *C. sempervirens*, and one species of each of *Juniperus* (*J. phoenicea*) and *Cedrus* (*C. atlantica*). The coniferous forests of the Palestine-Israel and Jordan coastal mountains contain: *Juniperus phoenica, J. oxycedrus, Cupressus sempervirens* and *Pinus halepensis* (Zohary, 1962; Al-Eisawi, 1996).

Though most of the Afro-Asian Red Sea coastal lands, including the southern part of Sinai, are located within the latitudes of the tropical hot climate (29° 50′N to 12° 35′N, 43° 3′E), yet their vegetation forms comprise coniferous forest essentially in the higher altitudes of the coastal mountains where temperature is very low and suitable for the growth. For example, in the southern mountains of Sinai the highest peak "St Katherine" is up to 2,641 m above sea level and the air temperature in winter is usually below zero. In Gebels Halal and Yalaq of the southern Sinai coniferous forest occurs in *Juniperus phoenicea* trees associated with: *Origanum isthmicus* (endemic in Gebel Halal) *Astoma seselifolium, Ephedra aphylla, E. campylopoda, Rubia tenuifolia* and *Cupressus sternbergia. Cupressus sempervirens* has been recorded by Migahid et al. (1959) in Gebel Halal, but neither Täckholm (1974) nor Boulos (1995, 1999) recorded it in the flora of Egypt. However, it is well known that few *Cupressus sempervirens* trees have been cultivated for centuries in the gardens of the Monastry in St Katherine in south Sinai. Thus, the *Cupressus* shrubs recorded by Migahid et al. (1959) might not be natural, but might also have been cultivated around the natural spring known as Ain[11] El-Goweirat of Gebel Hala. Today these plants are absent.

In the African Red Sea coastal mountains, the coniferous forests are not well developed. A few *Juniperus procera* trees have been recorded by Andrews (1950–1956) on the Red Sea coastal mountains of the Sudan*J. procera* and other coniferous trees are absent from the Egyptian Red Sea coastal mountains (Täckholm, 1956; Boulos, 1995; Zahran and Willis, 2008). However the situation is different in the Asian Red Sea coastal lands. Dense coniferous forests characterize the south western mountains of Saudi Arabia (the Asir). Allered (1968) estimated about 25 million acres of coniferous forest at the tops of these mountains. The dominant coniferous trees are *Juniperus procera* and *J. polycarpos* with abundant growth of *Cupressus sempervirens* (Migahid and Hammouda, 1978). The highest altitude (about 3,000 m) on Al-Sudda summit, 25 km south of Abha city, represents the water shed zone of these mountains. In this zone, *Juniperus* trees establish a closed canopy with little understorey. Trees up to 17–20 m high and 70–85 years old have been recorded (Zahran, 1982b). At the lower altitudes (< 2,000 m), *Juniperus* trees still predominate associated with trees and shrubs of other species such as *Olea chrysophylla, Pistacia palestina, Dodonaea viscosa, Commiphora* spp., and *Dracaena ombet*., woody herbs, such as *Psiada arabica, Eryops arabica, Lavandula dentata,* and *Euphorbia retusa*, grasses, such as *Themeda triandra,* and *Hyparrhenia hirta* and succulents such as *Euphorbia thi* and *Caralluma* spp. (*C. retrospeciens, C. pedicillata* and *C. sinaica*). In zones lower than 1,000 m, where there are noticeable change in climatic conditions, mainly an increase in air temperature, *Juniperus* trees disappear and *Acacia* spp. (*A. etbaica, A. nubica* etc.) dominate. The associates are: *Commiphora* spp. (*C. africana, C. opobalsamum, C. quadricincta*), *Grewia tenax, Olea chrysophylla, Ficus pseudosycomorus,* and *Adenium arabicus*. *Acacia asak, A. mellifera* and *A. etbaica* scrubland are abundant in these lower zones associated with *Themeda triandra, Artemisia judaica, Pulicaria crispa, Euphorbia cuneata, Scorzonera intricta, Aloe vera,* and *Euphorbia nubica*.

In the coastal mountains, of Yemen, Wood (1997) stated that *Juniperus procera* is the only native conifer. It grows on the altitude ranging between 1,800 and 3,300 m. Although *Juniper* woodland is probably the natural climax there, its thick vegetation, which is such a characteristic feature of the Asir Mountains of Saudi

[11] Ain is the Arabic for spring.

Arabia, is not recognized in Yemen. This presumably is due to cutting and overgrazing through many centuries. In fact, coniferous forest throughout the Afro-Asian Med. and the Red Sea coastal lands have been and still are exposed to uncontrolled exploitation for many purposes e.g. building constructions, ship building, furniture, railway sleepers, harbour works and pulp and paper production. Many areas have been unfortunately, deforested with subsequent soil erosion. Projects aiming at conservation, rehabilitation and afforestation of these precious coniferous forest are badly needed.

#### 3.3.8.4 Scrubland Vegetation

Scrubland is an arid land vegetation type formed mainly of woody trees, shrubs and undershrubs associated with some succulents, grasses and herbs, all are xerophytes. It is a characteristic ecological feature of the wadis and mountains of the Red Sea coastal deserts and the southern part of Sinai. Zohary (1962) recorded 60 tropical species of this vegetation from the Palestine-Israel coast. Boulos (1983) referred to a few medicinal woody trees and shrubs belonging to this vegetation form from North African countries.

The floras of the scrubland vegetation of the Afro-Asian Red Sea coastal lands comprise elements belonging to various tropical floras, most important among them are the Sudanian elements, such as species of *Acacia, Phoenix, Balanites, Dracaena, Moringa, Salvadora, Hyphaene, and Caralluma.*

The geographical distribution of scrubland vegetation form along the Afro-Asian Mediterranean and Red Sea coastal deserts and its altitudinal zonation on the coastal mountains has been discussed by many authors e.g. Tothill (1948), Kassas (1956, 1957, 1960), Ozenda (1958), Keay (1959), Hemming (1961), Zohary (1962), Greco (1966), Kassas and Zahran (1971), Pottier-Alapetite (1981), Boulos (1983, 1995, 1999, 2000, 2002, 2005), Konig (1986), Wood (1997), and Zahran and Willis (2008). These and other studies have clarified the relationship between climatic conditions and the latitudinal distribution of scrubland vegetation and its floristic elements. The altitudinal zonation pattern of this flora on the slopes of the mountains of the Red Sea also has been discussed. These are outlined below.

**A. Geographical distribution**

**a. *Acacia* spp.**

More than 10 species of *Acacia* are recorded in the Afro-Asian Mediterranean and Red Sea coastal lands, each has its own distribution pattern.

1. *Acacia raddiana* (= *A. tortilis* subsp. *raddiana*) is a woody tree (> 12 m high) that seems to have wide ecological amplitude. It grows everywhere in the Afro-Asian Med. and Red Sea coastal lands but with varying abundance.
2. *Acacia pachyceras* var. *najdensis* (= *A. negevensis*) is a tree 3–8 m high, restricted to the SW Med. coastal desert and south Sinai but absent elsewhere.

3. *Acacia tortilis* (= *A. tortilis* subsp. *tortilis*) is a tree up to 5 m high, dominant in the wadis of Afro-Asian Red Sea coastal deserts, common in Sinai and Palestine-Israel desert but absent from the North African Med. Coast.
4. *Acacia laeta* is a shrub or tree up to 5 m high, abundant in the Afro-Asian Red Sea coastal desert, common in the SW Asian Med. coast but absent from the North African Med. coast.
5. *Acacia ehrenbergiana* is a shrub 1.5–4 m high, common in the Afro-Asian Red Sea coastal deserts including south Sinai but completely absent from Afro-Asian Med. Coast.

   The other *Acacia* spp., are: *A. albida* (= *Faidherbia albida*), tree up to 18 m high, *A. etbaica, a* shrub or small tree, 4–6 m high, *A. mellifera, a* shrub or small tree, 6 m high, *A. oerfota* (= *A. nubica*)*, a* shrub or small tree, 1.5–4 high and *A. asak, a* shrub or small tree, 6 m high. All of threse are restricted in distribution to the Afro-Asian Red Sea coastal deserts but absent from the south Sinai as well as from the whole of the Afro-Asian Mediterranean coastal lands. *Acacia cyanophylla* is a tree 5 m high, recroded by Migahid and Hammouda (1978) in the Red Sea coastal mountains of Saudi Arabia but recorded neither from Egypt, nor from Yemen. *A. edgeworthii* is a shrub 1.5 m high recorded by Wood (1997) from the stony steppe between 200 and 400 m altitude from the Tihama mountains of the Yemen coast but recorded neither in Saudi Arabia nor from the African Red Sea coast. However, *A. albida* and *A. cyanophylla* are known to be cultivated in Yemen (Wood, 1997). Kassas (personal communication) informed the author that *A. cyanophylla* and *A. salegna* are cultivated in Egypt, both are not African species but introduced from Australia.

**b. *Other woody and succulent species***

Apart from the woody *Acacia* species, the scrubland vegetation of the Afro-Asian Red Sea coastal lands contains many other woody and succulent trees, shrubs and undershrubs, all are xerophytes. These species can be classified according to their geographical distribution into three groups: species restricted to the Afro-Asian Red Sea coastal deserts, species restricted to south Sinai, and (3) species growing in both regions.

1. Scrubland species restricted to the Afro-Asian Red Sea coastal deserts and absent from south Sinai are mostly woody plants, with a few succulents. The woody species include: *Commiphora africana* (a shrub or tree), *C. opobalsamum* (a strong smelling shrub or small tree), *C. quadricincta* (an armed shrub), *Dodonaea viscosa* (a shrub), *Dracaena ombet* (a small stout tree growing at high altitudes in the mountains), *Euclea schemperi* (a shrub or small tree), *Grewia villosa* (a shrub), *G. tembensis* (a shrub), *Sida ovata* (a shrub), *Cadaba farinosa* (a shrub) *C. glandulosa* (a shrub), *C. rotundifolia* (a shrub), *Zilla spinosa* subsp. *spinosa* (a shrub), *Boscia senegalensis* (a shrub or small tree), *B. angustifolia* (a shrub or small tree), *Aerva lanata* (a shrublet), *Maerua oblongifolia* (a shrub), *Solanum*

*incanum* (an undershrub), *S. careens* (an undershrub), *Abutilon bidentatum* (an undershrub), *A. longicuspe* (an undershrub), *Senna holosericea* (= *Cassia holosericea, an* under shrub) *Rhus abyssinica* (a tree or shrub), and *Olea europaea* subsp. *cuspidata* (= *O. europaea* subsp. africana, a tree or large shrub), *Jasminum floribundum* (a shrub or scrambler), *J. fluminense* (a climbing or scrambling, a shrub), *Sterculia africana* (a tree), and *Cordia sinensis* (= *C. gharaf,* a tree or shrub). The succulent species of this group include: *Euphorbia thi* (a shrub), *E. cuneata* (a shrub), and two *Caralluma* species, namely: *C. acutangula* and *C. retrospiciens* (both Cactus like bushes).

The flora of Yemen comprises the greatest number of *Caralluma* species (20 species) among the other floras of the countries of the Afro-Asian Red Sea coastal lands (Wood, 1997). The species recorded in the Tihama mountains include: *Caralluma acutangula* (= *C. retrospiciens*), *C. commutata, C. deflersiana, C. penicillata* (abundant in *Acacia* and *Euphorbia* scrub), *C. plicatiloba, C. quadrangular* (endemic to SW Arabian Peninsula) and *C. subulata* (endemic to SW Arabian Peninsula).

2. Scrubland species restricted to south Sinai include: *Capparis spinosa* var. *aegyptiaca* (a shrub), *Solanum sinaicum* (an undershrub), *Rhamnus lycioides* subsp. *graeca* (a shrub) and *R. disperma* (a shrub) and two succulent undershrubs: *Caralluma sinaica, C. europaea* (= *C. negevensis*).
3. Scrubland species growing in the Red Sea coastal deserts as well as in south Sinai include the following: *Ficus palmata* (*F. pseudosycomorus*, a shrub), *Maerua crassifolia* (a shrub or small tree), *Rhus tripartita* (a shrub), *Salvadora persica* (a shrub), *Pistacia khinjuk* (a tree), *Balanites aegyptiaca* (a tree), *Ziziphus spina-christi* (a tree), *Tamarix aphylla* (a tree or shrub), *Lycium shawii* var. *shawii* (a shrub), *Leptadenia pyrotechnica* (a shrub or small tree), *Maerua crassifolia* (a shrub or small tree), *Hyphaene thebaica* (a tree), *Phoenix dactylifera* (a tree), *Calotropis procera* (a shrub), *Retama raetam* (a shrub), *Launaea spinosa* (a shrub), *Moringa peregrina* (a tree or shrub), *Grewia tenax* (a shrub), *Aerva javanica* var. *javanica* (an undershrub), *Senna italica* (= *Cassia obovata*, an undershrub), *S. alexandrina* (= *Cassia acutifolia*, an undershrub) and *Cocculus pendulus* (a liane or shrub).

Species belong to the scrubland vegetation, both woody and succulents, have been also recorded in the Afro-Asian Mediterranean coastal lands. The flora of Palestine-Israel coastal lands comprises about 60 tropical species many of them belong to the scrubland vegetation form (Zohary, 1962). These include: *Acacia* spp. (*A. pachaceras* var. *najdensis* (= *A. negevensis*), *A. raddiana, A. tortilis and A. laeta*), *Calotropis procera, Ziziphus spina-christi, Balanites aegyptiaca, Sebestena gharaf,*[12] *Grewia villosa, Maerua crassifolia, Abutilon pannosum* (= *A. muticum*),

[12] Prof. L. Boulos informed the author that *Sebestena gharaf* could be *Cordia gharaf* (*C. sinensis*).

*A. fruticosum, Solanum incanum* and *Cassia obovata*. Pottier-Alapetite (1981) reported few tropical species from the flora of Tunisia, namely: *Acacia raddiana, Calligonum comsum, Capparis spinosa, Retama raetam* and *Tamarix aphylla*. Ozenda (1958) recorded a large number of tropical genera and species from the inland North African desert, far from the Med. Coast. Most of these are elements of the scrubland vegetation of the Red Sea coastal deserts such as: *Acacia albida, A. raddiana, Balanites aegyptiaca, Boscia salicifolia, Cadaba farinosa, C. glandulosa, Calligonum comosum, Caltropis procera, Capparis spinosa, C. galeata, C. decidua, Caralluma tombuctuensis, C. venenosa, Ficus carica, F. salicifolia, F. insens, Grewia populifolia, Hyphaene thebaica, Phoenix dactylifera, Leptadenia pyrotechnica, L. heterophyllum* (= *L. arborea*), *Lycium shawii* (var *shawii*), *Maerua crassifolia, M. angolensis, Olea laperrini* (endemic), *Panicum turgidum, Pistacia atlantica* (endemic), *Ziziphus mauritianus, Rhamnus lycioides, Salvadora persica, Tamarix aphylla* and *Zilla spinosa*.

Interesting notes on *Olea europaea* and *Caralluma* species are given by Zohary (1962) and Boulos (2000). *Olea europaea* subsp. *europaea* var. *europaea* is widely cultivated in all countries of the Afro-Asian Med. coastal lands, whereas *O. europaea* subsp. *europaea* var. *sylvestris* grows naturally on the North African Med. coast (Boulos, 2000). Zohary (1962) recorded O. *europaea* var. *oleaster* in the carob-lentisk maquis vegetation of the SW Asian Med. coast of Palestine-Israel, mentioning also that the genus *Caralluma* is represented by two related endemic species and one endemic variety. *C. negevensis* is a Saharo-Sindian species, *C. aaronis* is an Irano-Turanian species, while *C. europaea* var. *judaica* is Mediterranean. *C. maris-mortus* (= *C. sinaica*) is an endemic species. He added "All the species of *Caralluma* are without doubt relics of the Tertiary period during which time tropical vegetation prevailed over most parts of Palestine".

**B. Altitudinal zonation**

Altitudinal zonation is obvious in the scrubland vegetation of the Red Sea coastal mountains of Egypt, the Sudan and Saudi Arabia.

**(a) Red Sea coastal mountains of Egypt**

The range of mountains extending along the Egyptian Red Sea coast can be classified under two main sections: 1. the mountains of the northern section facing the western coast of the Gulf of Suez from Suez (27° 20′N) to Hurghada (27° 14′N) and the mountains of the southern section facing the western coast of the Red Sea proper from Hurghada to Mersa Halaieb (22°N) at the Sudano-Egyptian border. The mountains of the northern section are (from north southwards): Gebel Ataqa (817 m), Gebel Kahalya (660 m), Gebel Akheider (367 m), Gebel El-Galala El-Bahariya (700 m), Gebel El-Galala El-Qiblya (1,200 m), Gebel Abu Dokhan (1,705 m), Gebel Qattar (1,963 m), Gebel Shayeb El-Banat (2,187 m) and Gebel Umm Anab

(1,782 m). Gebel Shayed El-Banat is the highest of these peaks extending some 40–50 km to the west of Hurghada and representing the southern part of the Gulf of Suez and NE part of the Red Sea proper (Zahran and Willis, 2008).

The mountains of the southern section are the Gebel Nugrus group (24° 40′N to 24° 50′N) including: Gebel Migif (1,198 m), Gebel Zabara (1,360 m), Gebel Nugrus (1,504 m) and Gebel Hafafit (857 m), Gebel Samiuki group comprises three main blocks: Gebel Abu Hamamid (1,786 m), Gebel Samiuki (1,283–1,486 m), and Gebel Hamata (1,977 m) and Gebel Elba group on the Sudano-Egyptian border (22°N) and comprises Gebel Elba (1,428 m), Gebel Shindeib (1,911 m), Gebel Shindodai (1,526 m), Gebel Shillal (1,409 m), Gebel Makim (1,871 m) and Gebel Asotriba (2,217 m). Gebel Elba is particularly favoured by its position near the sea. The richness of the plant life of Gebel Elba area is so notable that it is considered to be one of the main phytogeographical regions of Egypt (Täckholm, 1974).

Plant life of the Egyptian Red Sea coastal mountains depends on a combination of factors: (i) expanses of water body traversed by wind before reaching the coast, (ii) distance from the shore line to the mountains and the topographic features of the stretch, (iii) altitudes and (iv) exposure. Plant life on mountains (scrubland vegetation) facing limited body of water of the Gulf of Suez (i.e. northern section), indicate an arid climate. By contrast, plant life on the Elba group indicate a less arid climate and present an example of a mist oasis. Altitudinal zones of vegetation are recognizable on the slopes of the higher mountains differences between the plants of seaward and of leeward slopes are obvious (Kassas and Zahran, 1971).

The effect of topography on precipitation is a universal phenomenon but is more pronounced near the coast. Though located in arid regions, the Red Sea coastal mountains, particularly the Gebel Elba group, can cause ample orographic precipitation. The northern blocks of mountains receive lesser water from orographic rain. Such differences in the amounts of rainfall have their effects on the plants of these mountains. In terms of the number of species recorded from the four mountain groups it was found that: 53 species from the Gebel Shayeb El-Banat group, 92 species from the Gebel Nugrus group, 125 species from the Gebel Samiuki group and 458 species from the Gebel Elba group. Though it is not the highest, Gebel Elba is the nearest to the sea (20–25 km) facing NE bend of the shore in such a manner that it faces in northward direction an almost endless stretch of water. In contrast, Gebel Shayeb El-Banat faces only a narrow stretch of water (the mouth of the Gulf of Suez). The amount of orographic rain is rather higher on Gebel Elba than that on Gebel Shayeb El-Banat.

In Gebel Elba, plant growth on the slopes of different exposures are obvious: north and east slopes have richer plant growth than south and west slopes. The scrubland vegetation on north slopes varies in relation to altitude, but on south slopes there are few differences except perhaps between the more or less barren slopes and the vegetated runnels that dissect them. Four main altitudinal zones are distinguished on the north and east slopes of Gebel Elba (Kassas and Zahran, 1971):

(a) a basal zone of *Euphorbia cuneata* scrub,
(b) a middle zone of *Euphorbia nubica* scrub,
(c) a high zone of *Acacia etbaica* scrub and,
(d) a tope zone with a patchwork variety of plants.

The altitudinal limits of these zones are not fixed by absolute levels but related to a combination of altitude, degree of slope, air temperature, wind velocity and other climatic and physiographic features. The fourth top zone comprises stands of *Dracaena ombet, Dodonaea viscosa, Delonix elata, Euclea schimperi, Rhus abyssinica, Ficus salicifolia* and *Pistacia khinjuk.* Within this zone, ferns and bryohytes also abound e.g. *Adiantum capillus veneris, Actiniopteris australis, Fumaria pallenscens, Riccia* spp., and *Mannia androgyne.* This may indicate that the plants growing at the top of Elba do not belong to the arid climate regions as they comprises a great variety of species of wet areas with high water requirements. On the south and west slopes of Gebel Elba, plants growing are mostly confined to runnels with an open scrubland type dominated by *Commiphora opobalsamum.* At the mountain base the runnels are mostly dominated by *Acacia tortilis* scrublands (for details please see section "In Egypt", II. Coastal mountains of Chapter 2).

**(b) Red Sea coastal mountains of the Sudan**

The Red Sea coastal mountains of the Sudan can be represented by the Erkwit Plateau (19°7′N, 37°20′E, Alt. 1,080 m). Erkwit is a deserted summer resort, lies at about 45 km to the south-west of Suakin on the Red Sea coast and about 30 km to the east of Sinkat on the inland plain at the edge of a steep escarpment dropping abruptly to the Red Sea plains. At its northern boundary are Gebel Nakeet (1,176 m) and Gebel Essit (1,143 m), these two mountains drop to Khor Dahand which separates Erkwit from the barren hills on the other side of the Khor. At the eastern boundary is Gebel Sela (1,273 m) the highest evergreen mountain of the district. At the southern boundary are Gebel Tatasi (1,190 m), Gebel Lagagribab (1,209 m), and Gebel Auliai (1,191 m) separated by Khor Amat from the barren mountain further south (Gebel Erbaba, 1,923 m). At the eastern boundary are Gebel Hadast (1,147 m) and Gebel Mastiokriba (1,113 m) which also drop to Khor that separates the Erkwit Plateau from the desert plain to its east (Kassas, 1956).

Erkwit receives a rainfall (218 mm) greater than the neighbouring areas, as compared to data from two stations: Suakin (181 mm) to the NE and Sinkat (127 mm) to the west. Suakin, a Red Sea port, represents areas with winter rainfall. Sinkat, which lies on the inland plain, represent areas with summer rainfall. Erkwit lies in between and receives both the summer and winter rainfalls. Tothill (1948) stated that the climate of Erkwit may be due to the happy combination of: latitude, situation and elevation. Its latitude is close to 19°N which divides the Sudan into an arid desert region to the north and a tropical continental region to the south (Ireland, 1948). In the north prevails the dry northerlies while in the south the southerlies bring summer rainfall. Erkwit enjoys the maritime modification of the Red Sea: as the continental northerlies pass over the warm water of the sea they absorb a considerable amount of moisture. Being situated on a westerly bend of the Red Sea, Erkwit is exposed to about 650 km of open sea in the direction facing the northerlies (the N.E. Trades). The moisture-laden wind meets no dissipating obstacle before impinging on the cool hills of Erkwit and orographic precipitation consequently occur. After passing over the Erkwit Plateau the northerlies resume their dry continental characteristics. During the winter months, the Erkwit Plateau is frequently swathed in clouds for weeks. This entails considerable dew precipitation which is more marked the higher the elevation, and which supplies the vegetation with a valuable water resources (Kassas, 1956).

The vegetation of Erkwit has a trizonal pattern: an arid, a transitional and a moist zone (Andrews, 1948). However, Kassas (1956) point out the features of five vegetation zones dominated and co-dominated by: *Maytenus senegalensis* (zone I), *Maytenus senegalensis – Euphorbia abyssinica* (zone II), *Euphorbia abyssinica* (zone III), *Dracaena ombet – Euphorbia abyssinica* (zone IV) and *Euphorbia thi* (zone V). Such zonation is actually a climatic phenomeon related to the wetness and dryness of the zones (for details of these five altitudinal zones please see section "In the Sudan" of Chapter 2).

The relationship between the local climate and vegetation is obvious in the Erkwit Oasis, deduced from the following (Kassas, 1956):

1. The total plant cover varies from 70% in particular habitats of the wettest zone I to only 5% in the driest zone (zone V),
2. The distribution of ferns, mosses and liverworts is limited to zones I and II evidence for interzonal differences in water resources available to the roots, Atmospheric moisture is much greater in zone I. gradually decreasing downwards in zones III, IV and V.
3. Vegetation is denser on the seaward slopes of the mountains than the leeward ones. This shows, again, that the atmospheric humidity (mist and wind-born water) is an effective water resource.
4. Generally, the conditions that made the limited mountainous area of Erkwit into an oasis with thick vegetation and a wide range of plant diversity amidst arid country are: firstly, the combination of summer rainfall of the territory to its west and winter rainfall of the Red Sea to its east; secondly, it receives sea-mists and wind-born moisture which face no obstacle until they meet the edge of the Erkwit plateau.

5. Local differences in physiographic factors, altitudes and distance from the edge of the escarpment cause the zonal pattern of the vegetation due to difference in the amount of available water.

**(c) Red Sea coastal mountains of Saudi Arabia**

The altitudinal zonation of the vegetation of the Saudi Arabian-Asian Red Sea coastal mountains is described in details by Konig (1986) in six locations: Tanuma, Gabal Sauda-ad-Darb, Slope of Hilly Tihama, Gabal Sauda, Al-Ulayya-Bisa, and draining runnels and wadis.

A short note on the zonation pattern of Al-Ulayya-Bisa region is given below.

The mountains of Al-Ulayya-Bisa region reach on altitude of 2,200 m. and the escarpment to the Red Sea is comparatively moderate so that rain shadow effects along the eastern slopes are not so marked (mean annual = 137 mm) as compared with that of Gabal Sauda, with an elevation up to 2,650 m, being 559 mm/year. Here, five zones of vegetation are recognized depending upon elevation: 2,200 m (*Juniperus* woodland), 2,000 m (*Olea* sclerophyllous scrub), 1,800 m (*Acacia gerrardi* xerophytic scrubland), 1,600 m (semi-desert) and 1,400 m (*Acacia-Commiphora* xerophytic scrubland). The *Juniperus* woodland of the higher altitude include: *Juniperus excelsa, Euryops arabicus, Euphorbia schimperiana* and *Cultia richardiana.* Below 2,100 m, *Acacia gerradii* becomes more characteristic of the vegetation, beginning with *Juniperus-Acacia gerrardii* woodland with *Cultia richardiana, Sageretia thea* ssp. *thea* and *Dodonea viscosa.* This community is replaced by *Olea-Acacia gerrarrdii* woodland below 2,000 m with characteristic species: *Acacia gerrardii, Olea europaea* spp. *africana, Barbeya oleoides, Pistacia chenensis* ssp. *falcata, Jasminum floribundum, Euclea schemperi, Ficus salicifolia, Psiada punctulata, Carissa edulis* and *Hypoestes forsskaolii.* With increasing distance from the watershed, only *Acacia gerrardii* xerophytic scrubland occurs, which is replaced below 1,850 m by semi-desert vegetation characterized by scattered dwarf shrubs such as: *Teucrium yemense, Campylanthus pungens, Pulicaria crispa, P. somalensis* ssp. *somalensis, Heliotropuim strigosum, Morettia canescens, Crotalaria emarginella, Salvia aegyptiaca, Lavandula pubescens, Aerva persica, Farsetia longisiliqua, Fagonia bruguieri, Indigofera spinosa, Periploca aphylla, Otostegia fruticosa,* and *Lycuim shawii,* in addition to grasses such as *Eneapogon schimperianus, Tetrapogon villosus, Danthoniopsis barbata* and *Eneapogon brachystachyus.* A particularly conspicuous species is the highly endangered *Dracaena ombet* found between 1,850 and 1,700 m where it grows in a number of places on the NW slopes. Below 1,600 m, *Acacia – Commiphora* xerophytic scrubland with *A. asak, A. hamulosa, A. tortilis, Commiphora myrrha, C. opobalsamum* and *Euphorbia cuneata* occurs.

# Chapter 4
# Climate–Vegetation and Human Welfare in the Coastal Deserts

## 4.1 Introduction

Deserts, coastal and inland, represent serious challenge for life on Earth. The lack of water makes survival difficult for both wildlife and people, but the desert has a fascinating allure and a rich wildlife. The unique biota of deserts are ancient civilization including fragile creatures that need to be protected for the welfare of desert peoples (Oldfield, 2004).

Deserts have attracted the attention of explorers for thousands of years, and still do. The ancient explorers include the Great historian Herodotus, who traveled through Egypt and Central Asia nearly 2,500 years ago and Alexander the Great of Macedon, who, shortly after, led a huge army from central Europe to become one of Egypt's Pharoahs. In the fourteenth century the great Muslim explorer Ibn Battouta set out to explore the known Muslim world, undertaking numerous journeys from his own country Morocco across the Sahara to Timbuktu located on the River Niger in central Mali and throughout Arabia, and India as far as China.

Ecologically, the desert is a unique ecosystem with two large components: the physical environment (abiotic) and the biological environment (biotic). The physical environment includes: atomospheric factors (climate and gases), topography, soil, ground water etc., whereas its biotic components are the green plants (producers), domestic and wild animals (consumers), microorganisms (decomposers) and man (the main beneficiary). The physical environment provides the ecosystem with renewable energies (solar and wind), raw materials such as gases, minerals, and living space that the biological community needs for growth and maintenance. All biotic and abiotic components of the ecosystem are so connected in their functions that it is difficult to describe the system by separate categories classified according to the roles these components play, all operate in harmony. The state of balance and stabilization of the desert ecosystem may continue forever unless it is exposed to external disturbing factors which can be natural (earthquakes, floods, volcanoes etc.) or man's overuses of the natural resources.

The different vegetation forms of the Afro-Asian Mediterranean and Red Sea coastal deserts are the outputs of the prevailing climates, with factors of the physical environment also important. These vegetation forms comprise many plant species

M.A. Zahran, *Climate–Vegetation*, Plant and Vegetation 4,
DOI 10.1007/978-90-481-8595-5_4, © Springer Science+Business Media B.V. 2010

adapted to live and reproduce under the environmental stress of the arid climate. These plants also provide the local people of these deserts with most of their daily needs of food, fodder for livestock, medicine, wood, fuel etc.

Now a question can be raised: what is the role that could be played by the natural vegetation growing under the arid climate of the Afro-Asian Med. and Red Sea coastal deserts for the welfare of the local people? The answer to this question will help greatly in the developmental plans of these coastal deserts. It is well known that the flora forming these vegetation forms have low water requirements and some of them are salt tolerant. At the same time, many of these plants have proved to have agro-industrial potential. Having these advantageous properties, these plants could be propagated under the arid climate to be introduced as non-conventional crops in coastal deserts using the available water, mainly ground and partly rainfall, for irrigation. The fruits, seeds, branches, bulbs, and tubers of many species are eaten by local people while others are components of traditional medicines and yet a third group are palatable by livestock. Also, the vegetative yields of some desert species, if scientifically managed, could be used as raw materials for strategic industries such as paper pulps, drugs, perfumes, fodder, etc. Apart from these valuable desert plants, renewable energies to run the development plans of coastal deserts could be secured from solar and wind energy of the prevailing climate.

To achieve our goal, three major issues will be discussed under three topics:

1. Degradation of desert vegetation;
2. Conservation of desert vegetation;
3. Vegetation and sustainable development of the deserts.

Degradation and conservation of vegetation are human activities but conducted for opposite purposes. Degradation leads to destruction of vegetation and removal of the only producer component of the desert ecosystem, resulting in excess areas of non-productive lands added to the present deserts (desertification). Conservation of vegetation, on the other hand, if scientifically oriented and well managed, will result not only in maintaining the natural vegetation in good condition but also in supporting its sustainable development, increasing land productivity aimed at securing the daily necessities for the livelihoods of local desert people.

## 4.2 Degradation of Desert Vegetation

Land[1] degradation is a huge problem in terms of the area affected with as much as 80% of arid, semi-arid and dry subhumid climatic zones suffering some degree of soil and/or vegetation degradation. Vegetation in rangelands is, by far, the most extensive land degradation (Dregne, 1997).

The drylands, which include coastal and inland deserts, occupy one third of the Earth's land surface but support less than one seventh of its population. They are

---

[1]Land here means all resources namely: soil, ground water, plants, animal, etc.

recognized as areas of declining productivity, deteriorating human conditions and settlement retreat: where environmental degradation calls for urgent remedial action (Mabbutt, 1978). However, the populations of these coastal and inland deserts are increasing considerably from about 628 million people in 1977 to 1,800 millions in 1994, i.e. threefold increase within less than 20 years. The requirements of these people increase at the same rate. The only alterative is the rational use of deserts natural resources in a sustainable manner (Dregne, 1997).

Low and uncertain rainfall is the major environmental problems in the drylands and this is not likely to be prevented by technological developments in the near future. The rainfall of drylands can vary greatly seasonally, with wide fluctuations over years and decades frequently leading to long periods of drought. Over time the, dryland (desert) ecosystem has become attuned to this variability in moisture levels with plants, animals, and microorganisms able to respond quickly to its presence or to with stand its absence. Scientifically and well managed irrigation method using the available water (mainly ground water, promises a direct remedy for rainfall deficiencies).

Mclillo and Salama (2008) reported that people have survived in dryland areas by adjusting themselves to the natural fluctuation of the arid climate. They have learned to protect the biological and economic natural resources of their lands (soil quality, fresh water supply, vegetation, domestic and wild animals, etc.) with age-old strategies such as adopting nomadic life-styles in their agricultural practices and in the raising of livestock. However, in recent decades these traditional strategies have become less practical with changing economic and political circumstances, population growth and a tendency towards more settled communities. When land managers cannot or do not respond with flexibility to climate variations, desertification can easily be the result.

The impacts of man the natural vegetation is very ancient, starting with primativ hunter-gatherers approximately 8000 B.C. before the Mesolithic period (Garrod and Bate, 1937). These activities did not affect vegetation to any great extent. With the development of sedentary farming, large stretches of woodlands were turned into arable lands, resulting in the disappearance of large areas of arboreal plant associations. In the Mediterranean region, for example, degradation of vegetated areas was very intense caused by the rooting out of plants covering hundred of square km in the midst of a potential woodlands which become bare rocky outcrops (Zohary, 1962). The primitive farmers cultivated these degraded areas with their crop plants.

The impacts of man on desert vegetation is manifested in various ways. Among these, there is the direct eradication of vegetation through excessive use, followed by spontaneous plant migration from adjacent areas. Man replaces natural habitats by artificial ones thus creating new vegetation types. Along with cultivated plants, man introduces (non-intentionally) weeds and adventitious plants which can adversely affect or even exterminate indigenous species. No doubt, man is responsible for remobilizing fixed sand dunes.

Destruction of vegetation by man is discussed by Zohary (1962) for the Palestine-Israel Med. Coastal lands, where there has been heavy damage of forest to obtain wood for charcoal production and other local needs as well as for occasional export to neighbouring countries. Grazing, though unable to bring about the total destruction of the indigenous vegetation, has led to the impoverishment or annihilation of the life-sustaining ecosystem.

Degradation of vegetation has been also discussed by Nahal (1997) who stated that in the arid, semi-arid and dry sub-humid climatic zones, the major causes of vegetation destruction are: the misuse, mismanagement, overexploitation and overgrazing of the renewable natural resources. These impacts have lead to the deterioration of fragile desert ecosystems, by changing the quality of its rangelands, loss of biodiversity, degradation of natural habitats, and the rarefaction or extinction of plant and animal species by the loss of ecotypes, races and varieties. Vegetation forms of the coastal and inland deserts, including those of the Afro-Asian and Red Sea coastal lands, are shrinking at an accelerating rate particularly in relatively highly populated areas.

Deserts, unlike the other biomes of the world, are expanding and are taking over areas of grasslands and woody dry savannah. This process, according to (Oldfield, 2004) is known as desertification or the diminution or disturbance of the biological potential of the land, leading ultimately to desert-like conditions. Overgrazing and excessive fuel-wood extraction in desert management results in loss of vegetation cover and soil erosion. Man-made deserts are usually barren and almost lifeless.

Invasion of plants to colonise new territories is a result of intentionally or unintentionally man's activities that may cause considerable disruption to native ecosystems such that it is biologically unsound to consider introduced species as adding to the biodiversity of the region (Drake et al., 1989). By introducing alien species human have strongly influenced the natural distribution patterns of many species with unfortunate consequences for many native communities.

In view of the alarming status concerning the ongoing processes of vegetation degradation, one might expect a large decline of biological diversity particularly in the coastal and inland deserts (Blondel and Aronson, 1999). Historical records, both palaeontologically and archaeologically, shows that human-induced decline of biological diversity started many thousands of years ago in most parts of the Mediterranean basin. Greater losses of the biota are to be expected if humanity continues its current unsustainable use of natural resources. Along the coastal lands of the Afro-Asian Med. and Red Sea coasts, man's impacts on the biodiversity by for example establishing new summer and winter resorts, has exacted a heavy and in some cases potentially catastrophic price on the biodiversity of these coastal lands. The loss of biodiversity is an irreversible consequence of environmental degradation.

Scientific evidence identifies degradation and desertification particularly in the coastal and inland deserts, as a serious and accelerated threat to human welfare at a time when an enormous increase in food production is needed. The problem is undeniable urgent because land reclamation costs rise steeply as degradation proceeds. Man rather than climate is the chief agent of degradation and desertification. Tolba (1997) as he seeks to wrest a living from the fragile desert ecosystem under harsh and unpredictable climatic condition and a variety of social and economic pressures, thus there is misuse or overuse of the land. Too frequently people have acted in this way because no other alternatives are apparent. Although man's impacts are the agent of land degradation, but at the same time people are also its victims.

To achieve the goal of reducing land degradation through dryland development, planners should address three areas (Tolba, 1997): social, economic and political development with emphasis on issues such as poverty, food housing, employment, health, education, population pressure and demographic imbalance, conservation of natural resources with emphasis on water, energy, soil, minerals, plant and animal resources, and sustainable environmental use by emphasizing soil fertility, and preventing soil loss, pollution and deforestation.

Revegetation of endangered areas of the coastal and inland deserts largely with xerophytes, halophytes and psammophytes could be an appropriate solution. Woody shrubs and undershrubs have been successfully utilized to stabilize dunes, and building ground. Once established, such areas can be a source of much-needed fuel. With good management, grazing will be possible, the roots, young shoots and leaves of the cultivated woody species could provide valuable industrial materials such as waxes, gums, medicine, fiber, edible oils, etc. (Schechter, 1978).

## 4.3 Conservation of Desert Vegetation

Conservation, biologically, means protection, preservation and careful management of natural resources and of the environment.

For the preservation of natural resources for present and future generations, arid land countries should adopt a well-defined national policy and strategy on nature conservation and management for keeping their ecosystems running.

The poorest communities, mostly in arid lands, are those most vulnerable to ecological disasters. Ecosystem exhaustion inexorably brings famine, disease and mass migration. Converting environmental conservation policies into developmental opportunities should be the objective of all governments (IUCN, 2008).

The desert ecosystem and its natural vegetation are in need of conservation attention worldwide. Changes in traditional land use and uncontrolled developmental pressures can have detrimental effects on this fragile ecosystem and although its plants (and animals) are adapted to survive in harsh conditions, they are also vulnerable to human over-exploitation. The fine balance between man and the desert ecosystem is easily upset particularly when increasing numbers of people need to make a living from marginal land: the carrying capacity of the desert vegetation will not be enough leading to overgrazing and overuse and consequently destruction of vegetation (Oldfield, 2004). Action to conserve habitats and the biota of desert ecosystems need to be planned and carried out at a local level with the full involvement of those who depend on the land for their livelihoods. Regional and international agreements are also important to provide a framework for conservation of habitats and biodiversity of the desert ecosystem for the sustainable use of its components and the fair and equitable sharing of the benefits arising from the use of genetic resources. Many of the desert countries of the world have drawn up plans which identify priorities for species and habitat conservation and detail how these will be undertaken. It is also planned to conserve species which are threatened at a national level. The plan also outlines the development of national legislation to protect species and to forbid exploitation of threatened species.

To achieve goals related to biological conservation and hence sustainable development of desert ecosystems, a useful approach is to analyse the feedback mechanisms that keep ecosystem running. Studies of positive or negative feedback cycles at local or regional levels may be crucial long term strategy for planning new management policies. For the ecosystem. Blondel and Aronson (1999) believe that conservation of vegetation relies on two strategies. The first is to assist threatened plant species to survive and the second is to protect habitats against destruction. The best strategy for the long-term protection of biological diversity is the preservation of natural communities, so called in situ, preservation.

A multifaceted approach to biodiversity conservation, including ecological function and socio-cultural values as a strategic choice for both study and management. From the ecological and conservation points of view, biodiversity fall within three general categories (Lister, 2008). The most prevalent focus on ecological structure (form), followed by a focus on function (ecological processes) and then an implicitly value-oriented focus on wealth or richness (resources).

There are two general classes of roles for biodiversity in ecosystem: ecosystems: stability and ecosystem function. That biodiversity is connected to ecosystem stability is an old and dominant thems in ecology. The conventional generalization that there is an inherent "balance" or equilibrium in nature, is linked to successional theory: as an ecosystem becomes more diverse during succession, it is believed that they become more stable. The maintenance of ecosystem stability has been frequently advocated as a basis for conservation. The direct association of diversity with stability of the ecosystem came about with the postulation that stability is imparted by increasing the number of links in the ecological food web.

The role of biodiversity in ecosystem function is obvious. The essential processes of living system (nutrient cycling, carbon and water cycling, productivity etc.) are certainly dependent to some degree on the diversity of games, species, populations, communities, and landscapes whose structures and composition perform these functions. The diversity of functions themselves is undoubtedly critical to the maintenance of ecosystems. All species play a small but significant role in ecosystem function, and if conservation emphasized only those species considered keystone species, it would be a serious mistake (Erhilich and Erhilich, 1981).

Equally important in the conservation plans for desert vegetation is maintaining constant contact with local people in order to: identify their needs, involve them in decision-making processes, and strengthen their capacity of action by means of governance systems that establish clear rights with regard to the use and ownership of natural resources. It is also necessary to identify economic opportunity that compensate the role of these people in the maintenance of environmental services, to create a diverse and competitive local economy based on product quality and to educate and enable local people to ensure the financial viability of conservation projects (IUCN, 2008). Taking these aspects into account will promote a sense of ownership among local desert communities enabling them to share responsibility in the conservation of natural resources, and also create jobs and income-generating opportunities that can benefit local communities and improve services such as water, energy, fodder, food etc. These are vital steps to guide conservation practice into, a rout towards the sustainable development.

For the Afro-Asian Med. and Red Sea coastal lands that have been subjected to long periods of human disturbance, there exist two possible alternative to address the continued degradation of vegetation and consequently the whole ecosystem. These are: restoration or rehabilitation (Aronson et al., 1993). Restoration and rehabilitation of ecosystem both aim at recreating self-sustaining ecosystems characterized by autogenic succession in plant and animal populations and sufficient resilience to repair themselves following natural or human perturbations. Restoration of an ecosystem aims at the complete return to a pre-existing state in terms of species composition, that is, the re-establishment of all indigenous species and the elimination of all exotics. In fact, this is virtually impossible. In contrast, rehabilitation, concentrates on repairing damaged or blocked ecosystem functions by raising ecosystem stability and productivity. It is primarily practiced in economically disadvantaged regions where human populations still depend on local resources for their livelihood and survival such as desert ecosystems. However, it is also highly pertinent to conservation aims since increasing productivity and profitability for local people in these ecosystems should ultimately reduce pressure on natural resources particularly plant resources. Biodiversity is in fact a key component of sustainable development of the ecosystems.

## 4.4 Sustainable Development of the Deserts

### *4.4.1 Introduction*

Sustainable development is defined as development that meets the needs of the present without compromising the ability of future generations to meet their own needs. It is a development strategy that managers all assets, natural resources and human resources as well as financial and physical assets, for increasing long-term wealth and well-being (Anonymous, 1987). Pearce et al. (1990) consider that sustainable development is a value-laden word implying change that is desirable and a *vector* of desirable society objectives the elements of which might include:

1. increase in real income capita;
2. access to natural resources;
3. a "fairer" distribution of income.

It is a path towards social justice and environmental protection. Sustainable development rejects policies that support current living standards by depleting the productive base, mainly plant and other biotic and abiotic resources of the ecosystem leaving future generation with poorer prospects and greater risks than today. SD, thus, promotes harmony between man and nature (Anonymous, 1987). Sustainable development can be realized through three main elements: social equity, economic efficiency and environmental conservation (Kassas, 2004).

It is worth stating that all religions (Islam, Christianity, Judaism, Buddhism etc.), consider that the environment is the source of life and the store of the resources of nature; accordingly we must protect, conserve and develop its assets and prohibit their abuse and destruction. This is obvious in the idea of the revival and restoration or recovery of lands through agricultural activities. Prophet Muhammed says "On Doomsday if anyone has in hand a sapling of a palm he should plant it". Such a positive attitude towards green plants, the only producers in all natural ecosystems, is admirable. Thus, from a religious point of view, man has to take the correct measures to conserve vegetation for the sustainable development of his own environment, i.e. for his own benefit and for the betterment of life for the future generations.

The sustainable development of the Afro-Asian Mediterranean and Red Sea coastal deserts is described here under four major issues: alternative developmental plans, sustainability: a challenge towards a better future, natural resources and plants with promising potentialities.

## *4.4.2 Alternative Developmental Plans*

Conservation and sustainable development of coastal deserts should ideally be directed towards a common goal: the rational use of the natural resources of the desert ecosystem to achieve the highest quality of living for the local people (Anonymous, 1987). However, environmental stress is usually the expected result of the growing demand on scarce resources particularly in the desert ecosystems of the arid lands. Within any ecosystem each species of plants and animals, exists as a population, the growth or decline of which depends on the capacity of the systems to provide its requirements. These must limit population growth of all plants and animals as well as people. The environmental limits to growth determines the *carrying capacity* for any species which may be at: subsistence level, security level or optimum level (Dasmann et al., 1960). The optimum level is, in fact, the normal objective for human societies, their domestic animals and their crops. These levels are controlled by the major factors affecting the ecosystem, namely: climate, soil, water and the complex of biotic factors.

From the ecological point of view, planning for sustainable development of coastal and inland deserts should be oriented to enhance the goals of development and anticipate the effects of development activities on the natural resources, both renewable and non-renewable, and processes of the environment. Dasmann et al. (1960) proposed six alternative options for the developmental plans of any particular ecosystem:

1. The desert can be left without man's interference, i.e. in a completely natural state and reserved for scientific and educational uses, watershed protection and/or for its contribution to landscape stability;

2. It can be developed as a national park (protectorate) with the natural landscape remaining largely undisturbed, to serve as a setting for outdoor recreation and ecotourism;
3. It can be used for limited and controlled harvest of its wild vegetation or animal life, but maintained for the most part in a wild state, serving to maintain landscape stability, support certain kinds of scientific or educational uses, provide for some recreation and tourism, and yield certain commodities;
4. Desert vegetation can be used for more intensive harvest to use its biomass as raw materials for various industries or pasture for livestock or intensive wildlife culling. In this case its value as a "wild" area for scientific study diminishes; its value for tourism and outdoor recreation may decrease but is not lost; its role in landscape and watershed stability is changed,
5. The wild vegetation and animal life of the deserts having been removed in part, the area can be intensively utilized for cultivation of pasture or farming crops,
6. The wild vegetation and animal life having been almost completely removed, the desert can be used for intensive agricultural, industrial and/or transportation purposes.

If one of the first three options is selected, choice remains, to a great extent, to switch to other uses as to use the land for any of the latter three objectives. Selection of the fourth option reduces the possibility of restoring the land to any of the first three categories but does not eliminate restoration completely. If one of the fifth or sixth options is selected, any shift to the other options, within a reasonable period of time would be difficult and costly. To keep the balance of the desert ecosystem its sustainable development has to depend mainly on its biotic and abiotic components. Introduction of foreign elements from outside it must not be haphazard, and scientific and restricted measures should be considered to avoid the disturbance of the balance of the ecosystem. That is the correct base upon which the decision makers could depend to select the correct choice.

### 4.4.3 Sustainability: A Challenge Towards a Better Future

The sustainable development of the Afro-Asian Med. and Red Sea coastal lands is the means of enhancing the well-being of both present and future generations who can enjoy the benefits of the continued use of the natural resources. However, activities for the implementation of the proposed developmental projects in these coastal areas require an understanding of the types and distribution of their habitats, climatic factors, and the natural biota (Anonymous, 1982). The building of a sustainable future requires new information, strategies, techniques, approaches and links that will all act together to develop the comprehensive and holistic management and decision-making processes implicit in the concept of sustainability (Kumar et al., 1993).

Coastal deserts, such as those of the Afro-Asian Med. and Red Seas, are important areas characterized by rich natural resources of economic values and biological diversity that are, therefore, regarded as sources of income to a wide range of local people. However, several factors such as industrial development, tourism villages and urbanization, exert continuous pressure on these natural resources leading to a breakdown of the ecological functions of the coastal desert habitats. Pollutants, such as oil etc., have had adverse effects on natural biota, especially sedentary species such as coral bivalves and mangrove vegetation. Pollutants may also threaten various industries, especially those depending on sea water such as desalination plants (Al-Sanbouk, 2002). An integrated coastal management programme could be effective in enhancing the sustainable development of these coastal deserts. It will provide the government(s) with coordination mechanism and processes to resolve conflicts in land use harmonizing environmental, economic and social activities for the success of the proposed plans.

To ensure that the concept of sustainability becomes embedded in decision-making processes that affect coastal areas, there are seven guiding principles that have to be considered. These principles provide guidelines for the integrated management and development of the coastal deserts (Anonymous, 1998a):

1. Understand that management of renewable natural resources is of strategic importance for social and economic development and is cost-effective in the long-term;
2. Recognize that sustainability requires the need to maintain the integrity of coastal systems and that this implies limits to the use of resources generated by the systems;
3. Understand that the carrying capacity of the natural vegetation for animal feeding and human uses is variable but not infinite;
4. Develop integrated management strategies that allow multiple use of natural resources in which complementary activities are integrated and conflicting activities are avoided or reconciled,
5. Establish protected areas;
6. Ensure good co-ordination in coastal management activities and involve local people to ensure effective management and equitable socio-economic development;
7. Accept that coastal planning management should not be fixed, but an ongoing process, with modifications made in the light of updated information and changing human need.

## *4.4.4 Natural Resources*

The natural resources of the coastal and inland deserts to be discussed here are: climate resources and land resources.

### 4.4.4.1 Climate Resources

The atmosphere with its components, gases and climate elements, is essential to all life; a particular combination of gases, temperature, sun light, wind forces, precipitation, humidity, evaporation etc. are, actually, the main factors. The atmosphere not only supplies the different ecosystems of the globe with renewable resources of essential materials, conditions and forces that maintain all living organisms, but it acts as a "filter" through which sunlight and radiant energy reaches the earth's surface, and it provides, as a result, an insulating or layer without which variations between day and night temperatures will be too extreme for the survival of any known forms of life.

All climatic elements collectively play an effective role in the sustainable development of the Afro-Asian Mediterranean and Red Sea coastal lands. Brief notes on air temperature, rainfall and energies (wind energy and solar energy) are given below.

**A. Temperature and rainfall**

Air temperature of the coastal lands of both Mediterranean and Red Seas are suitable all year around for the growth and reproduction of all plant species adapted to live there. The amounts of annual rainfall vary considerably, being, relatively, higher along the Mediterranean. As described in Chapter 1, the annual rainfall of the Med. coasts, which occurs mainly in winter, is, to some extent, reliable enough for certain kinds of rain-fed agriculture. Some of the rain percolates downwards and is usually stored in the shallow layers of the soil and could easily be pumped for irrigation during the dry (summer) season. Kassas (1970, 1972a) described a successful example and infers that the western Med. coastal land of Egypt (Mareotis coast) was, and is still, a promising agricultural and horticultural region. During the Graeco-Roman period the cultivation of vineyards flourished and good wine was produced (Mareotis wine) and stored. By the tenth century, this coastal belt gradually declined and the vineyards were replaced by desert. "It is unlikely that there have been major climatic changes during the last 2,000 years that could have caused the deterioration of the area" (Kassas, 1972a). However, according to De Cosson (1935), early in the twentieth century some attention was given to the Mareotis region. The extension of a railway westward of Alexandria to Mersa Maturh and the plantation of vine (*Vitis vinifera*), fig (*Ficus carica*), date palm (*Phoenix dactylifera*) olives (*Olea europaea*), carob (*Ceratonia siliqua*) almond (*Prunus amygdalus*) and Pistacio (*Pistacia vera*) under mainly rain-fed irrigation was successful (Kassas, 1972b). At present, apart from the above mentioned fruit trees and according to Ayyad (1983), the main land uses of the Mareotis is grazing. Limited rain-fed farming is also carried out during winter. The stored underground water, in shallow wells and cisterns is used during summer. The main annual crop is barley (*Hordeum vulgare*).

Along the Red Sea coastal land, however, where the climate is arid or extremely arid, the annual amount of rainfall is too small to secure the least amount of freshwater for rain-fed agriculture particularly in the littoral salt marshes. In these coastal saline strips, sea water or its derivatives may be an alternative source to grow highly salt-tolerant species (halophytes) with economic potentialities. Landwards, in the coastal montane countries, the situation is different. The vegetation of these areas enjoys a reasonable amount of rainfall both in winter (Med. affinity) and summer (tropical-monsoon affinity). Construction of dams across the large wadis of these montane Red Sea countries is very helpful in storing rainfall water for agricultural and domestic purposes. Here, xerophytes of economic values could be propagated and introduced as non-conventional desert crops.

**B. Energies**

Non-renewable energy resources (fossil fuels) provide about 95% of the commercial energy used worldwide. The North-African Middle East coastal and inland deserts provide a considerable part of the world's oil and gas reserves. The world will run out of affordable supplies of these non-renewable energies unless there is a radical change in the consumption patterns or we find some other major renewable energy resources. According to Cunningham and Saigo (1992), the developed countries which have only 20% of the total population of the world consume about 66% of all commercial energy. In addition, energy consumption is rising rapidly in the developing countries and in 2020 is expected to be 4 times that of 1980.

According to Anonymous (1980), the woody trees and shrubs which are the main source of energy for the local people of the coastal and inland deserts, the other renewable sources are: geothermal, and the wind and sun. The latter two are considered here.

**(a) Solar energy**

The earth receives solar energy as radiation from the sun and the amount greatly exceeds mankind's use. This resource has a familiar daily as well as seasonal variations, and is significantly affected by weather. It has a relatively low intensity, with a peak of about 1 km/m$^2$ at sea level. Every country has access to the resource, to different degrees. The applications of solar energy are quite diverse, including direct thermal, electric power generation using thermodynamic cycles, and direct conversion to electricity with photovoltaic (PV) systems (Anonymous, 1994). Solar energy results from the process of continuous nuclear fusion in the sun, which is, actually, the ultimate source of all known energies (El-Qattamy, 1975). Now, as reserves of oil and gas are dwindling, and as constraints on the use of coal are growing and the future of nuclear power is in doubt, solar energy is making a comeback to both new and familiar forms (Brown, 1981).

Apart from green plants which capture the sun's energy through the process of photosynthesis, new technologies permit solar energy to be harnessed in innumerable ways. It can be captured through such devices as rooftop collectors, photovoltaic cells and buildings incorporating solar architecture.

All natural ecosystems operate with solar energy, but no two ecosystems (or countries) are precisely the same in this respect. The desert ecosystem does, in fact, receives the greatest amount of sun energy as compared to other ecosystems of the earth. The richest desert is the Empty Quarter of Arabia followed by the Sahara of North Africa (Table 4.1).

**Table 4.1** Solar energies received by five deserts (After El-Saiegh, 1976)

| Deserts | Area ($km^2$) | Heat energy of sun | |
|---|---|---|---|
| | | $kW/h/km^2 \times 10^6$ | $kW/h \times 10^9$ |
| 1. Sahara, Africa | 7,770,000 | 2,300 | 17,871,000 |
| 2. Arabian Peninsula | 1,300,000 | 2,500 | 3,250,000 |
| 3. Middle and Western Australia | 1,550,000 | 2,000 | 3,100,000 |
| 4. Kalahari, Africa | 518,000 | 2,000 | 1,036,000 |
| 5. Mogif, South California | 35,000 | 2,200 | 77,000 |
| *Total or mean rate* | 11,173,000 | Rate = 2,260 | 25,344,000 |

El-Qattamy (1975) stated that changing the total amount of sun energy (25,344,000 kW/h $\times$ $10^9$) reaching five deserts (Table 4.1) into electric energy could produce about 130,000 $\times$ $10^{12}$ kW/h, equivalent to 3,000 (three thousands) times the amount of electric energy needed for the world.

This indicates the importance of the deserts as major receivers of solar energy. Comparisons can be made with the clean space of the high levels of the atmosphere where there are neither absorbing nor reflecting materials the sun's rays. In these clear and clean areas insolation about 1,353 $W/m^2$ whereas in the desert is about 800 $W/m^2$. In the areas where there are clouds and/or water vapour, insolation decreases by about 30% due to absorption. In addition to clouds, dust, smokes and other air pollutants reflect radiation.

Latitudinally, insolation decreases gradually from the equator northwards and southwards. The best areas for the utilization of sun energy are situated between latitudes 40° north and south of the Equator. The deserts located between these latitudes receive the highest amounts of solar energy (El-Saiegh, 1976).

The above discussion shows that the deserts represent a very suitable area for the efficient utilization of solar energy as a cheap, clean and renewable natural resource. Solar energy could be used to obtain amounts of the badly needed fresh water by desalinating sea water. According to El-Saiegh (1976), 300 $km^3$ of sea water could provide $36 \times 10^7$ $m^3$ of fresh water/year for domestic purposes, or $50 \times 10^7$ $m^3$/year for agricultural and industrial purposes. Solar energy could also be used to drill for ground water. Securing adequate amounts of fresh water is the first requirement for developmental programmes in the deserts upon which depends the following:

1. Agricultural schemes,
2. Establishment of new factories;
3. Establishment of new villages and settlements,
4. Availability of enough fresh water for domestic uses and attracting more people to live there.

In addition to desalination of sea water and pumping ground water, solar energy could be used to produce electric power, operation of sun furnaces, etc.

**(b) Wind energy**

About 2% of the sun's energy striking the earth ultimately results in winds, in two major ways (Chiras, 1991). First, sunlight falls unevenly on the earth heating some areas more than others. Warm air rises, and cooler air flows in from adjacent areas. The earth's principal circulation pattern develops as warm air near the equator rises, drawing cooler polar air towards the tropics. The earth's rotation then causes air to circulate clockwise in the Northern Hemisphere and antichlockwise south of the equator. The second major wind-flow pattern results from the unequal heating of land and water. Air over the oceans is not heated as much as air over the land. Therefore, cool oceanic air often flows landward and replaces warm rising air.

The potential of wind energy is enormous (Cunningham and Saigo, 1992; Miller, 1997). Tapping the globe's windiest spots could provide 13 times the electricity now produced worldwide. Wind energy could supply 20–30% of the electricity of many countries. However, "Today, wind generated electricity accounts for only a tiny portion of the world's enormous energy needed" (Chiras, 1991).

Our question is: what are the advantages and disadvantages of the utilization of wind energy? Wind energy is a clean renewable source of electricity: harvesting it takes limited areas of land and is safe to operate. Moreover, wind technologies do not preclude other land uses. Wind farms, for example, leave ample space to be grazed by animals or be planted with various crops. "Wind power is an unlimited source of energy at favourable sites, and large wind farms can be built in only 3–6 months. This system emits no carbon dioxide nor other air pollutants during operations. They need no water for cooling and their manufacture and use produce little water pollution" (Miller, 1997). The same author stated that the costs of producing electricity in wind farms is about half that of a new nuclear plant and is competitive with coal. Expert opinion is that by the middle of the twenty-first century, wind power could supply more than 10% of the world's electricity. However, wind power can be used only in areas with sufficient winds, e.g. sea shores and high mountains, both occur along the Afro-Asian Med. and Red Sea coasts. The wind does not blow all the time, so backup systems and storage facilities are needed. Storage technologies seem to be one of the major weaknesses. In addition, wind generators can be noisy and may impair both television reception and telecommunications. A survey of wind patterns in Egypt identified the Red Sea and the Med. coastal deserts as priority sites for wind farms and a large wind farm is being established at Zaafarana on the Red Sea coast.

#### 4.4.4.2 Land Resources

Ecologically, the word land infers to all renewable and non-renewable natural resources above-and under the soil surface including: groundwater, soil, natural vegetation, wildlife, microorganisms, and man. The first three resources are discussed below.

**(a) Groundwater**

Water, which provides life for man, animals, plants and microorganisms, is scarce in deserts. Precipitation (rainfall) is low and is not reliable for developmental projects. Water is not only a matter of quantity, but also a matter of quality.

In view of the great scarcity of surface water supplies, or even their complete absence throughout much of the time, groundwater assumes a vital importance in the deserts. Groundwater is more reliable than rainfall in that it can be drawn upon throughout the year and is usually less affected by a relatively dry period, even if for successive years (Dixey, 1966). The Bedouins inhabiting the deserts, e.g. Med. and Red Sea coastal deserts, are accustomed to sinking wells by simple equipment in suitable places to depths of as much as 100 m. However, greater amounts of water supplies may be obtained by means of bore-holes which can be constructed under difficult conditions of hard rocks or loose sand, and which can be drilled to greater depths. This, however is beyond the capacities of the local people.

Dixey (1966) warned that groundwater could be misused by over pumping or by extracting water over successive years at high rates. The result is a gradual lowering of the water table in the vicinity of the bore-hole and the reduction and eventually exhaustion of the supply. In the parts of the Med. and Red Sea coastal deserts directly affected by sea water (littoral salt marshes), the groundwater is in contact with salt water of the sea. In this case, over-pumping leads to a rise of salt water into the wells with consequent pollution of the supply, rendering it unfit for agriculture or other domestic use. In addition, due to the distribution of oil fields along parts of the Red Sea coastal deserts, e.g. in the Egyptian section, oil pollution may be expected and the water of the wells will be unfit for all uses.

Since ground water is replenished by rainfall, which is very scarce in coastal deserts, it might be expected that the volume stored (mostly non-renewable) is decreasing, a factor that decreases its economic potentiality as a permanent source of fresh water. Supplies of the required quantities of fresh water are essentially to initiate development programmes. The most reliable way is the desalination of sea water which can be maintained along the whole stretches of the Afro-Asian Mediterranean and Red Sea coastal deserts. Solar energy is a cheap and effective option for the power required for sea water desalination. Treated drainage water could then be used to establish forests for wood production.

**(b) Soils**

Soils of the coastal and inland deserts accumulate soluble products of weathering in the upper part of the soil profile; these products are present as calcium carbonates and soluble salts (Jewitt, 1966). Soil texture varies widely from heavy clay to coarse sands, reflecting the influence of parent material, although most have one horizon of heavy texture. Dregne (1976) stated that the soils of arid lands (deserts) fall into two main orders:

*aridisols* or essentially desert soils and *entisols* or alluvial soils and soils of sandy and stony deserts. Aridisols are mineral soils distinguished by the presence of horizons formed under recent conditions of climate or those of the earlier periods. The surface stratum, the epipedon, is usually light-coloured and there may be a salt or clay stratum near the surface. The salt layer is well represented in littoral zones directly affected by sea water. "Most of the time when temperatures are favourable for plant growth, aridisols are dry or salty with consequent restrictions on plant growth" (Zahran and Willis, 2008). However, such growth restriction may not apply to the salt-tolerant plants (halophytes) of the Afro-Asian Med. and Red Sea coastal deserts, which can withstand saline conditions. Many of these plants are of economic value and could play a considerable role in sustainable development programs. The entisols, the most common type along the Red Sea coastal desert, are also mineral soils with little or no development of horizons.

The soils of the Afro-Asian Med. and Red Sea coastal deserts, like those of the other deserts, contain low levels of organic matter, are slightly acidic to alkaline in reaction at the surface, show accumulation of calcium carbonate within the topmost 1.5 m (5 ft) layer, have weak to moderate profile development, are of coarse to medium texture and have low biological activity (Dregne, 1976). In the upstream parts of the wadis of the Red Sea coastal deserts, the soils show a surface layer of stones, pebbles or gravel that constitutes a desert pavement from which the fine particles have been removed by the action of wind and/or water.

Generally, the soil of the Afro-Asian Med. and Red Sea coastal desert is suitable for different types of agriculture if fresh water is available.

**(c) Natural vegetation**

Many species of plants that supply 90% of the world's food, fodder, fibre, drugs etc. were domesticated from wild plants found in the tropics (Miller, 1997). Existing wild plants, many of them still unclassified and unevaluated, remain interesting to plant ecologists, agronomists and genetic engineers for developing new crop strains; some of them may become important sources of food, livestock fodder, clothes, rubber, dyes, paper pulp, drugs, perfumes, etc.

Wild plants are vital components of ecosystems. Plants supply animals with food, recycle nutrients essential for agriculture and help to produce and maintain fertile soil. They also produce oxygen and carbon dioxide in the atmosphere, moderate earth's climate, help regulate water supplies and store solar energy as chemical energy. Moreover, green plants may help to remove poisonous substances, and make up a vast gene pool of biological diversity.

Ecologists and conservationists believe that hundreds of wild species are disappearing at an alarming rate, and preserving plants is a must for their actual or potential usefulness as renewable natural resources. At the same time, wild species have an inherent right to exist: it is ethically wrong to cause the extinction of any plant species.

According to Miller (1997), extinction of plants (and animals) is a natural process. As the planet's surface and climate have changed over its 4.6 billion years of existence, species have disappeared and new ones have evolved. However, since the dawn of agriculture, about 10,000 years ago, the rate of species extinction has increased sharply as human settlements have expanded worldwide. It was estimated that during recent centuries at least 4,000 species became extinct mostly because of man's activities, and the number is rising. The greatest threat to most natural plants is destruction, fragmentation and degradation of their habitats especially in the coastal and inland deserts.

The natural wealth of the flora of the Afro-Asian Med. and Red Sea coastal deserts, the producers of their main ecosystems (mangrove, reed, swamps, salt marshes, sand dunes, desert plains, wadis and mountains), comprises hundreds of annual and perennial species. These plants could be considered the backbone for the sustainable development of these coastal deserts. Annual species, such as *Zygophyllum simplex* and *Asphodelus tenuifolius*, usually cover wide areas between perennial trees and shrubs during the wet seasons. Such seasonal green cover is usually used as natural range for the local domestic and wild animals. However, the permanent framework of the different vegetation types is formed mainly of perennial halophytes and xerophytes. Apart from being adapted to live under salinity and/or aridity stress, these plants have various economic potentialities. Taking these advantages into consideration, these plants could be propagated in these coastal deserts depending upon the available water. To highlight their economic value, the vegetative yields of the plants to be cultivated there, could be used as materials for certain industries, e.g. livestock fodder, paper, drugs, oil, etc.

Economically, the flora of the Afro-Asian Med. and Red Sea coastal deserts can be categorized under six main groups:

1. Fodder producing plants,
2. Drug producing plants,
3. Fiber producing plants,
4. Oil and perfume producing plants,
5. Wood and fuel plants,
6. Food plants.

A considerable number of these economically important plant species have more than one usage. For example *Acacia* app. and *Balanites aegyptiaca* (xerophytes) as well as *Avicennia marina* and *Rhizophora mucronata* (mangroves, halophytes) are valuable as fodder, drug and wood producers.

Water is the limiting factor; without it no development could be implemented. Accordingly, for the shortage of fresh water, development programmes along whole stretches of the Med. and Red Sea coastal deserts will depend upon saline and/or brackish waters. A key role could be played by halophytes in the sustainable environmental development of these coastal deserts. About this Shay (1990) stated that salt-tolerant plants may provide a logical alternative for many development programmes, and saline farmland could be used without remedial measures. Groundwater too saline for irrigating conventional crops can be used to grow salt-tolerant plants. This means that the Afro-Asian Med. and Red Sea Coastal deserts may serve as new agriculture lands using different concentrations of sea water and ground saline water. These plants may be grown also in salt affected lands unsuitable for conventional crops, to produce food, fuel, fiber, resins, essential oils and pharmaceutical materials. Accordingly, saline agriculture is essentially needed and domestication of salt-tolerant plants (halophytes) currently growing in saline soil and introduce them as non-conventional crops may have potential promise.

## *4.4.5 Plants with Promising Potentialities*

### 4.4.5.1 Introduction

Of the enormous areas distinguished as extreme arid, arid and semi-arid, only about one tenth is under cultivation, either as irrigated or rainfed. By far the greater part of the arid environment is unused except by grazing and browsing animals and other limited daily uses by the local inhabitants (Bedouins), as far as can be seen at present, these lands are unlikely to be used for large scale developmental programs. Rainfed agriculture is subject to drought and partly as a consequence people are tending to concentrate on those areas where rainfall is more reliable. It seems that great areas of the arid lands that have been degraded may be less heavily used in the future and, thus may have the opportunity to recover (Grove, 1985).

For the future it might be suggested that economic xerophytes and halophytes will mainly be needed for desert areas where people in arid countries are congregate.

Nabhan and Felger (1989) stated that it is ironic that much of the modern agricultural developmental projects in the deserts of the arid land countries, e.g. Tushki project in the Western Desert of Egypt, depend on temperate or tropical crop species not adapted to the harsh desert climate with its extreme temperatures, very low rainfall, low soil moisture, low humidity and high evaporation. These plants require large amounts of irrigation water as well as micro-environmental modification to be economically productive. These crops are good yielders per unit area given groundwater for irrigation, but they are costly both economically and energetically. In addition, irrevocable groundwaters depletion is becoming a common tragedy in the deserts worldwide. Borlaug (1983) stated that we have the most to gain in the desert crop production by the utilization of plant species better adapted to environmental stresses in marginally productive arid zones. He sees this gain primarily accomplished by incorporating into conventional crops hardy genes from wild relatives adapted to live under desert environmental stress.

Plants with promising potentialities in the Afro Asian Med. and Red Sea coastal and inland deserts are described here under four headings: currently used native plants, proposed exotic shrubs, palm trees and case Studies.

#### 4.4.5.2 Currently Used Native Plants

The major vegetation forms of the Afro-Asian Mediterranean and Red Sea coastal lands (see Chapter 3) comprise many species adapted to live under aridity and/or salinity stresses and in the same time proved to have promising potentialities for the sustainable development of these coastal lands. They can be used directly and indirectly. All of their parts (roots, stems, branches, leaves, flowers, fruits, seeds, barks etc.) could be used directly as food, fodder, fuel, timber, drug and as raw materials for various industries e.g. paper, hydrocarbons, oils, perfume, drugs, resins, dyes etc. Honey production is the indirect use of these plants supplying honey bees with nectar all the year around.

**(a) Direct uses**

Le Houerou (1985) throws light on a number of useful plant species of potential economic values, as fodder and fuel, naturally growing in countries of the Afro-Asian Mediterranean and Red Sea coastal lands. The selection of species is based on the view to their possible use in revegetation programmes as well as to their adaptation to meet virtually all environmental stress within these coastal deserts. Le Houerou (1985) grouped forage plants under three groups: perennial grasses e.g. species of *Agropyron, Elymus, Bromus, Cenchrus, Cynodon, Dactylis, Digitaria, Hordeum, Lasuirus, Lolium, Oryzopsis, Panicum, Phalaris, Sporbolus* and *Stipa*, perennial and annual fodder legumes, e.g, species of *Astragalus, Hedysarum, Lotus, Medicago, Melilotus, Onobrychis, Trifoluim, Trigonella, Vicia, Acacia, Prosopis,* and *Ceratenia,* and fodder species from other families e.g. species of *Periploca* (Asclepiadaceae), *Opuntia* (Cataceae), *Atriplex* and *Haloxylon* (Chenopodiaceae), *Morus* (Moraceae), and *Calligonum* (Polygonaceae) etc. The fuel species are grouped under two groups: native and introduced species. The native fuel species include: *Pinus halepensis*, *Cupressus sempervirens*, *A. atlantica*, *Tamarix aphylla*, *T. stricta, Populus alba* and *P. euphratica*. The introduced species include: the Australian phyllodineous species namely: *Acacia saligna*, *A. cyclops*, *A. salicina* and *A. ligulata*, the arid zone *Eucalyptus* species namely: *E. astringens, E. brackusayi, A. camaldulensis, E. oleosa, E. gomphocephala, E. microtheca, E. occidentalis, E. salmonophloia* and *E. toquata* and two *Casuarina* species: *Casuarina stricta* and *C. cunninghamiana.*

Zohary (1962) classified the flora of Palestine-Israel Med. coastal land under 5 groups: food, industrial, pasture and honey plants. Among the food plants, the leaves and stems of some species (e.g. *Rumex roseus, Chenopoduim spp., Diplotaxis acris, Sisymbrium irio, Malva rotundifolia Lactuca cretica, Silybum marianum* etc.) can be used as greens and salads, tubers or bulbs of (e.g. *Erodium hirtum, Astoma Seselifolium, Hordeum bulbosum, Allium ampeloprosum* etc.) can be used as vegetables, spices and condiments can be furnished by *Laurus nobilis*, *Capparis spinosa, Majorana syriaca*, species of *Teucrium* etc.; edible fruits can be collected from several plant species (e.g. *Ochradenus baceatus*, *Crataegus azarolus, Pyrus syriaca, Prunus ursina, Prosopis farcta, Ceratenia siliqua, Nitraria retusa, Salvadora persica, Ziziphus spina-christi, Z. lotus, Arbutus andrachne* and *Lycuim* spp), numerous legumine fruits can be used as pulses (e.g. species of *Astragalus, Vicia, Lathyrus, Pisum, Lens* etc.).

The industrial plants of the flora of Palestine-Israel coastal area contain: fiber and wicker plants such as species of *Arundo, Salix, Typha, Arundo, Phragmites, Scirpus, Cyperus papyrus*, and *Juncus*, oil plants such as species of *Brassica, Sinapis, Calepina, Eruca* and *Ricinus*; tannins can be obtained from species of *Quercus, Pinus, Pistacia*, *Arbutus* and *Rhus*. There are only a few plants which are used for the extraction of essential oils, e.g. *Majorana syriaca* and *Thymus capitatus*. In addition, there are few wild species known for having been used in the past in vegetable dyes, are of no present value because of the introduction of synthetic dyes. Among these dye plants are: *Reseda luteola* (orange), *Tephrosia apollina* (blue), *Echium italicum* roots (red), *Phelypaea lutea* (yellow) and the indigo plants (*Indigofera* spp). Soap plants, mainly belong to the Chenopodiaceae, have been widely used in the past and are still used today in desert regions (Zohary, 1962). Pieces of *Haloxylon* spp., *Anabasis* spp., *Salsola* spp. Could be used as sources of potassium for the soap manufacture.

There are many medicinal plants naturally growing in Palestine-Israel coastal area, used chiefly in folk medicine. A considerable number of these plants are widespread in the drug markets of Cairo, Damascus, Beirut, Jerusalem, Hebron, and Gaza, while others are commercially rare, although they are used extensively by people who collect them. Examples are: *Polygonum equisetiforme, Ochradenus baccatus, Papaver rhoeas, Peganum harmala, Balanites aegyptiaca, Ruta bracteosa, Thymelaea hirsuta, Eryngium spp., Plumbago europaea, Calotropis procera, Alkanna strigosa, Calamintha incana, Majorana syriaca, Teucrium polium, Achillea santolina, A. fragrantissima, Inula viscosa, Artemisia herba-alba* and *Helicophyllum crassipes.* Some herbs are used as panaceas for healing all kinds of illnesses, while others are specific to particular ailments.

Other native medicinal plants included in older or modern pharmacopoeias are: *Chenopodium ambrosoides, Laurus nobilis, Glycyrrhiza glabra, Althaea officinalis, Malva silvestris, Foeniculum vulgare, Ammi visnaga, A. majus, Nerium oleander, Salvia sclarea, S. triloba, Hyoscyamus muticus, Datura stramonium, D. metel, Valeriana dioscoridis, Citrullus colocynthis, Matricaria chamomilla,* etc.

Boulos (1983) describes the medicinal values of more than 300 species of the vascular plants naturally growing in North African countries. The Compositae contains the highest number (42 species) of drug producing plants followed by Labiatae (29), Legumenosae (28), Gramineae (13), Cruciferae (12), Euphorbiaceae (10), Cucurbitacea (7), and Cupressaceae (6). The other families contains from 1 to 5 medicinal species.

Hundreds of palatable grasses and leguminous species in the local flora of Palestine-Israel coastal land have been the source of pasture for cattle and sheep from ancient times to the present day (Zohary, 1962). Among the most valuable native perennial grass species are: *Dactylis glomerata, Hordeum bulbosum, Lolium perenne, Oryzopsis holciformis, O. coerulescens, O. miliacea, Phalaris bulbosa, Andropogon distachyus, Arrhenatherum palaestinum, Brachypodium pinnatum* and *Bromus syriacus.* The rainy (winter) season is characterized by a set of grazing plants, the most abundant of which are: *Avena sterilis, A. barbata, A. wiestii, Bromus scoparius, B. sterilis, Cutandia memphitica, Hordeum murinum, Koeleria phleoides, Aegilops ovata, A. peregrina, A. bicornis, Brachypodium distachyum, Phalaris brachystachys, P. minor* and *Lolium gaudini.* Leguminous plants which grow mainly in winter, are important component of natural pastures in the months of February and March. They include for example many species of *Trifolium, Lathyrus, Medicago, Pisum, Astragalus* and *Vicia.*

Improvement by propagating palatable xerophytes and halopytes could provide an inexhaustible source for the development of impoverished ranges. Such plants include *Atriplex halimus, Kochia indica, Colutea istria, Crotalria aegyptiaca, Stipa* spp., *Aristida spp., Panicum turgidum, Pennisetum dichotomum* etc.

**(b) Indirect use: honey production**

The estimated world's annual production of honey, according to official statistics reaches millions of tonnes. The three major exporters are China, Mexico and Argentina, all of which include subtropical arid regions (Crane, 1985). Such areas can be important for honey production: when skies are clear, energy from the sun is freely transmitted to plants, and some of this energy is converted into sugar which plants sectrete as nectar. If sufficient bees are present, this nectar can be converted into honey and harvested.

Plants supply all bees with all their food resources, i.e. nectar, honey dew and pollen, and also the propolis they use in building their nests. Bees wax, venom and bee milk (used for feeding the young) are synthesized by adult worker bees, beeswax largely from carbohydrates in nectar and honey, and venom and bee milk largely from proteins in pollen. Immature bees cannot be reared without pollen, or as adequate pollen substitute provided by the beekeepers.

Using bees to crop the land is an additional way of getting a harvest. If the honey sources are plants that yield seed or fruit, the bees may increase yields by their pollinating activities.

The eleven vegetation forms of the Afro-Asian Mediterranean and Red Sea coastal lands are rich with nectar plants and plants that produce a lot of pollen and hence provide suitable areas for bee-keeping and honey production all year around. Zohary (1962) stated that wild plants in the Palestine-Israel SW Mediterranean coast are the primary sources for bee keepers. There are hundreds of excellent honey plants supplying nectar, pollen, or both to the hive. According to the seasonal distribution of flowering plants, there is possibility of obtaining 3–4 honey crops during the year. Spring is the most fruitful season for the growth of therophytes (ephemerals, annuals and biennials) as well as of the blossoms of cultivated orange trees. Among the winter honey and pollen plants are: *Senecio vernalis, Diplotaxis erucoides, Sinapis* spp., *Raphanus raphanistrum, Maresia pulchella* (and numerous other species of Cruciferae), *Amygdalus communis, Trifoluim* spp., *Medicago* spp. (and other annual leguminous species), *Euphorbia cybirensis, Ridolfia segatum, Styrax officinalis, Echium judaeum* and representatives of other families. In early summer bees harvest nectar from *Daucus maximus, D. aureus, Ammi visnaga, Centaurea* spp., *Carthamus tenuis, C. glaucus, Cichorium* and *pumilum,* whereas *Prosopis farcta, Thymus capitatus* and many species belong to family compositae and others supply nectar to bees in mid summer. Of the late summer plants the most important species are *Inula viscosa, Rubus sanctus, Lythrum salicaria* and *Lippia nodiflora* (all are hydrophytes). This means that continual rotation of hives is needed to take the advantage of the seasonal distribution of flowering making the Palestine-Israel Med. coastal land one of the most favoured areas for yielding multi-seasonal honey crops. This advantage is, fortunately, a characteristic feature of the Afro-Asian Mediterranean and Red Sea coastal lands.

Through the mangrove project supervised by the Ministries of Agriculture and Environment in Egypt (2003–2006), bee hives have been established in sites of mangrove vegetation along the Red Sea coast of Egypt. High quality and large quantities of honey are currently produced all year around. This can be considered a management tool to prevent heavy cutting and grazing of the mangrove trees. The local inhabitants are gaining a lot from honey production.[2]

---

[2]The author of this book was the Executive Manager of that project.

#### 4.4.5.3 Proposed Exotic Shrubs

In a world of rapidly increasing human population, it is a dream either to make the desert permanently arable or to give it a natural vegetation cover. The best way to achieve this would be a "forest" which reproduces itself quickly and provides useable wood, edible fruits, and forage for domestic animals, and at the same time provide a place for relaxation and enjoyment. A forest that gives the region economic and ecological stability will guarantee, by careful use and management, the survival of its inhabitants for a very long time (Steinen, 1985).

Many trees and shrubs naturally growing in the different parts of the arid and semi-arid regions of the world have high economic values but, unfortunately, are not native to the Afro-Asian Med. and Red Sea coastal lands. A proposal to introduce and propagate some of these species in the coastal lands ought to be considered. There are at least three species of these exotic valuable plants, namely: Jojoba, Jatropha (both oil-producing plants) and guayle (a rubber producing plant).

1. **Jojoba**

   Jojoba (*Simmondsia chinensis*) is a shrub or small tree of SW North America. It has edible seeds containing a valuable oil (40–60% liquid wax of high value). This oil is similar to sperm-whale oil and has an expanding list of uses, from an engine lubricant to cosmetics. Jojoba wax is valuable for its stability, purity simplicity, and lubricity and can be modified by partial dehydrogenation to produce a variety of soft white waxes and creams for use in industry. Unfortunately, the remaining residue after wax extraction cannot be used as animal feed due to the presence of an unusual toxin (Wilson and Witcombe, 1985).

   The attraction of this crop for people in the warm arid regions of the world is its potential to grow and produce good yields with comparatively little water. It grows in native stands where rainfall is less than 120 mm/year and it can be also grow on saline soils with saline water. Once established the plant can have a net positive photosynthesis with low water potentials. However, it grows best and produces highest yields at precipitation levels of between 380 and 500 mm/year.

   Most interest in this plant has been shown by the USA, Israel, Egypt, Mexico and Australia with both large and small farming organizations involved in the development of this crop. Whether these plantations are profitable depends on predictions of the future price of jojoba oil (jojoba does not yield seed until five years after planting) and on whether claims for the seed yield of the varieties planted are fulfilled. The seed yield of plants grown from seed may vary from 0.6 to 5.4 kg/bush. Many of the areas where jojoba was first planted may be unsuitable, since it is now realized that jojoba is more sensitive to cold than was commonly thought (the temperature should not fall below –4° C), and, although established plans are very drought tolerant, an adequate supply of water

(600–750 mm/year) is often necessary during the first two years of establishment. In addition jojoba is not free of disease problems and is attacked by many different pathogens, e.g. *Phytophthora* and *Pythium* species and a member of insects.

In conclusion it is unlikely that the "miracle" plant jojoba will enable many people to get rich quickly. Research is needed on the selection and breeding of jojoba for productivity, cold resistance, early maturation, disease resistance, hermaphroditism and an upright shape that facilitates mechanical harvesting. Harvesting by hand is very expensive and it is considered that jojoba can only capture the high-volume low-cost lubricant market if its plantations are harvested by machines.

**2. *Jatropha curcas***

*J. curcas* is a large shrub 3–4 m high belong to family Euphorbiacea. It has ovoid, black fruit 2.5 cm long that breaks into three 2-valved cocci; seeds ovoid-oblong, dull brownish black. It flowers in hot and rainy seasons, and sets seeds in winter when it is leafless.

*J. curcas* is native to tropical America but is now widespread throughout India. It is an oil plant and can be a suitable source for biomass energy especially in rural areas. Its oil can be transformed into diesel engine fuel (Khoshoo and Subrahmanyam, 1985).

High quality oil with a calorific value of 9,470 kcal/kg can be extracted from the seeds of *J. curcas,* which compares favourably with 10,600 kcal/kg for gasoline and 6,400 for ethanol. The oil is readily soluble in diesel oil and gasoline and can be used in fuel mixtures for gasoline engines. The substitution of diesel oil is of prime importance to developing countries and a diesel engine performance test (the Yanmer SA 70-L diesel engine) indicates that the engine performance and fuel consumption are very similar for *J. curcas* and diesel oil when the same engine is used. If a *Jatropha*-oil industry is developed, it would essentially be an agro-industry because vegetable oil seeds would be used as the raw material for an alternative fuel.

Other uses of this oil are as an illuminant, for making soaps and candles, and for its medicinal properties. The seed cake can be used as manure and the leaves are used as feed for raising eri-silk worm.

The plant can grow from the seed, or from stem cuttings buried in the soil which soon develop roots. The mean annual yield of air-dried seeds from a 5 year old tree is 2–3 kg.

*J. curcas* has a number of advantages: – the plant is adaptable to a wide range of soil types, including those of arid regions; it grows quickly and is easily cultivated; the processing for oil extraction is very simple; the energy balance appears to be favourable; it is available for direct practical use in the rural areas; no engine structure improvements are necessary and it can be grown under conditions which offer no competition to food or animal feed production.

**3. Guayule (*Parthenium argentatum*)**

The principle source of natural rubber is *Hevea brasiliensis,* a large tree, which meets nearly one-third of the world's rubber demand, the remainder being met by synthetic products of petroleum-based chemicals. However, following the oil crisis, the price of synthetic rubbers is less competitive on account of its high cost and non-availability of the peterochemicals. There is an ever-increasing demand for natural rubber on account of its elasticity, resilience, and low heat production. The projected international demand for natural rubber is expected to outstrip the *hevea* rubber production, leading to a worldwide shortage. This shortage cannot be met by *hevea* rubber because of limitations in extending plantations. World attention has been drawn to a search for alternative sources of natural rubber and guayule is currently considered the best contender, since it can be grown in dry, arid, semi-arid and non-agricultural lands.

Guayule *(Parthenium argentatum)* is a native to north-central Mexico and south-west USA, where it grows in poor arid, and semi-arid, desert land. It occurs in drylands over an area of 276,700 $km^2$ and occupies favourable sites on limestone hillsides at altitudes of 1,200–2,100 m in regions with less than 400 mm rainfall (Khoshoo and Subrahmanyam, 1985). It needs a great deal of sunlight and low night temperatures and has been cultivated in many parts of the arid lands. The plant has been grown from seed for three successive generations. Seeds are obtained from April to October but the best seeds are available during dry hot months from mid-April to mid-June. They are sown outdoors in raised nursery beds during October–January. One month old seedlings in polythene bags or earthern pots are transplanted directly in the ground. The seedlings are best transplanted during the winter months (November–February).

Guayule can also be propagated by shoot cuttings. Vegetative propagation by apical shoot cuttings has been successful when undertaken in July or August. It starts flower production in March when 3–4 months old and continues until November. Flowering and growth cycle in old plants start again in March. A 2-year-plant normally produces some 10% rubber by dry weight; some varieties yield as much as 25% and, with chemical stimulants, rubber production can be increased at early stages of growth to 30%. Guayule rubber is found not in a specialized lactifer system, but in the parenchyma of stems and roots as latex particles similar in size to those obtained from *Hevea.* For this reason it cannot be "tapped" but must be extracted from the tissues, and because it contains no natural antioxidant it degrades rapidly in contact with air, so the plant must be processed within a few days of harvesting (Hall, 1985).

Whether or not any country can establish guayule as an economically viable crop, however, depends on many factors. First, and foremost, it needs the price of oil to increases to be competitive with synthetic rubbers. Guayule can be a locally produced source of polyisoprene rubbers.

It seems likely that in coming decades there will be markets for all the natural rubber than can be produced, whether *Hevea,* guayule, or other plants. There is a continuing rise in world consumption of rubber, and natural rubber is still preferred in many applications. *Hevea* can be only cultivated in a limited tropical zone, which makes it vulnerable to political, economic, or biological problems. There may also be justification for the cultivation of guayule outside the realm of conventional economics, because of the need to stabilize desert margins, find crops adapted to desert environments, and provide jobs and incomes for desert dwellers where farming conventional crops is risky or impossible.

Other important products of guayule under investigation are the terpenoid constituents present in the stem and leaf resins. Numerous novel sesquiterpene esters and sesquiterpene lactones have been identified from guayule, other species of *Parthenium* and F1 hybrids (Rodriguez, 1977). Recently, oxygenated pseudoguaianolides were found to be very effective in deterring feeding by phytophagous insects (Isman and Rodriguez, 1983). Since at the present time the only source of natural rubber is the tropical tree. *Hevea,* the development of guayule and hybrids as desert hydrocarbon crops for the arid zones of the world in necessary and timely (Rodriguez, 1985).

#### 4.4.5.4 Palm Trees

Although palms are very conspicuous when they occur in arid and semi-arid areas, the scientific literature concerned with better utilization and development of plant resources of such areas pays relatively limited attention to them. This apparent neglect may be explainable, in part, by two related factors. First, palms are not true xerophytes and could not survive in most arid or semi-arid areas without the presence of underground water sources. Secondly, because the palms exhibit a scattered and highly variable distribution, and form a typical vegetation associations when they do occur, they are unrelated to the characteristic climatic climax vegetation formations (Johnson, 1985).

The Afro-Asian Mediterranean and Red Sea coastal lands are characterized by the growth of two palm species: date palm (*Phoenix dactylifera*) and doum palm (*Hyphaene thebaica*). Date palms commonly occurs along the stretches of the coastal lands of both seas, whereas doum palms only occur in the southern sections of the Afro-Asian Red Sea coastal lands but are absent from those of the Mediterranean. However, both species occur in the remote desert oases of North Africa. A brief accounts of these two palms and their potential economic uses are given below.

**(a) *Phoenix dactylifera***

What is probably the most familiar and most valuable tree of the desert is the date palm. It ranks among the three most important of the world's economic palms, in company with the coconut and oil palm. Thus as far as arid and semi-arid land palms are concerned, the date palm is a developed tree crop and may serve as an example of what potentially could be achieved with other species.

*P. dactylifera* is a large dioecious palm reaching 30 m in height. It has a relatively thick trunk covered with persistent leaf bases. Suckers occur at the base of the tree. The leaves are pinnate, averaging 3–6 m in length and form a crown of more than 100 leaves. A new leaf is produced approximately every month. Upon reaching sexual maturity at about 5 years, the female palm blossoms once each year. Fruits require some 6 months to ripen. An average of a dozen inflorescences are borne on a single tree. The date palm fruit is cylindrical in shape, about 5 cm long and 2 cm wide, and contains a single hard seed. Annual fruit yield per adult tree varies from as little as 5 kg to more than 100 kg.

The date palm thrives in the hot, dry climates of subtropical and tropical regions, providing there is an abundant supply of water. One decided advantage possessed by the date palm is that its needs can be met by brackish water without adverse effects. It can also withstand cold temperatures down to –7°C. Essential to high fruit production is the absence of rainfall during the period of pollination (FAO, 1982).

The cultivation of date palm evolved in oases and elsewhere in combination with other crops. Thus, there is a long tradition of inter-cropping with citrus, figs and other perennial crops; within sparsely planted palm groves, annual crops such as wheat, and beans, are grown as well as forage grasses for livestock.

Date varieties are classified as being soft, semi-dry or dry on the basis of moisture content; Medjool, Deglet Noor and Thoory are examples, respectively. A dry, fresh ripe date contains 75–80% sugar and is a good source of iron and potassium. Dates eaten with dairy products make an acceptable diet and some desert groups subsist on that combination for months at a time. Dried dates can be stored almost indefinitely. Whether fresh or dried, dates are consumed raw, chopped and fried in butter, boiled and then fried, preserved whole, and made into preserves, syrup or date butter. Macerated pulp can be made into a beverage with water or milk and drunk as is or allowed to ferment. Date wine can be made into vinegar or distilled to produce spirits. Cull dates commonly are utilized for the latter (Johnson, 1985).

Livestock can be fed date pulp and softened or ground seeds on a dry weight basis, date seeds contain 20.64% starch, 4.38% sugars, 6.43% protein and 9.2% oil. Date seed oil is of good quality, but does not occur in sufficient quantity to justify commercial extraction.

Date palm growing today is still heavily concentrated in the North African and Near East region. Currently Iraq and Saudi Arabia are leading producers, each with 15.2% of the total. They, together with Egypt and Iran, account for 56.7% of the world's dates.

Literally hundreds of domestic uses are reported, with the greatest number occurring in remote parts of the date-growing areas of the Old World. In brief, the tree can be tapped for sap which can be drunk fresh, reduced to palm sugar or fermented into palm wine. When a date palm is felled, the palm heart is extracted and consumed. The leaves are used for thatch and to construct fences. To both Jew and Christian, the date palm leaf has ritual significance. Leaflets are woven into mats, baskets, fans, etc. and fiber extracted from the leaves makes a strong rope. The midribs serve to make crates, simple furniture, and chicken coops or can be burned as fuel. Leaf bases also are used as fuel and as floats for fishing nets. The trunk can be cut into boards and employed in construction as rafters, for walls, and to fashion shutters and doors. Young green dates are strung as necklaces by children; fresh flower heads are distilled into tarah water for flavouring sherbet; and date seeds used in traditional medicine. Finally, the date palm is an excellent provider of shade in the hot deserts and is greatly appreciated for its landscape value (Johnson, 1985; Amer and Zahran, 1999).

**(b)** ***Hyphaene thebaica***

The doum palm shares the desert oasis habitat of subtropical and tropical Africa with the date palm. But unlike the date, it is known in the wild and its natural range includes Africa, the Middle East and West India. It has a relict distribution which is somewhat uncertain because there are about 40 species of *Hyphaene. Hyphaene* is a unique in the Palm in having a dichotomously branching stem, present in most species.

Because its presence is most often linked to poorly drained soils with a high water table, the doum palm is an indicator species in Africa: it can occur as a solitary tree, or may form pure stands. The palm reaches a height of 10 m under favourable conditions and branches to form 4–16 crowns of fan-shaped leaves. It is dioecious and bears large, smooth brown fruits composed of a juicy pulp and a very hard endosperm. Each tree produces about 50 kg of fruit per year. Baboons and elephants eat the fruit. In Egypt, the doum palm has been cultivated since ancient times and has long been considered a sacred tree, symbolizing masculine strength. The most important product is the fruit, a common wild fruit of the Middle East. The pulp is edible and is described as having a taste suggestive of gingerbread or carob pods. Doum fruits have been found in many Egyptian tombs. In early times the fruits were made into cakes. At present in Egypt they are sold dried, and are reconstituted and eaten as a paste. The palm heart is also edible.

A doum fruit weighs an average of 20 g. Fruits of the sweet type are composed of, by weight, 22–30% exocarp, 39–42% mesocarp and 34–44% endosperm; those of the bitter type of 30–41% exocarp, 23–27% mesocarp and 34–44% endocarp (Fanshaw, 1966). When mature the endosperm has the hardness of vegetable ivory. For that reason the seed has been used for centuries to carve trinkets and other small objects, and in Egypt early in the present century was exploited commercially to make buttons. The endosperm contains about 10% oil (Eckey, 1954).

The fan-shaped leaves are used widely for thatch and to weave mats, baskets, and bags. Fibre is extracted from the leaves and made into rope. Camels eat young leaves of the palm. Doum wood is strong and durable, and has utility for posts, beams and can be hollowed out for water pipes. It has a chocolate-brown color streaked with black and makes attractive furniture. The fruit pulp and roots are employed in folk medicine.

#### 4.4.5.5 Case Studies

For a constructive and continuous role that could be played by the natural vegetation in the sustainable development of coastal and inland deserts, enough drought resistant and/or salt tolerant plant species with economic potential should be available all year around. Actually, the required quantities of such valuable plants probably cannot be secured from the present vegetation cover. It is necessary to widen the areas covered by these plants by introducing them as non-conventional crops to be cultivated under aridity and/or salinity stress. Available rainfall and ground water have to be used, on a scientific bases, for irrigation. Sea water neat or diluted could be used also to propagate highly salt tolerant plants e.g. mangrove species. Fortunately, successful field experiments have been conducted to propagate some of these species to produce food, fodder, fiber, drugs, perfumes, oils, fuel, timber etc. in the coastal and inland deserts of the arid region of the world. The species tested included: trees, shrubs, undershrubs, grasses, sedges, rushes etc. Case studies world wide indude: Zohary (1962), Malcolm (1972), Walsh et al. (1975), Dregne (1977) Mann (1978), Anonymous (1980, 1997, 1998b), Biswas and Biswas (1980), Schechter and Galai (1980), Zahran and Abdel Wahid (1982), Zahran et al. (1983), Ahmed and San Pietro (1986), Barrett-Lennard et al. (1986), Goudie (1990), Lieth and Al-Masoom (1993a, b), Clough (1993), Zahran (1993b, 2004, 2007), Al-Zayani (1993), Pessarakli (1993), Ben Haider (1994), Abdel Razik (1994), Ashour et al. (1997), Khalifah (2000), Abou Deiah (2001) and Saenger (2001).

The results of studies carried out on representative species of relevant coastal vegetation forms are discussed under five groups: livestock fodder plants, fiber-producing plants, drug-producing plants, wood-producing plants and oil-producing plants.

**Group I: Livestock Fodder Plants**

The floristic compositions of the natural vegetation forms of the Afro-Asian Mediterranean and Red Sea Coastal lands are rich and characterized by considerable number of palatable species belonging to Gramineae, Leguminosae, Chenopodiaceae Cruciferae, Compositae, Nitrariaceae, etc. These palatable plants can be considered as reliable local natural range plants when their total vegetative yields are high enough to maintain continuous fodder supply for the local livestock. Unfortunately, this is not ascertainable under the arid and semi-arid climate of these coastal lands. As rainfall is generally low, the density and biomass of range plants is not enough to supply the requirements of the livestock all the year around. It will also not be possible to get homogeneous natural vegetation with palatable plants only, because other unwanted and/or poisonous weeds will be present. The alternative and promising solution is to propagate certain palatable xerophytic and/or halophytic species proven to have high nutritive values under the dry or saline stress of these coastal deserts. Short accounts on the propagation of representative xerophytes fodder-producing and halophytes: (*Panicum turgidum, Kochia* spp., *Atriplex canescens* and *Leptochloa fusca*), as well as a case study of the sustainability of range lands are described below.

**Case Study 1. *Panicum turgidum***

*P. turgidum* is a common xerophytic fodder grass that predominates in both arid and hyperarid coastal and inland areas. It is a highly palatable grass: its seeds are used as human food. In addition, *P. turgidum* is an effective binder of sand and could be used to stabilize dunes. *P. turgidum* extract is used in nomadic medicine as a voluntary agent for removing white spots on the eye (Boulos, 1983). The powder from its underground stems is used in healing wounds. Unfortunately, this multi-use drought resistant grass is usually overgrazed beyond its capacity to remain vigorous. However, it exhibits high growth rates in late spring and early summer months from buds and rhizomes hidden underground soil surface and are thus partly protected from grazing (El-Kabalawy, 2004).

Batanouny et al. (2006) conducted germination and growth experiments on this xerophytic palatable grass to study its reproductive capacity.

**a. *Germination experiment***

Germination of intact and dehusked seeds of *P. turgidum* has been tested under different temperatures in light and dark. Under constant temperature, intact *P. turgidum* seeds did not germinate at 15°C, but the germination percentage increased up to 37% at 35°C. Temperature between 20 and 30°C showed higher values of germination under dark conditions (31.7%) than those under light (21.7%). Alternating temperatures improved germination percentage reaching 51% under 10–20°C increased to 84% under 20–30°C.

In contrast, germination of dehusked seeds was favoured by light. Germination was highest 46 and 43% under constant temperature of 35°C in light and dark 10 and 25°C, decreased to 51% when germination was tested under 20 and 30°C.

**b. *Growth experiments***

**(i) By transplants**

This experiment was conducted in a newly reclaimed area at El-Noubarya, Western Desert (148 km south of Cairo) along the Cairo-Alexandria desert road. Transplants obtained from a *Panicum turgidum* community in the nearby area were planted every week into holes (20–30 can deep) to discover the best time of the year for propagation. This turned out to be April or October, with sprouting percentages of 55% (April) and 50% (October).

**(ii) By seeds**

Fully mature seeds collected from natural stands of *Panicum* sown and irrigating daily for seven days and then every other day for a month showed successful germination. Five days after sowing, there was 10% seedling emergence gradually with time to 85% after 25 days. Four-month old *Panicum* plants produced dense culms (mean 3,450 culms, mean height 107 cm). Though successful, Batanouny et al. (2006) did not recommended propagation by seed for dry-land agriculture, instead proposing that seed be densely sown in nurseries and the seedlings transplants for propagation. Success would be maximized if the transplants were kept moist until planting in the field, and the upper parts of their culms dipped before transplantation to reduce transpiration and promote tillering.

**Case Study 2. *Kochia* Forage Halophytes**

To introduce *Kochia* plants as fodder-producing halophytes that can be propagated under saline and/or arid conditions, establishment experiments were conducted in salt-affected lands of Saudi Arabia and Egypt.

In Saudi Arabia a field experiment was carried out in the Bahra area of the Red Sea coast (midway between Jeddah and Mecca) where the shortage of green fodder, especially in summer, is a problem (Zahran, 1986).

The salt-affected land near a saline artesian well (4,851 ppm total soluble salts) dominated by *Suaeda monoica* (a salt-tolerant perennial shrub) was cleared of its natural vegetation. The sandy to sandy-loam soil was alkaline (pH = 8.1, 8.9) with the highest amounts of solouble salts (EC = 12 mmhos $cm^{-1}$) in the surface layer, decreased down the profile.

Seeds of *K. indica* (obtained from Egypt) and *K. scoparia* (obtained from Texas, USA) planted not too deep (ca. 1 cm) and irrigated with saline well water at weekly intervals showed high percentages, with most of seeds germinating after 4–10 days. Growth of the plants continued normally, even though after 100 weeks the soil salinity had increased substantially (e.g. from 12 to 20 mmhos cm in the surface layer). By the end of May, the plant cover was more than 80%. The plants attained maximum vegetative growth during summer (July–August), with *K. indica* producing more (mean fresh weight = 8.5 kg per bush) than *K. scoparia* (5.6 kg) plant became brownish-yellow during September, and were dry by October–November, but produced profuse amounts of seeds, which were collected in December.

In natural stands, both *Kochia* species germinate in February, but if seeds are also germinated during summer, this will result in two vegetative crops each year, increasing the economic value. The potential was confirmed by sowing *K. indica* in June and by October the vegetative yield of green forage was as high as 55.3 t $ha^{-1}$. These results emphasize the potential importance of *Kochia* plants; they are salt-tolerant, drought-resistant, rich in nutritive substances, palatable to live stock, rabbits, and poultry, and poultry, and can be planted at any season of the year.

The promising results obtained from the propagation experiment with *K. indica* in Saudi Arabia led to a second phase to study the effects of fertilizer treatment, salinity stress, and irrigation intervals on the vegetative yields of these plants. This was carried out on a piece of saline land in Damietta on the Med. coast of Egypt, where the soil is saline, poor, infertile, and non-productive. Conventional crops usually cultivated in the other part of the Damietta Governorate fail to grow in this soil because of its high salt content. The irrigation water for the experiment was supplied from the Om Dingel Canal with high salt content ($EC = 3.7$–$5.3$ mmhos cm$^{-1}$) relative to water from the Damietta Branch of the Nile ($EC = 0.33$–$0.41$ mmhos cm$^{-1}$). The salinity of the drainage water is higher, $EC = 5{,}590$–$5{,}630$ $\mu$mhos cm$^{-1}$ (Zahran et al., 1992).

The natural vegetation of the area is halophytic and includes *J. acutus, J. rigidus, K. indica, Zygophyllum aegyptium, Spergularia marine, Beta vulgaris, Cyperus rotundus, Tamarix nilotica, Cynodon dactylon, Mesembryanthemum crystallinum,* and *Melilotus indicus.*

There was a highly significant increase in productivity with increasing doses of both N and P fertilizers reaching about 6, 15, and 22.5 t ha$^{-1}$ for the three successive harvests. Yields increase with age, but there was no effect of an irrigation interval of 20 as opposed to 30 days.

Increased soil salinity adversely affected but this effect decreased with plant age until in 7 months old plant the yield was higher (9.8 t ha$^{-1}$ dry matter) in the high salinity plots than that in the low-salinity plots (8.7 t ha$^{-1}$ dry matter). This probably indicates that salt tolerance increase with age: older plants are more tolerant to increased soil salinity than are younger plants.

**Case Study 3. *Atriplex canescens***

Growth and forage yield of the fodder shrub: *Atriplex canescens* was tested in the salt-affected land of the delta of Wadi Sudr in Sinai focussing on the effect of three types of soil amendments: farm yard manure, town refuse and elemental sulphur (El-Housini et al., 2004). The soil was sandy-clay, calcareous (56% $CaCO_3$) with an EC of = 8.5 mmhos/cm and pH of 7.9. The sole source of irrigation was underground brackish-saline water containing about 4,500 ppm dissolved salts. Seedlings raised from seeds at 7-month old and at 30–40 days intervals according to seasons for two years. The results elucidate that:

1. *A. canescens* can be propagated as a forage halophyte in salt affected land.
2. Biomass production increased by increasing amounts of manures.
3. Total carbohydrates and protein content increased with increasing levels of manures, but crude fibre content decreased-probably as a consequence of increasing amount of protein. Ash content showed no clear trend.

**Case Study 4. *Leptochloa fusca***

The growth of the fodder halophyte kallar grass (*Leptochloa fusca*) in both coastal and inland salt-affected lands of Egypt was experimented with Ashour et al. (2002).

The field trials used local water (diluted sea water and brachkish drainage water) for irrigation, and biomass production and chemical composition were evaluated. On the eastern coast of the Gulf of Suez (Sinai), four dilutions of sea water were tested (5,000–20,000 ppm) on root stumps irrigated weekly or twice weekly and harvested every 2 months. Biomass was the highest (39.18 t/ha/year fresh and 18.5 t/ha/year dry) when irrigated with 10,000 ppm sea water, decreasing with more saline water. Increasing salinity increased ash content, soluble carbohydrates, crude protein and crude fat and decreased crude fibre content.

Near lake Qaroun of Fayium Depression, Western Desert, the weekly or twice-weekly irrigation used agricultural drainage water with (pH = 7.4, EC = 2.8 mmhos/cm$^{-1}$), again on transplanted stumps and harvested at 2-month intervals. Total yield was 23.6 t/ha/year (fresh weight) and 14.4 t/ha/year (dry weight). The chemical composition of *L. fusca* of this experiment were: ash (14.7%), fat (1.83%), proteins (9.2%), carbohydrates (42.9%) and fibre (31.4%). Other elements determined were: potassium (1.02%), sodium (2.14%), calcium (0.54%), magnesium (0.4%) and phosphorus (0.16%). Thus, kallar grass (*L. fusca)* is therefore a useful halophyte rich in nutritive value, which could be propagated in salt-affected lands of the inland and coastal deserts irrigated with diluted sea water or brackish drainage water. The grass tolerates salinity up to 20,000 ppm, but high yield was obtained when it was irrigated with 10,000 ppm sea water. It is easily propagated through root stumps and produces during summer when traditional green fodder is unavailable.

**Case Study 5. Sustainability of Range Lands**

**a. *General Remarks***

In arid countries, the pressure of land-use, overgrazing, over cutting, cleaning of agriculture etc., coupled with extreme climatic aridity and uncertainty of rainfall, has resulted in an advanced stage of desertification. Natural vegetation is currently regressing at a rate of 1–2% annually (Le Houerou, 1973). El-Kady (1987) stated that the intelligent use of rangeland in the deserts for optimum yields requires a detailed knowledge of what and how much the land can produce under given circumstances, i.e. knowledge of the biological potentialities. It also requires management of land so as to insure continuity (sustainability) of optimal yields by conserving existing resources of soil, water, vegetation and wild animals, and monitoring the condition of each element. Monitoring will allow for prediction of crucial long-term effects of man-made manipulation. Sound management including land protection usually results to an increase in density, cover and vigour of the vegetation and consequently better soil stabilization (Halwagy, 1962).

According to Ellison (1960), when left to fallow, dryland ecosystems disturbed by overgrazing or stressed by drought, may recover. Recovery tends to advance at a slow pace because of the low productivity except in years of above-average rainfall. Monitoring of changes to predict long-term effects of management manipulations is of crucial significances in the formulations of suitable management plans (Ayyad, 1983). Such monitoring can be viewed at three levels. The first is concerned with ecosystem components and human impacts on them, looking at changes in soil characteristics, the abundance of biodiversity (plants, animals, and microorganisms) above and below soil and the roles these play in energy flow and nutrient cycling. These can then be used to construct simulation models for predicting future changes at the ecosystem levels (El-Kady, 1987). The second level includes monitoring changes in the patterns of vegetation composition and of physiographic using vegetation mapping (remote sensing and aerial photographs) followed by ground truthing. The third level includes the composition of salient features of large areas in successive years to evaluate landscape changes.

**b.** ***Grazing Experiments***

The effect of protection and controlled grazing on the vegetation composition and productivity as well as the rate of consumption of phytomass by domestic animals was assessed by El-Kady (1980), during three successive years 1977–1979 in the non-saline depression of El-Omayed, 80 km west of Alexandria in the Western Mediterranean Coastal desert. The climate here is arid with mild winters and warm summer (UNESCO, 1977).

Five grazing treatments were studied, one outside fenced plots with *ad lib* grazing (about 6 livestock per 10 ha) and four in fenced plots representing 50, 50, 25 and 0% of the *ad lib* level. Fencing started at various times. Changes in the density, cover, frequency, phytomass and the phenological sequence of species were recorded and compared to those of the same species outside the fenced plots. Grazing animals in each of the two 50%-grazing plot were observed in detail to see what and how frequently they ate. A consumption index was then calculated to assess the relative preference of different plant species by domestic animals. The following results have been obtained:

1. Seven perennial species constituted the major part of the animal diet in the controlled grazing plots: *Asphodelus microcarpus, Echiochelon fruticosum, Thymelaea hirsuta, Plantago albicans, Crucianella maritina, Helianthemum lippii* and *Gymnocarpos decander*, beside one annual species (*Rumex pictus)* during growing season. Outside the fenced plots, *A. microcarpus, E. fruticosum* and *T. hirsuta* were the main species in the diet. Relative preferences, changed from season to.
2. Remarkable increases were recorded in total density and cover of perennials, in frequency and presence of animals, and in phytomass as a result of protection and controlling grazing. Some plants exhibited negative responses and the productivity of most species was more pronounced as the degree of controlled grazing increased especially with an initial period of full protection. Thus, partial protection and

controlled grazing may provide outcomes than full protection. Light nibbling and removal of standing dead biomass by domestic animals can promote vigour and growth of defoliated plants, and the availability of nutrients (especially nitrogen) can be enhanced by the passage through animal guts and out as faeces.

3. The consumption by domestic animals was about 20 and 40% of the net primary production of shoots in the controlled grazing plots and the unprotected area, respectively.
4. The phytomass and necromass provided amounts of total digestable nutrients ranging from 520 g/animal/day to 960 g/animal/day in the controlled grazing plots, but only half these amounts in the free-grazing area where protein contents were 10% less. These amounts were estimated to be far short of their requirements of animals in the free grazing area, and scarcely adequate for those under the rotational system with 50% grazing pressure.
5. To maximize productivity, adequate amounts of supplementary feed rich in protein should be supplied.
6. Similar studies need to be conducted in other rangelands of coastal and inland deserts to provide information necessary for the protection, recovery, management and consequently sustainability of these rangelands.

**Group II: Fibre Producing Plants**

The natural flora of the Afro-Asian Mediterranean coastal lands compreises a considerable number of xerophytes, psammophytes and halophytes that could be considered as fibre producing plants. Among these are: *Ammophila arenaria, Calotropis procera, Leptadenia pyrotechnica, Caralluma spp., Dracaena ombet, Desmostachya bipinnata, Stipa tenacissima, Juncus rigidus, J. acutus, Gossybium arboraem, Thymelaea hirsuta, Lygeum spartum, Phragmites australis* etc. All of these plants grow under aridity and/or salnity stresses, some are dominant and wide spread. A plan aiming at the sustainable utilization of these plants by introducing them as non-conventional fibre producing crops in these coastal deserts could be encouraged. The vegetative yields of these plants could be used as raw materials for various fibre industries. Representative species are given below.

**Case Study 1. *Stipa tenacissima***

Nadji et al. (2009) isolated soda lignin, dioxane lignin and milled lignin from alfa grass (*Stipa tenacissima)* growing abundantly in the western countries (Morocco, Algeria and Tunisia) of the North African Med. coastal lands (see Chapter 3). The total antioxidant capacity of the lignins was comparable to commercial antioxidants commonly used in thermoplastic industry. *S. tenacissima* could easily propagated by seeds under rainfed irrigation to get the biomass production needed.

**Case Study 2. *Thymelaea hirsuta***

El-Ghonemy et al. (1977), Shaltout (1983) and Bornkamm and Kehl (1990) reported that *Thymelaea hirsuta* is a perennial evergreen shrub 40–200 cm tall. It is a circum Mediterranean in distribution but of minor importance along the European Med. coastal belts. It is a common xerophyte in the North African and SW Asian coastal lands where it inhabits sandy formations, rocky ridges and coastal plains. *T. hirsuta* seems intolerant of the increased aridity because the gradual landward (southward) decrease in rainfall, rise in temperature and evaporation and reduction in soil moisture contents are associated with a progressive decline in density and vigoure of *T. hirsuta.* The southern most limit of its presence is 70–75 km from the Med. coast of Egypt (Zahran and Willis, 2008).

Anatomical studies and fibre length measurements show that it can be considered as a renewable source of the raw material in the fibre and/or rayon industries. It contains a considerable number of relatively long fibres (3,000–6,000$\mu$) which these industries prefer (Zahran and Boulos, 1974).

**Case Study 3: *Phragmites australis***

*P. australis* is a robust, erect reed swamp perennial grass; with culms up to 2.5–6 m high. It is the most widely distributed flowering plant in the world, found throughout North and South America, Europe, Asia, Africa and Australia.

Ecologically, *P. australis* is classified as a semi-aquatic plant that grows in various habitats and in fresh and brackish waters. It is tolerant to saline conditions, but its performance decrease with increasing salinity (Batterson and Hall, 1984).

*P. australis* has several uses for man and animals. It is a high quality livestock forage during early growth readily eaten by cattle, goats, sheep, horses and donkeys. It can be cut for hay. It is an excellent soil stabilizer because of the fact that it can propagate both vegetatively and by means of seed and that its deep root spreads rapidly, thus it is so useful for soil conservation and for the stabilization of sediments. *P. australis* is used also in folk remedies for mammary carcinomast and leukaemia. The biomass of *P. australis* can be used as an energy source via methane emission, as building material for houses, rafts, and mats, and a source of cellulose in the paper and textile industries (Eid, 2008).

*P. australis* is a relatively good seed producer, with up to 1,000 fertile fruit per inflorescence. However, vegetative propagules are the most important means by which the plant spreads. Colonies expand peripherally by lateral rhizome growth, typically subterranean (Marks et al., 1994).

**Case Study 4. *Juncus rigidus and J. acutus***

*J. rigidus* is a densly tufted salt-tolerant rush with slender pungent nodeless leafy shoots (culms) more than one metre high developing every year from sympodial creeping rhizomes. These rhizomes extend horizontally exploiting wide areas and can produce dense plant growth within a few years. *J. acutus*, has the same morphology and is also salt tolerant but its rhizomes grow in a particular way that give rise to a great number of green culms forming circular patches of various sizes.

*Juncus* plants have many uses both in old and new. *Juncus* culms are used by local people to make the best quality mats, and its seeds are used in oriental medicine as a diuretic and a remedy for diarrhaea etc. These seeds are rich in fatty acids especially palmitic, oleic, linoleic, lauric, myrestic, stearic and linolenic acid. There are 48–52 different glycerides of the oils extracted from the seeds of *J. rigidus,* suggesting the possibility of using them as potential oil sources for various purposes. However, the most important industrial use of *Juncus* plants is the use of clms as raw materials for the paper industry (Boyko, 1966). This was confirmed by Zahran et al. (1972) who found that the fibre length was 407–2,421μ with an average width of 16μ (ratio = 92.7:1) a cell walls 6–5μ thick. The percentage of longer fibres (> 1,000μ) is greater than 50%. These characteristic are preferred in the paper industry.

Chemical analysis and a pilot experiment carried out in one of the paper mills in Egypt proved that *Juncus* culms contain low ash (6.5%), low lignin (13.3%), high alphacellulose (39.8%) and high yields of unbleached pulp (36.8%). Depithed unbleached pulp of *Juncus* (grade index = 73%) stronger are much stronger than that of rice straw (grade index = 24%) and bagasse (42%) compared to that of imported softwood long fiber unbleached kraft pulp (grade index = 100%). Also, the bleaching of *Juncus* pulp gave brightness of 76 photo volts and good strength properties. In addition, *Juncus* pulp can be used alone to produce paper while the other local raw materials (e.g. rice straw, bagasse) should be mixed with wood pulp before producing paper. The countries of our area are importing wood pulp to produce good quality paper and/or to improve the strength properties and grade index of paper produced from their local raw materials. The amount of paper and/or wood pulp imported from abroad is sharply increasing every year. A promising alternative is to depend upon local raw materials proved to produce good quality paper such as *Juncus* plants. Accordingly, if sufficient amounts of *Juncus* culms can be made available to paper mills, we can reduce the quantity of paper and/or paper pulp imported. However, large-scale economic production of paper entails large-scale production of *Juncus* plants from natural population. This will hopefully, realized through the cultivation of *Juncus rigidus* and *J. acutus* in the salt affected lands of these countries. Fortunately, and according to Zahran (1993b), both *Juncus* species were successfully grown experimentally in saline soil near Lake Manzala and in calcareous, soil near Mariut Lake of the Med. coastal land of Egypt. Thus, *Juncus* halophytes are plants of agro-industrial and economic potentiatities that could play an effective role in the sustainable development of the salt affected areas of the Afro-Asian Med. and Red Sea coastal deserts.

**Group III: Drug-Producing Plants**

Plants are the amazing sources of chemical compounds and have always played a major role in the treatment of diseases. Plants are, considered Nature's Green Pharmacy providing drugs to maintain the health of humans and animals.

Prehistoric man used plants to battle diseases, induce hallucinogenic experiences and to word off evil spirits. The advent of western medicine diminished the importance of herbal medicine (Boulos, 1983).

In the countries of the Afro-Asian Mediterranean and Red Sea coastal lands, the use of herbs for therapeutic purposes dates back to times immemorial. In the deserts of these countries where people live in tribes far away from each other, there is almost no proper medicinal care, and this function is fulfilled by particular people using folk medicine formulae based on the crude materials from their local environment. "Plants are, actually, used as drug with slight or almost no change leading in most cases to satisfactory results" (Boulos, 1983).

Medicinal plants constitute a considerable proportion of the flora of the eleven vegetation forms of the Afro-Asian Med. and Red Sea coastal lands, all are used in folk medicine. Unfortunately, most of them are seriously threatened due to over use since there is no alternative. Thus "there is a great need to provide a framework for the conservation and sustainable use of these valuable renewable natural resources" (Hammouda, 2005). The following pages throw light on representative medicinal plants naturally growing in the coastal lands.

**Case Study 1. *Salsola tetrandra***

*S. tetrandra* (Chenopodiaceae) is a halophyte growing naturally in coastal salt marshes of Afro-Asian Mediterranean coastal lands. It tolerates soil salinity up to 2.5%. Chemical screening of *S. tetrandra* indicates that it contains: alkaloids, glycosides, saponins, tannins, sterols and organic acids. Bioassays using extract of this halophyte have determine its action on the intestine, uterus, heart, blood pressure and respiration, anthelmintic properties and toxicity (Zahran and Negm, 1973).

Its action on blood pressure and respiratory movements was examined in dogs anesthetized by rectal administration of pentobarbitone sodium (Nembutal), recording blood pressure from the carotid artery and respiratory movement via a sphygomograph connected to chest movement. The effect on the heart was tested on perfused heart.

When *S. tetrandra* alcoholic extract was injected (in different doses) intravenously into the femoral vein of dogs, there was a sudden transitory rise in the blood pressure following by a gradual full to the original level. There was no noticeable effect on respiratory movements. The extract markedly decreased the force of cardiac contraction of the heart.

Toxicity experiment of 4% alcoholic extract in water on mice and rabbits showed a minimum lethal dose of 0.4 g/kg body weight. The alcoholic extract inhibited uterine activity at all stages of the sexual cycle, and relaxed the intestinal musculature indicating its possible antispasmodic and anticonstipation value.

There are also some indications of anthelmintic activity of *S. tetrandra* alcoholic extract, hence it depressed intestinal taenia, and then a purgative can be used to expel the depressed parasite.

The therapeutic possibilities of *S. tetrandra* as an intestinal antispasmodic and anthelmintic drug indicates that wide areas occupied by the plant along the Mediterranean coastal land may be of economic importance. To be quite sure of a permanent supply of *S. tetrandra*, it is necessary to protect and develop the wild vegetation and also to transplant it to new non-productive salt-affected areas of the Afro-Asian Mediterranean coastal lands.

**Case Study 2. *Calotropis procera***

*C. procera* (Oshar, Asclepiadaceae) is a xerophytic shrub or small tree 3–5 m high, with the softwood covered with thick cork-like bark, light-brown in colour. The leaves are light green, simple, large and broad, up to 25 cm long. The flowers are green on the outside, inside pink. Fruits are large (15 cm across), smooth, apple-like, green in colour, and spongy. When mature the fruit open and reveals the seeds which are packed into a compact core and covered by long silky hairs, facilitating their dispersal by wind (Täckholm, 1974; Springuel, 2006).

*C. procera* has various uses including ropes from its inner bark. However, its most valuable component is the milky latex, used to treat many human and animal disorders. In the same time, the latex cause serious inflammation to eye that may lead to blindness. Decoction of bark and latex used in veterinary medicine, anti-leprosy for scabies. Powered dried leaves vermifuge in small doses; dry leaves are smoked as cigarettes for asthma, cataplasm (poultice) of fresh leaves for sunstroke. The latex is applied to teeth to loosen them, and also for toothache. The leaf extract is a cardiotonic, the root an emetic, and expectorant; root bark is used for dysentery, elephantiasis, syphilitic ulcers, stomachic, diaphoretic (Boulos, 1983).

*C. procera* is regenerated by seeds which are easily germinated without treatment and can be planted directly into moist ground. Seedlings grow rapidly with sufficient watering.

**Case Study 3. *Cassia senna***

*C. senna (Senna alexandrina,* Senameki, Leguminosae*)* is a xerophytic glabrous undershrub, multi-stemmed reaching a height of 1–1.5 m and width of 2 m. The pale green stems are erect and densely branched at 20–30 cm above the ground. Leaves are compound with 3–7 pairs of elongating grayish-green mucilaginous leaflets having a peculiar odour and sweetish taste. Yellow recemed flowers, fruits are flat pods green when young and changed to yellow-brown when mature, each fruit contains six seeds. Root system is extensive in the ground.

*C. senna* is used in traditional medicine and has a commercial value. It is well known as a purgative and is used in the treatment of influenza, asthma and nausea. “Infusion of powdered leaflets and pods is a popular laxative and purgative” (Boulos, 1983).

*C. senna* is usually propagated by seeds and germination occurs without any treatments. However, to enhance germination scarification is recommended to break the outer cover of the seeds either by scratching or by putting the seeds in hot water for one day.

**Case Study 4. *S. Cymbopogon schoenanthus***

*C. schoenanthus* (Halfa Barr, Gramineae) is a perennial xerophytic aromatic, stout, densely tufted grass about 1 m high, with filiform narrow leaves and shallow fibrous roots. For its medicinal value, the grass is highly threatened by over exploitation. It grows in alluvial deposits of desert wadis, in rocky habitats as well as in sandy soil. A high temperature is very suitable for its growth and it tolerates a prolonged dry period, it is a true xerophyte.

Springuel (2006) stated that *C. schoenanthus* (halfa barr) was known in ancient Egypt as one of the ingredients for making the famous *Kyphi,* a scent free from oil and fat. It is used intensively in indigenous medicine as a diuretic, a painkiller in cases of colic and an antipyretic. Pharmaceutically, this grass is used in preparing the drug Proximol. Halfa barr is a healthy and refreshing hot drink especially popular in Upper Egypt. "Infusion of plant diuretic, emmenagougue, astringent, carminative, sudorific, antirheumatic, cataplasm for wounds of camels. Infusion of flower febriuge" (Boulos, 1983).

The seeds of *C. schoenanthus* are very small and difficult to germinate due to their long dormancy: they need to be in the soil for at least 8 months. Thus, it is preferable to germinate this grass by root cuttings. Once established, halfa barr will self-propagate when moisture is available.

**Case Study 5. *Origanum syriacum* subsp. *sinaicum***

*O. syriacum* subsp. *sinaicum* (margoram, za'atar, bardagoosh, Labiatae) is an endemic tomentose herb or undershrub. Its stem (40–90 cm) is erect much branched, the leaves are broadly ovate, entire and palmate-veined, verticillasters 2–8 flowered in dense spik-like infloresences, or often in panicles, the corolla is lilac. It grows in rock habitat of the Sinai mountain. It is thought to be the tree Hyssop of the Bible (Hammouda, 2006). This medicinal plant is highly threatened because it is heavily collected to prepare herbal tea for various medicinal uses. There is a great need to conserve its natural growth and to propagate it both in situ and ex-situ.

Za'atar has a long history as a medicinal and flavoring herb. It has been used in folk medicine to treat cold, coughs and gastrointestinal problems. It is also used as antibacterial, antifungal, and anti-rheumatic. The essential oil produced from *Origanum* (oreganol oil) is extracted from its leaves and flowering branches. The oil is rich in a long list of minerals including: calcium, magnesium, zinc, iron, copper, boron, and manganese in addition to vitamins.

*Origanum* is propagated by seed, division and basal cuttings (Hammouda, 2006). Seed germination (in a green house) usually takes place within two weeks and the seedlings are planted into pots when they are large enough to handle and later planted out into permanent positions in early summer. Division propagation usually takes place either in March or October. Very easy, larger bits, can be planted out directly into their permanent position in late spring or early summer.

Basal cuttings is also easy way to propagate this medicinal plant. Shoots with plenty of underground stems are collected when they are about 8–10 cm above the ground during June, placed in individual pots and kept in light shade in a cold frame or green house until they are well rooted.

**Case Study 6. *Salvadora persica***

*S. persica* (Arak, Salvadoraceae), the tooth brush tree, is a xerophytic tree or shrub. It has opposite leaves and small greenish-white flowers in rich terminal panicle.

Fruits are small white or pale purplish, globose drupe of pungent taste. The stem is heavily branched, and both stems and branches are whitish and bear numerous leathery leaves with a strange smell similar to mustard hence the English name: mustard tree. The roots are extensively branched from the base of the trunk and are very long spreading both horizontally and vertically, and deeply penetrating the soil.

The tooth-brush tree grows abundantly in certain wadis of the Red Sea Coastal Desert and Sinai. For its dense growth, one of the wadis of the Red Sea Coastal Desert is called "Wadi Arak". *Salvadora* is an evergreen plant that keeps its leaves even during the prolonged dry season.

*S. persica* is very well known in Egypt, Saudi Arabia and throughout the Middle East because of its use in traditional medicine. It is used to treat gonorrhea, spleen, boils, sores, gum diseases, headache, stomach pain and respiratory disorders. The leaves, roots, bark and flowers contain an oil that is diuretic. The powered leaves are mixed with millet flour and honey and made into small balls to be taken every morning for 40 days as an antisyphilitic. The fruits are edible, stomachic, carminative, febrifuge, fortify the stomach and bring good appetite (Boulos, 1983). The powered bark of arak is used in the treatment of snake and scorpion bites. The most important use of arak since ancient times is to clean the teeth. Its young stems and lower parts of the stems close to the roots are used as a tooth-brush.

Propagation of *S. persica* can be by seeds that germinate easily without treatment. However, due to its low seed production, it is preferable to propagate it by cuttings that are obtained from naturally growing plants cuttings are taken from the base of the stem close to the root. Propagation by tissue culture is also recommended (Springuel, 2006).

**Case Study 7. *Solenostmma arghel***

*S. arghel* (argel, hargal, Asclepiadaceae) is an evergreen erect perennial xerophytic undershrub 0.6–1.0 m in height. It is a blue-green, finely velvety-pubsecent plant with elliptical lanceolate Leaves. Flowers are white and occur in axillary umbels, fruit ovate, smooth, very hard of dark purple colour. *S. arghel* is a typical evergreen xerophytic undershrub that extends vegetative activity for at least three years of a rainless periods. In dry periods some individuals may shed their leaves and even some branches, but others remain evergreen throughout the vegetative periods. Reproductive activity is high in the third year from September to June. Flowers start to appear in September and fruits in April-June. Ripe fruits open and seeds are released (Hamed, 2005).

*S. arghel* is a threatened plant due to its heavily collection from the wild to be sold as a folk medicine in the *Attarin* shops. Propagation is successful by seeds which germinate under a wide range of temperature: 20–40°C in soil. The most suitable temperature for its germination is 35°C. This may indicate its specific adaptation to high temperature. Pre-sowing treatment of seeds with growth stimulators, e.g. Thiourea 5, increases the percentage of germination up to 92.5%. Survival of seedlings is better when seeds are germinated in pots or plastic bags rather than sown directly. The seedlings can be planted out after 3–4 months in any season except winter. Seeds can also be sown directly into the ground, preferably sandy soil at the beginning of summer (May–June). Daily watering is required until the first seedlings appear, and then watering can be reduced to 3 times/week and 2 times/week when the seedlings are 2 months old. The germination percentage, is generally, low (about 30%) when the seeds are sown directly in the ground (Springuel, 2006).

**Group IV: Wood Producing Plants**

Among the eleven vegetation forms of the Afro-Asian Mediterranean and Red Sea coastal lands there are five forms with wood producing trees and shrubs: (mangrove vegetation, broad-leaved evergreen forest, stunted woodland, coniferous forest and scrubland). The following are three case studies throwing light on representative trees and shrubs that could be considered for wood production and, planted in the Afro-Asian Med. and/or Red Sea coastal lands.

**Case Study 1**

Seven trees and shrubs of the forest of the Palestine-Israel Mediterranean coastal land have been propagated (Zohary, 1962). These are: *Pinus halepensis, Ceratonia siliqua, Cupressus sempervirens, Tamarix aphylla, T. gallica, Quercus ithoburensis* and *Pistacia atlantica.*

1. *Pinus halepensis* is a drought-resistant tree that attains a height up to 10 m. It had been successfully and extensively cultivated on the high planteaus for shade, timber and cover. It grows fairly well in the western Negev desert under 100–200 mm annual rainfall. It is the only pine species that can withstand severe climatic conditions.
2. *Ceratonia siliqua* tree has been cultivated through the ages because of the forage yielded by its pods. Its successful growth is very promising, especially in the Mediterranean foothills. *C. silique* is drought-resistant.
3. *Cupressus sempervirens* is a drought-resistant tree but less important in forestry. It is cultivated in both Sinai and the Palestine-Israel mountains.
4. *Tamarix aphylla* is a common forest tree in the arid zone. It is planted in the Negev desert where it thrives well on sand dunes under extreme conditions of low rainfall, but only when it is widely spaced; otherwise it soon impoverishes the water resources and stop growing. It is indigenous only in the wetter wadis of the southern Negev.
5. *Tamarix gallica* occurs in a variety of forms some of which are resistant to wind and sea spray such as extent that they can be used successfully as windbreaks. It is planted on dunes of the Negev desert, where it similarly requires relatively high amounts of soil moisture. Both *Tamarix* species are easily propagated by cuttings.
6. *Quercus ithaburensis* is a fast growing oak particularly on the Mediterranean mountains. It is a broad-leaved tree for quite promising afforestation.
7. *Pistacia atlantica* is a very promising both as forest tree and as a stock on which *P. vera,* a fruit tree, can be successfully grafted. It grows under rather poor rainfall conditions and is therefore suitable for arid zones.

**Case Study 2. *Balanites aegyptiaca***

*Balanites aegyptiaca* is a drought-resistant thorny shrub or tree with a height between 6 and 12 m. It has an edible drup fruit of about plum-size, green turning yellow. It is abundant tree in the scrubland vegetation form of the Afro-Asian Red Sea coastal lands, and can continue bearing fruits during periods of drought (Täckholm, 1974; Anonymous, 1998b).

The *Balanites* tree is relatively deep-rooting with a strong tap root and is semi-deciduous dropping some but not all, of its leaves during dry season. Natural regeneration may be either through seedlings or by root suckers. As most desert trees, it grows slowly. It begins to fruit at about 4–7 years old and fruiting period depends on the local climate. In the most favourable conditions trees bear fruit and flowers simultaneously. Fruiting can be heavy with yields of between 45 and 200 kg. It is a multi-use tree. The young leaves, fruits, thorns are eaten by livestock; ripe fruit pulp edible, seed kernel is used for making bread and soap and contains 30–58% of an edible oil (Zachom oil), the residue, which contains 50% protein, is used in cooking and soap making. The medicinal uses of *B. aegyptiaca* are many (Boulos, 1983). Fruit and kernel extracts are lethal to water snails and are used in the control of schistosomiasis, they are a mild laxative, the leaves clean malignant wounds, the bark is a fumigant to heal circumcision wounds; the root extract is used for Malaria.

The wood of *B. aegyptiaca* is heavy, hard and strong and is widely used to make domestic utensils, small farm tools, furniture and specialist goods. It is considered a good fuel, and produces high quality charcoal. The wood gives out considerable heat with very little smoke, making it very suitable for use as a firewood inside buildings. The shell of the kernel is also a fuel source. The wood provides also timber for construction (Ayensu, 1983).

Germination of *B. aegyptiaca* seeds collected from Egypt's desert was tested in pots in a nursery (450 $m^2$ area) in Wadi Allaqi, Eastern Desert, Egypt and establishment of 500 *Balanites* seedlings was demonstrated. The drip irrigation method was used for watering the seedlings. During the year of the experiment 300 seedlings survived and showed healthy growth.

**Case Study 3. *Silviculture and Rehabilitation of Mangrove Forest***

Mangrove plants are multi-purpose trees and shrubs inhabiting the swampy shorelines of the tropical seas and oceans. Quite apart from their noticeable role in protecting the shorelines against erosion, all parts are useful to man. Wood and fuel can be obtained from their stems and branches, livestock fodder from their leaves and fruits, food for people from their fruits drugs from their leaves, fruits and roots.

The Afro-Asian Red Sea coastal belts are characterized by the growth of two mangrove species: *Avicennia marina* and *Rhizophora mucronata.* Human interference is causing deterioration of these mangrove forests and therefore field experiments were conducted to propagate the two species along the Red Sea coast of Egypt aiming at the afforestation of new areas and rehabilitation of deteriorated sites.

An ecological survey included the inspection of mangrove stands to collect fruits and propagules for transplantation. Seedlings either germinated from seeds in nurseries or collected from nature were transplanted into expansion areas of established mangrove stands, or into empty sites, or into rehabilitation sites of degraded mangrove stands. Of course, sea water was the only source of irrigation. About 3,000 *Avicennia marina* seedlings collected from natural stands and more than 4,000 nursery-reared seedlings of *Rhizophora mucronata* from the nursery were transplanted into established or new area, and 5,000 seedlings into rehabilitation sites. So far, the growth of more than 90% of the *A. marina* and *R. mucronata* plants seems normal, and are gaining vigour and size (Anonymous, 2006).

**Group V: Oil Producing Plants**

The world-re-knowned olive trees and shrubs (*Olea* spp., family Oleaceae) and castor oil herbs and shrubs (*Ricinus* spp., family Euphorbiaceae) grow naturally in the Afro-Asian Med. and Red Sea coastal lands. For their high economic values, wide areas of the North African and SW Med. coastal lands are cultivated with *Olea europaea* as well as with *Ricinus communis* for the commercial production of oils from their seeds (Zohary, 1962; Branigan and Jarrett, 1975; Danin, 1983; Boulos, 1983, 2000; Dallman, 1998; Blondel and Aronson, 1999). Apart from these two plants, the flora of the eleven vegetation forms of the Afro-Asian Med. and Red Sea coastal lands comprises a considerable number of multi-purpose species which, in addition to their economic potentialities as fodder, drug, fiber or wood producing plants, can also produce volatile and/or stable oils. Among these are species of: *Achillea, Acacia, Artemisia, Balanites, Capparis, Cyperus, Cymbopogon, Juncus, Moringa, Origanum, Solanum, Teucrium* etc. (Boulos, 1983; Anonymous, 1998c; Zahran, 1993b, 2004; Batanouny, 2005, 2006). However, the determination of the economic potentialities of these native species as oil-producing plants needs further studies. There are two exotic species (*Simmondsia chinesis* and *Jatropha curcas*) which proved to have high economic potentialities as oil producing plants have been introduced as non-conventional crops to be cultivated in certain parts of the Afro-Asian Med. and Red Sea coastal lands. These two case studies are discussed below.

**Case Study 1: Jojoba (*Simmondsia chinensis)***

*Simmondsia chinensis,* Jojoba, pronounced as hoh-hoh-bah, to an evergreen xerophytic shrub that is endemic species in an extensive arid areas of the Sonoran Desert in Arizona, California, New Mexico and numerous Western states of USA between 25 and 31°N covering about 120,000 square miles (193,200 km$^2$) (Gentry, 1958; Anonymous, 1975; Rodriguez, 1985; El-Hadeedy, 1984).

*S. chinensis* is a diocious woody shrub that live up to 200 years. *Jojoba* tolerates extreme desert temperatures; daily summer highs of 40–43°C in the shade are common in its habitats. A 30 mm of rain a year are enough to support large stands of productive Jojoba bushes. *Jojoba* would impose only a small strain on water supplies, which are now so heavily exploited by the propagation of conventional crops having high water demand. A truly drought-resistant desert shrub, Jojoba grows well under marginal soil and moisture condition. It is, actually, the proper plant to be cultivated in deserts for sustainable development. Apart from that, Jojoba shrubs seems very tolerant to saline and alkaline soils and saline irrigation water, the causes of barrenness in many arid areas. It can grow in soils with salinity of up to 7,000 mg/l and with careful management it may be able to make these barren deserts productive (Anonymous, 1975; El-Hadeedy, 1984).

Perhaps Jojoba's greatest attribute for arid lands is that it requires little or no water during summer. To produce new growth, set flowers and produce seeds, Jojoba needs moisture during winter and spring. The seeds usually mature in the late spring, and moisture is unnecessary. This schedule fits well with the desert environment of the Afro-Asian Med. and Red Sea coastal lands where winter is the wet season. Water is needed for Jojoba shrubs when it is most available ad not needed when it is scarce. This schedule contrasts with most other crops which survive drought only if they receive moisture during the hottest and dry months.

The economic value of Jojoba shrubs stems from its oil-producing seeds up to 50% of the weight a colourless, odourless oily liquid with unusual properties that is commonly referred to as "Jojoba oil". This oil is chemically liquid wax made up of non-glycerides esters with a narrow range of chemical composition; the esters are almost entirely composed of straight chain acids and alcohols, each with 20 or 222 carbon atoms and one unsaturated bond. Jojoba oil is, unique: an unsaturated, liquid wax that is readily extractable in large quantities from a plant source. Waxes of this type are difficult to synthesize commercially; the only other natural source is the sperm whole, an endangered marine species.

Jojoba waxes are used in many industries for a wide variety of applications in lubricants, paper coatings, polishes, electrical insulation, carbon paper, textiles, leather, precision casting and pharmaceuticals. "Jojoba oil is potentially useful for all these products" (Anonymous, 1975).

Jojoba oil does not change in composition as the seed matures, nor does it change during storage. Seeds analyzed 25 years after harvest show no change in easter composition. Thus, dried seeds can be stored without deterioration or chemical change.

Jojoba oil is preferable as lubricant due to the fact that it may be made to react with sulphur to yield a stable product containing a relatively large amount of sulfur, which works well as a lubricant additive. This oil could be used in cars and trucks (Sample, 2003). Sperm whale oil is widely used in lubricants. The composition and physical properties of Jojoba are close enough to sperm oil to suggest the use of Jojoba oil as a substitute for most of sperm oil. In fact, Jojoba oil has several advantages over the similar product from the sperm whale (Anonymous, 1975):

1. It has a mild, pleasant odour,
2. It contains non-glycerides and very little besides the liquid wax,
3. It requires little or no refining to prepare it for most lubrication purposes,
4. It is a vegetable product that can be produced in many resource-poor countries.

The Jojoba liquid wax has many other applications (El-Hadeedy, 1984) as: a component of hair oil, shampoo, soap, face cream, sunscreen compounds, coating for some medicinal preparations, stabilizer for penicillin products, inhibitor to growth of tubercle bacilli, cooking oil, low caloric additive for salad oil, printing ink composition, chewing gums, surfactants, detergents, driers, emulsifiers, resins, plasticizers, polishing wax (for floor, furniture and automobiles), protective coating for fruits, paper containers cosmetic for lipsticks etc. Thus, Jojoba shrub may be considered, "Green Gold" (El-Mogy, 2009). Pilot experiments carried out in the deserts of USA, Israel and Egypt have shown that Jojoba shrubs can be easily cultivated vegetatively by stem cuttings as well as sexually by seedlings produced from seeds (Anonymous, 1975; El-Hadeedy, 1984). Such success have encouraged private companies in Egypt and Israel (*Jojoba Israel*) to cultivate wide areas of Jojoba shrubs. In Israel the cultivation of Jojoba plant started in the mid 1960s whereas in Egypt it started in 1994. In both countries production of Jojoba seeds is so high and commercial that both companies have established mills to produce Jojoba oil for local consumption and for export as well (El-Mogy, 2008).

**Case Study 2. *Jatropha curcus***

*Jatropha* is a genus of approximately 175 succulent species belonging to the Euphorbiaceae. The name is derived from a Greek origin (Jatros = Physician = doctor and trophe = nutrition), hence the common name physic nut.

*Jatropha curcus* is a succulent shrub or tree native to central America naturalized in many tropical and subtropical areas including: India, Africa and North America. It is a deciduous plant with smooth gray bark which excludes watery and sticky latex when cut. The leaves are simple, broad, glabrous, and deeply palmately lobed with a long petiole. Ciliate glands usually represent the stipules. The flowers are monoecious, greenish-yellow in terminal long, peduncled, paniculate cymes. The fruit is usually a three chambered schizocarpic capsule splitting into three one seeded cocci. The exocarp remains fleshy until the maturation of the seeds.

*J. curcas* may produce more than one crop during a year, or flower and fruit continually under irrigation if the soil moisture and temperatures are suitable. The seeds are ovoid-oblong and black in colour and become mature when the capsule changes from green to yellow after 2 months of fruit splitting. Each fruit bears three seeds. Seed yields of *J. curcas* shrubs range from 1,500 to 2,000 kg/ha (Dar, 2007).

*J. curcas* grows almost anywhere – even on gravelly, sandy and saline soils (Reyadh, 2006). It can thrive also on the poorest stony soil and even in the crevices of the rocks. The leaves shed during winter months form a mulch around the base of the plant, enhancing the activity of earthworms around the root-zone which improves soil fertility.

*J. curcas* comes from the hot tropical and subtropical regions, it can tolerate extremes of temperature but not frost or inundation: a rising water table for a considerable period can kill the plant.

The non-edible vegetable oil of *J. curcas* seeds has the requisite potential of providing a promising and commercially viable alternative to diesel oil since it has desirable physicochemical and performance characteristics. Cars can run on *Jatropha* oil without much change in design (Reyadh, 2006). The oil content is

35–40% in the seeds and 50–60% in the kernel. The oil contains 21% saturated fatty acids and 79% unsaturated fatty acids. There are some chemical elements in the seed which are poisonous and render the oil not suitable for human consumption.

*Jatropha* seed oil has a very high saponification-value wax and is being extensively used for making soap in some countries. The oil is also used as an illuminant because it burns without emitting smoke.

The latex of *J. curcas* contains an alkaloid known as "jatrophine" which is believed to have anti-cancer properties. It is also used as an external application for skin diseases and rheumatism and for sores on domestic livestock.

Other uses of *J. curcas* parts are:

1. The tender twigs of the plant are used for cleaning teeth,
2. The juice of the leaves are used as an external application for piles,
3. The extract of the root has been reported to be an antidote for snake-bites,
4. A dark blue dye used for colouring cloth and fishing nets are produced from the bark,
5. Leaves are used as food for the tussore silkwarms,
6. Leaves are used for fumigating houses against bed-bugs,
7. Seeds are considered to be anthelimintic,
8. The ether extract of the plant shows antibiotic activity against *Styphyllococcus aureus* and *Escherichea coli* (Reyadh, 2006).

*J. curcas* seed oil contains several toxic compounds including lectin, saponin, aphorbol esters and a trypsin inhbitor, a family of compounds known to cause a large number of biological effects such as tumor promotion and inflammation. It is therefore necessary to find feasible routes for detoxifying of the oil (Haas and Mittelbach, 2000).

*J. curcas* could be propagated on a mass scale both by seeds and vegetatively by stem cuttings. For commercial cultivation it is normally propagated by seeds. Before sowings, the seeds are soaked in cow dung solution for 12 h and kept for another 12 h. Hot humid climate is preferable for good germination. Germinated seeds are sown in bags of soil and farm-yard manure. Seeds or cuttings can be directly planted in the main field, but pre-rooted cuttings in bags and transplanted into the field may give better results.

Recognizing its economic value and importance in sustainable development of deserts, the Ministry of Agriculture and Land Reclamation in Egypt introduced *J. curcas* to marginal and desert areas using treated waste waters for irrigation. This has established man-made forests in many parts of Egyptian desert where *J. curcas* is successfully growing with other trees e.g. species of *Acacia, Cupressus, Casuarina, Eucalyptus, Pinus,* and *Morus*. Drip and modified surface irrigation methods are followed.

# Bibliography


Abdel Ghani, M. M. (1993). Habitat features and plant communities of the Holy places, Makkah, Saudi Arabia. *Feddes Repertor.*, 104: 417–425.

Abdel Ghani, M. M. (1996). Vegetation along a transect in the Higaz mountains, Saudi Arabia. *J. Arid Environ.*, 32: 289–404.

Abdel Rahman, A. A., Shalaby, A. F., Balegh, M. S. and El-Monayari, M. (1965). Hydroecology of date palms under desert conditions. *Bull. Fac. Sci.* Cairo Univ., 40: 55–71.

Abdel Razik, M. S. (1994). *Avicenea marina. Al-Qurm. General Study and Propagation Experiments in Qatar.* Center for Scientific and Applied Research, Doha, Qatar: 142pp in Arabic + 18pp in English.

Abou Deiah, A. B. (2001). *Fodder Plants for Deserts and New Lands.* Publ. Desert Res Center, Cairo: 123pp (in Arabic).

Abu Al-Izz, M. S. (1971). *Lanforms of Egypt.* Translated by Dr. Yusuf A. Fayid, The American University in Cairo Press, Cairo, Egypt: 281pp.

Adams, R., Adams, M., Willens, A. and Willens, A. (1978). *Dry Lands: Man and Plants.* The Architectural Press, London: 152pp.

Ahmed, A. M. (1983). On the ecology and phytosociology of El-Qaa plain, South Sinai, Egypt. *Bull. Inst. Desert-Egypte*, 33(1–2): 281–314.

Ahmed, A. M. and Mounir, M. M. (1982). *Regional Studies on the National Resources of the NW Coastal Zone*, Egypt, US National Science Foundation, Oklohoma State University and the UNEP and Remote Sensing Center, Academy of Scientific Res. and Tech., Cairo, Egypt. 195pp.

Ahmed, R. and San Pietro, A. (eds.) (1986). *Proceedings of Prospects for Biosaline Research, USA – Pakistan Bio-saline Research Workshop,* Karachi, Pakistan: 584pp.

Ahmed, A. M. and Nassar, Z. M. (1999). Chemical composition of some natural range plants from the north western coast of Egypt. *Proceedings of the 6th National Conference on Environmental Studies Research*, Ain Shams University, Cairo: 551–568.

Al-Eisawi, D. (1996). *Vegetation of Jordan.* UNESCO, Cairo Office: 284pp.

Al-Hubaishi, A. and Müller-Hohenstein, K. (1984). *An Introduction to the Vegetation of Yemen: Ecological Basis, Floristic Composition, and Human Influence.* GTZ, Eschborn, Germany: 209pp + 95pp in Arabic.

Ali, S. F. and Jafri, S. M. H. (1976). *Flora of Libya*, Vols. 1–24. Department of Botany, El Faateh University Tripoli, Libya.

Al-Kholy, A. A. (ed.) (1972). *Aquatic Resources of the Arab Countries.* Science Monograph Series, Arab League Education Cultural and Scientific Organization (ALESCO), Muscat, Oman: 452pp.

Allan, T. and Warren, A. (1993). *Deserts: The Encroaching Wilderness.* Mitchell Beezley, IUCN, London: 176pp.

Allered, B. W. (1968). *Woodlands in Saudi Arabia.* FAO Report, Rome: 17pp.



M.A. Zahran, *Climate–Vegetation*, Plant and Vegetation 4,
DOI 10.1007/978-90-481-8595-5, © Springer Science+Business Media B.V. 2010

Al-Sanbouk, A. (2002). *Integrated Coastal Zone Management*. PERSGA'S Newsletter (16), Saudi Arabia: 13pp.

Al-Sodany, Y. M., Shehata, M. N. and Shaltout, K. H. (2003). Vegetation along an elevated gradient in Al-Jabal Al-Akhdar, Libya. *Ecol. Mediterr.*, 29(Fascicule 2): 125–138.

Al-Zayani, A. K. (1993). Development of the coastal environment by planting mangrove plants: Prospects of Gulf Region. *Proceedings of the Symposium on Desertification and Land Reclamation in the Council of Arab Gulf States*, Bahrain University: pp. 1–24 (in Arabic).

Amer, W. M. and Zahran, M. A. (1999). Plam trees in Egypt. *Proceedings of an International Conference on Date Palms*, Assiut University, Assiut, Egypt: 171–189.

Andrews, F. W. (1948). The vegetation of the Sudan. In: Tothill, J. D. (ed.), *Agriculture in the Sudan*. Oxford University Press, London: pp. 32–61.

Andrews, F. W. (1950–1956). *The Flowering Plants of the Sudan*, Vols. I–III. Sudan Gov., Khartoum: 237pp. (1950), 485pp. (1952), 597pp. (1956).

Anonymous (1960). *Climatic Normal's of Egypt*. Ministry of Military Production, Meteorological Department, Cairo: 237pp.

Anonymous (1975). *Products from Jojoba: A Promising New Crop for Arid Lands*. National Academy of Sciences, Washington, DC: 30pp.

Anonymous (1977). *Annual Environmental Report*. General Directorate of Meteorology, Ministry of Defense and Aviations, Kingdom of Saudi Arabia. Part I. Surface Climatologically Report: 35–90.

Anonymous (1980). *Firewood Crops, Shrubs and Tree Species for Energy Production*. Natural Academy of Science, Washington, DC: 237pp.

Anonymous (1982). *Sinai Development Study Phase I. Draft Final Report*. Vol. VII, Environment, Dame, and Moore, Chapter 4, Climatology and Meteorology: 16pp and Chapter 7, Terrestrial Ecology. 48pp.

Anonymous (1987). *Our Common Future*. World Commission on Environment and Development, Oxford University Press, Oxford, UK: 400pp.

Anonymous (1993). *Climatic Normal's of the Northern Section of the Yemen Republic*. Civil aviation and meteorological authorities, computer Department, sheets of Alhodaydah, means of 1980–1992.

Anonymous (1994). *New Renewable Energy Resources. A Guide to the Future World Energy*. Council, Kogan Page Limited, London: 391pp.

Anonymous (1997). *Proceedings of an International Seminar for Sustainable Development of Egypt's Red Sea Coast*. Mercarb Eddy, Intern. Inc., GEF, Egypt, Cario.

Anonymous (1998a). *The Regional Impacts of Climatic Changes. An Assessment of Vulnerability*. Intergovernmental Panel on Climatic Changes, UNEP, WMO, Cambridge University Press, Cambridge: 517pp.

Anonymous (1998b). *Cultivation and Use of Balanites aegyptiaca in Wadi Allaqi* (Lake Nasser Area, Eastern Nabia). Unit of Environmental Studies and Development, South Valley University, Aswan Egypt: 18pp.

Anonymous (1998c). *Preliminary Coastal Zone Management Action Plan for the Egyptian Red Sea*. Draft plan Report 3. Tourism Development Authority/EEAA, Red Sea Governorate, Egypt & GEF: 113pp.

Anonymous (2006). Assessment and Management of Mangrove Forest in Egypt for Sustainable Utilization and Development ITTO/MALR/MSERS/EEAA. Progress Report No. 6.

Anonymous (2008a). Climate change. *Scientific American*, June 2008, 21P.

Anonymous (2008b). *Lake Hula, The Columbia Encyclopedia*, 6th ed. Columbia University Press, New York: 6pp.

Aronson, J., Floret, C., Le Floch, E., Oralle, C. and Pontanier, R. (1993). Restoration and rehabilitation of degraded ecosystems. I. A view from the south. *Restor. Ecol.*, 1: 8–17.

Ascherson, P. and Schweiafurth, G.. (1889a). Illustration de La Flora d'Egypte. Mom. *Inst. Egypte*, 3(1): 25–260.

Ascherson, P. and Schweinfurthe, G. (1889b). *Supplement a l'illustration de la flore d'Egypte*, 2: 745–810.

Ashbel, D. (1945). *Hundred Years of Rainfall Observations (1845/5–1944/5)*. Mimeographed, Hebrew, Jerusalem, Israel: 217pp.

Ashbel, D. (1951). *Regional Climatology of Israel Jerusalem*. Meteorology Department Hebrew University, Hebrew: 244pp + 17pp.

Ashour, N. I., Serag, M. S., Abdel Halem, A. K. and Mekki, B. B. (1997). Farage production from three grass species under saline irrigation. *J. Arid Environ.*, 37: 299–307.

Ashour, N. I, Serag, M. S., Abdel Haleem, A. K., Mandoura, S., Mekki, B. B. and Arafat, S. M. (2002). Use of Kallar Grass (*Leptochloa fusca L.*) Kunth. In saline agriculture in arid lands of Egypt. *Egypt. J. Agron.*, 24: 63–87.

Aucher-Eloy, P. M. R. (1843). *Relations de voyages en Orient de 1830 a 1838, revueses annotees par mons*. Le Comte Jaubert. Paris, Roret: 775pp.

Auder, J., Cesar, J. V., Lebrun, J. P. (1987–1989). *Les Plantes Vasculaires de la Republique de Djibouti*, Vols. 1 & 2. CIRAD, Department d' Elevage et de Médecine Veterinaire, Farnce: 968pp.

Ayensu, E. S. (1979). Plants for médicinal uses with reference to arid zone. *Proceedings of Arid Land Plant Resources Conference on Texas Tech University*, Lubbock, TX: pp. 177–178.

Ayensu, E. S. (ed.) (1983). *Firewood Crops, Shrub and Tree Species for Energy Production*, Vol. 2. National Academy of Science, Washington, DC.

Ayyad, M. A. (1973). Vegetation and environment of the western Mediterranean coastal land of Egypt I. The habitat of Sand dunes. *J. Ecol.*, 61: 509–523.

Ayyad, M. A. (1976). Vegetation and environment of the western Mediterranean coastal land of Egypt IV the habitat of the non-saline depressions. *J. Ecol.*, 64: 713–722.

Ayyad, M. A (1983). Some aspects of land transformation in the western Mediterranean desert of Egypt. *Adv. Space Res.*, 2(8): 192.

Ayyad, M. A. and Ammar, M. T. (1974). Vegetation and environment of the western Mediterranean coast of Egypt. *Ecology*, 55: 511–524.

Ayyad, M. A. and El-Bayyoumi, M. A ( 1979). On the phytosociology of sand dunes of the western desert of Egypt. *Vegetatio*, 31(2): 93–102.

Ayyad, M. A. and El-Ghareeb, R. E. M. (1982). Salt marsh vegetation of the western Mediterranean desert of Egypt. *Vegetatio*, 49: 3–19.

Ayyad, M. A. and El-Ghareeb, R. E. M. (1984). Habitat and plant communities in the NE Desert of Egypt. *Communication in Agrisciences and Development Research*, College of Alexandria, University of Alexandriey, 7(6): 1–34.

Ayyad, M. A. and Ghabbour, S. I. (1986). Hot deserts of Egypt and the Sudan. Chapter 5. In: Evenari, M., et al. (eds.), *Ecosystems of the World*, Vol. 1B. Hot Deserts and Arid Shrublands. Elsevier, Amsterdam: pp. 149–202.

Ayyad, M. A. and Ghabbour, S. I. (1993). Dry coastal ecosystems of eastern north Africa. Chapter 1. In: Van der Mearel, E. (ed), *Ecosystems of the World 2B. Dry Coastal Ecosystems: Africa, America, Asia and Oceania*. Elsevier, Amsterdam: pp. 1–16.

Baeshin, N. A. and Aleem, A. A. (1987). Littoral vegetation at Rabegh (Red Sea Coast), Saudi Arabia. *Bull. Fac. Sci. KAU Jeddah*, 2: 123–130.

Banoub, M. W. (1979). The salt regime of Edku Lake (Egypt) before and after the construction of Aswan High Dam. *Arch. Hydro.*, 85: 392–399.

Bagnouls, F. and Gaussen, H. (1957). Les Climates Ecologiques etleur classification. *Ann. Geogr.*, 66: 193–220.

Baletto, E. and Casale, A. (1991). Méditerranéen insect conservation. In: Collins, N. A. and Thomas, J. A. (eds.), *The Conservation of Insects and Their Habitats*. Academic Press, London: pp. 121–142.

Ball, J. (1912). *The Geography and Geology of South Eastern Desert*. Egyptian Survey Department, Cairo: 394pp.

Ball, J. (1916). *The Geography and Geology of West Central Sinai*. Egyptian Survey Department, Cairo: 219pp.

Ball, J. (1939). *Contribution to the Geography of Egypt*. Survey and Mines Department, Cairo: 308pp.

Ball, M. C., Cowan, I. R. and Faquhar, G. D. (1983). Maintenance of the leaf temperature and optimization of carbon gain I. Relation to water loss in a tropical mangrove forest. *Aust. J. Plant Physiol.*, 15: 263–270.

Banaja, A. A., Beltagy, A. I. and Zahran, M. A. (1990). *Red Sea, Gulf of Aden and Suez Canal: A Bibliography on Oceanographic and Marine Environmental Research*. ALECSO–PERSGA, UNSCO, Belguen: 198pp, 48pp (in Arabic).

Barrett-Lennard, E. G., Malcolm, C. V., Stern, W. R., and Wilkins, S. M. (eds.) (1986). *Forage and Fuel Production from Salt Affected Wastelands*. Elsevier, Amsterdam: 459pp.

Barry, R. G. (1974). The world hydrological cycle. In Charley, R. J. (ed.), *Water, Earth and Man*. Metheum, London: pp. 11–29.

Batanouny, K. H. (1973). Habitat features and vegetation of deserts and semi-deserts in Egypt. *Vegetatio*, 27(4–6): 181–199.

Batanouny, K. H. (1987). Current knowledge of plant ecology in the Arab Gulf countries. *Catena*, 14: 291–316.

Batanouny, K. H. (1996). Medicinal plants is North Africa: as endangered components of biodiversity. In Batanouny, K. H. and Ghabbour, S. I. (eds.), *Proceedings of the Workshop on Arid Lands Biodiversity North Africa*. IUCN, SDC, Cairo: pp. 103–110.

Batanouny, K. H. (2005). *Encyclopaedia of Wild Medicinal Plants in Egypt I. Solanum nigrum, Tecrium polium, Pluchea discoridis*, and *Solenostemma arghel*. The Palm Press, Cairo: 81pp.

Batanouny, K. H. (2006). *Encylopaedia of Wild Medicinal Plants in Egypt II. Cappairs spinosa*, Cyperus rotundus, *Capparis sinaica, Acacia nilotica* and *Origanum syriacum*. The Palm Press, Cairo: 84pp.

Batanouny, K. H. and Baeshin, N. A. (1982). Studies on the flora of Arabia II. The Medina-Badr road, Saudi Arabia. *Bull. Fac. Sci. King Abdul Aziz Univ.*, Jeddah, Saudi Arabia, 6: 1–26.

Batanouny, K. H. and Ghabbour, S. I (eds.) (1996). Arid Lands Biodiversity in North Africa. *Proceedings of the Workshop ZUCN, SDC*, Cairo: 134pp + 5pp (in Arabic).

Batanouny, K. H. and Shams, K. A. (2006). Monogrph on *Capparis spinosa*. In: Batanouny, K. H. (ed.), *Encyclopaedia of Wild Medicenal Plants in Egypt 2: Capparis spinosa, Cyperus rotundus, Capparis sinaica, Acacia nilotica and Origanum syriacum*. The Palm Press, Cairo: pp. 3–22.

Batanouny, K. H., Zayed, K. M., Emad, H. M. and Kabeil, H. E. (2006). Reproductive ecology of *Panicum turgidum* Forssk. *Taeckholmia*, 26: 63–88.

Batterson, T. R. and Hall, D. W. (1984). Common reed. *Phragmites australis* (Cav.) Trin Ex. Stend. *Aquatics*, 6(2): 16–20.

Baumann, H. (1993). *The Greek Plant World in Nyth, Art and Literature*. Timber Press, Portland.

Ben Haider, B. M. (1994). *Mangroves (Al-Qerm). Development in UAE*, 1st ed. Nadwat Al Thaqafa and Al-Ouloum, Dubai: 53pp (in Arabic).

Birks, H. H., et al. (2001). Palaeolimnological responses of nine North African Lakes in the CASSARINA Project to recent environmental changes and human impact detected by plant macrofossil, pollen and faunal analysis. *Aquatic Ecol.*, 35: 405–430.

Birks, H. H. and Birks, H. J. B. (2001). Recent ecosystem dynamics in nine North African lakes in the CASSARINA Porject. *Aquatic Ecol.*, 35: 461–478.

Biswas, R. M. and Biswas, A. K. (eds.) (1980). *Desertification*. Pergamon Press, Oxford: 523pp.

Blondel, J. and Aronson, J. (1999). *Biology and Wild Life of the Mediterranean Region*. Oxford University Press, Oxford, UK: 313pp.

Bodenheimer, F. S. (1957). The ecology of manuals in the arid zone. In: *Human and Animal Ecology. Reviews of Research.* Arid Zone Res. 8: 100–137.

Boissier, E. (1867–1888). *Flora Orientalis* 5 Vols. and suppl. Basileae et Genevae, H. Georg. 1017, 1159, 1033, 1276, 868, 466pp. t5pls + map.

Booth, F. E. M. and Wickens, G. E. (1988). *Non-Timber Uses of Selected Arid Zone Trees and Shrubs in Africa*. FAO Conservation Guide. 19, FAC, Rome.

Borlaug, N. E. (1983). Contribution to conventional plan breeding to food production. *Science*, 219(4585): 689–694.

Bornkamm, R. and Kehl, H. (1990). The plant communities of the Western Desert of Egypt. *Phytocoenologia*, 19(2): 144–231.

Bornmuller, J. (1898). Ein Beitrag Zur Kenntniss de Flora Van Syrien und Palastina, Verhandl. *K. K. Zoolog – Bot. Ges. Wien.*, 48: 544–653.

Bornmuller, J. (1912). Zur Flora Van Palaestina. *Ungar. Bot. Blatter II*, 1/4: 3–12.

Boulos, L. (1960). *Flora of Gebel El-Maghara, North Sinai*. General Organization for Government Printing Office, Ministry of Agriculture, Cairo, Egypt: 24pp.

Boulos, L. (1971). The Flora of Libya. *Proj. Mitt. Bot. Staatssmml. Municher*, 10: 14–16.

Boulos, L. (1972). Our present knowledge on the flora and vegetation of Libya. *Webbia*, 26: 365–400.

Boulos, L. (1975). The Mediterranean elements in the Flora of Egypt and Libya. In: *La Flora du Bassin Mediterranean essai de Systematique Synthetique*. CNRS, Paris: pp. 119–124.

Boulos, L. (1977). A check list of the Libyan Flora. 1. Introduction. *Adiantaceae-Orchidaceae. Public Cairo Univ. Herb.*, 718: 115–141.

Boulos, L. (1979a). A check list of the Libyan Flora 2. *Salicaceae* to *Neuradaceae*. *CODEN: CNDLAR*, 34: 21–48.

Boulos, L. (1979b). A chick list of the Libyan Flora: 3 Compositae. *CODEN: CNDLAR*, 34: 307–332.

Boulos, L. (1983). *Medicinal Plants of North Africa*. Reference Publications, Inc., Algonac, MI: 286pp.

Boulos, L. (1989). Egyptian desert plants with promising economic potential. *Arab Gulf J. Sci. Res.*, 2: 41–108.

Boulos, L.(1995). *Flora of Egypt Chicklist*. Al-Hadara Publishing, Cairo, Egypt: 283pp.

Boulos, L. (1997). Endemic Flora of the Middle East and North Africa. In: Barakat, H. N. and Hegazy, A. K. (eds.), *Reviews in Ecology, Desert Conservation and Development. A festschrift for Prof. M. Kassas on the Occasion of 75th Birthday*. IDRC/CRDI, UNESCO/South Valley University, Cario: pp. 229–260.

Boulos, L. (1999). *Flora of Egypt: Volum I (Azollaceae–Oxalidaceae)*. Al-Hadara Publishing, Cairo, Egypt: 419pp.

Boulos, L. (2000). *Flora of Egypt: Volume 2 (Geraniaceae–Boraginaceae)*. Al-Hadara Publishing, Cairo, Egypt: 352pp.

Boulos, L. (2002). *Flora of Egypt: Volume 3. (Verbenaceae–Compositae)*. Al-Hadara Publishing, Cairo, Egypt: 373pp.

Boulos, L. (2005). *Flora of Egypt – Volume 4. Monocotyledons (Alismataceae–Orchidaceae)*. Al-Hadara Publishing, Cairo, Egypt: 617pp.

Boulos, L. (2009). *Flora of Egypt checklist*. Revised Annotated Edition, Al-Hadara Publishing, Cairo: 410pp.

Boyko, H. (1949). The climax vegetation of the Negev with special reference to arid pasture problems. *Palest. J. Bot. Reh. Ser.*, 9: 17–35.

Boyko, H. (1966). Basic ecological principles of plants growing by irrigation with highly saline or sea water. In: Boyko, H. (ed.), *Salinity and Aridity*. Dr. W. Junk Publisher, The Hague, Netherlands: pp. 131–200.

Bradely, R. S., Diaz, H. F., Eischeid, J. K., Janes, P. D., Kelly, P. M. and Goodess, C. M. (1987). Precipitation fluctuations over Northern Hemisphere land areas since the mid-19th century. *Science*, 237: 171–175.

Branigan, J. J. and Jarrett, H. R. (1975). *The Mediterranean Land*, 2nd ed. Mac Donald & Evans, London: 628pp.

Broom, J. (2008). The ethics of climate change. *Sci Am*: 69–72.

Broussonet, P. M. S. (1795–1801). *Botanical Explorations in Morocco*. Specimen Studied and Published by Willdenow, Desfontaines, Gouan and Cavanilles.

Brown, L. R. (1981). *Building a Sustainable Society*. W.W. North Comp., New York: 433pp.

Brown, A. F. and Massey, R. E. (1929). *Flora of the Sudan*. Sudan Government, Khartoum: 502pp.

Butzer, K. W. (1960). On the Pleistocene Shorelines of the Arab Gulf – Egypt. *J. Ecol.*, 68: 626–637.

Butzer, K. W. (1976). *Early Hydraulic Civilization in Egypt: A Study in Cultural Ecology*. University of Chicago Press, Chicago, UK.

Butzer, K. W. and Twidale, C. R. (1966). Deserts in the past. Chapter VII. In: Hills, E. S. (ed.), *Arid Lands. A Geographical Appraisal*. Methuen & Co. Ltd., UNESCO, Paris: pp. 127–144.

Chapman, V. I. (1960). *Salt Marshes and Salt Deserts of the World*. Hill, London: 392pp.

Chapman, V. J. (1964). *Coastal Vegetation*, Pergaman Press, London; 262pp.

Chapman, V. J. (1974). *Salt Marshes and Salt Desert of the World*, 2nd ed. Cramer, Lehre: 392pp (Complemented with 102pp).

Chapman, V. J. (1975). *Mangrove Vegetation*. Cramer, Lehre: 425pp.

Chapman, J. V. (ed.) (1977). *Ecosystems of the World. I. Wet Coastal Ecosystems*. Elsevier, Amsterdam: 428pp.

Cheylan, G. (1991). Patterns of Pleistocene turnover, current distribution and speciation among Mediterranean mammals. In: Groves, R. H. and Di Castri, E. (eds.), *Biogeography of Mediterranean Invasions*. Cambridge University Press, Cambridge, UK: pp. 227–262.

Cheylan, M. and Poitenvin, F. (1998). Conservazione Di rettili; anfibi. In: Monbaillui, X. and Torr, A. (eds.), *L' a Gestione Degli Ambients Costieri Einsulari del Mediterranco*. Ediziona Del Sole, Alghero: pp. 275–336.

Chiras, D. D. (1991). *Science: Action for a Scientific Future*, 3rd ed. The Benjamin Cummings Publishing Company Inc., New York: 549pp.

Churchill, D. M. (1973). *The Ecological Significance of Tropical Floras of Southern Australia*, Vol. 1. Special Publications Geological Society, Australia: pp. 49–86.

Cloudsley-Thompson, J. L. (1978). The future arid environment. In Mann, H. S. (ed.), *Arid Zone Research and Development. Proceedings of an International Symposium Arid Zone Research & Development*. Arid Zone Research Association, Jodhpur, India: pp. 469–474.

Clough, B. F. (1993). Constants on the growth, propagation and utilization of mangroves in arid regions. In: Lierh, H. and Al-Masoom, A. (eds.), *Towards the Rational Use of High Salinity Tolerant Plants*, Vol. 1. Kluwer Academic Publishers, London: pp. 341–352.

Cohen, C. R. (1980). Plate tectonic medal for the Oligo-Miocene evolution of the western Mediterranean. *Techtonophysica*, 68: 283–311.

Committee on Jojoba Unlization (1975). *Products from Jojoba: A Promising New Crop for Arid Lands*. Natural Academy of Sciences, Washington, DC: 30pp.

Covas, R. and Blondel, J. (1998). Biogeography and history of the Mediterranean bird fauna. *IBIS*, 140: 395–407.

Crane, E. (1985). Bees and honey in the exploitation of arid lands. Chapter 12. In: Wickens, G. E., Gooden, J. R. and Field, D. V. (eds.), *Plants for Arid Lands*. George Allen of Unwin, London: pp. 163–175.

Cufodontis, G. (1961–1966). Enumeratio plantarum Aethiopiae Spermatophyta, Sequentiae. *Bull. Jardin Bot. Nat. Belg.*, 31: 709–772; 36: 1059–1114.

Cunningham, W. P. and Saigo, B. W. (1992). *Environmental Sciences: A Global Concern*. Wm. C. Brown Publishers, Dubuque, IA: 632pp.

Dafni, A. and O'Toole, C. (1994). Pollination syndromes in the Mediterranean generalization and peculiarities. In: Arianoutson, M. and Groves, R. (eds.), *Plant–Animal Interactions in Mediterranean-Type Ecosystems*. Kluwer Academic Publishers, Dordricht: pp. 125–135.

Dallman, P. R. (1998). *Plant Life in the World's Mediterranean Climate*. University of California Press, Los Angeles: 257pp.

Danin, A. (1969). A new Origanum from Isthmic Desert (Sinai). *Origanum isthmicum* sp. *Israel J. Bot.*, 18: 191–193.

Danin, A. (1972). Mediterranean elements in rocks of the Negev and Sinai Deserts. *Notes Roy. Bot. Gard. Edinburgh*, 31: 537–440.

Danin, A. (1981). Weeds of eastern Sinai coastal area. *Willdenowia*, 11: 291–300.

Danin, A. (1983). *Desert Vegetation of Israel and Sinai*. Cana Publ. House, Jerusalem, Israel: 148pp.

Danin, A., Orshan, G. and Zohary, M. (1975). The vegetation of northern Negev and Judaean desert of Israel. *Israel J. Bot.*, 24: 118–172.

Danin, A., Shimda, A. and Listen, A. (1985). Contributions to the Flora of Sinai. III Checklist of the Species collected and recorded by the Jerusalem team 1967–1982. *Willdenowia*, 15: 255–322.

Dar, W. D. (2007). *Research Needed to Cut Risks to Biofuel Farmers*. Science and Development, New York.

Dasmann, R. E., Milton, J. R. and Freeman, P. H. (1960). *Ecological Principles for Economic Development*. John Wiley & Sons, London: 235pp.

Davies, J. L. (1972). *Geographical Variation in Coastal Development*. Oliver & Boy d, Edinburgh, UK: 204pp.

De Cosson, A. (1935). *Mareotis*. Country Life London, London: 219pp.

Deil, U. and Muller-Hohenstein, K. (1983). Zur Pflanzenwelt des Jemen-am beispiel sukkulenter Euphorbien–Jamen – Report. *Mitt. D. Deutsch Jemenitschen Ges.*, 14(2): 12–16.

Delile, A. R. (1809–1812). *Description de l' Egypte*. Histories Naturelle, Vols. I & II. Imprimrerie Imperiale, Paris: pp. 44–82, 145–300.

Desfontaines, R. (1798–1800). Flora Atlantica Sive Historia Plantarum Quae in Atlante Agrotunesiana et Algeriense Crescent. *J. Redoute et Marechal,* Paris: 9929pp, 444pp, 458pp, 261pp.

Di Castri, F. (1981). Mediterranean type shrublands of the world. In: di Castin, F., Goodal, D. W. and Specht, R. L. (eds.), *Collection Ecosystems of the World*, Vol. II. Elsevier, Amsterdam: pp. 1–52.

Di Castri, F. (1991). An ecological survey of the five regions with a Mediterranean climate. In: Groves, R. H. and Dicastin, F. (eds.), *Biogeography of Mediterranean Invasions*. Cambridge University Press, London: pp. 3–16.

Dixey, F. (1966). Water supply, use and management. Chapter V. In: Hills, E. S. (ed.), *Arid Lands. A Geographical Appraisal*. Methuen & Co., London and UNEESCO, Paris: pp. 7–102.

Drake, J. A., Mooney, H. A., di Castrif, F., Groves, R. H., Kruger, E. J., Rejmanek, M. and Williamson, M. (1989). *Biological Invasions: A Global Perspectives*. John Wiley, Chichester.

Drar, M. (1936). Enumeration of the plants collected at Gebel Elba during two expedition. *Fouad I Agricultural Museum Technology and Science Series,* Vol. 149. Ministry of Agriculture, Cairo, Egypt: 123pp.

Draz, O. (1956). Improvement of animal production in Yemen. *Bull. Inst. Desert d'Egypt*, 6: 69–95.

Draz, O. (1965). *Rangeland Development in Saudi Arabia (in Arabic)*. Riyadh University, Riyadh, Saudi Arabia (non-published report).

Draz, O. (1969). *The Hema System of Range. Reserves in the Arabian Peninsula: Its Possibilities in Range Improvement and Conservation Projects in the Middle East*. FAO/PL: PFC/B, Roma: 11pp.

Dregne, H. E. (1976). *Soils of Arid Regions*. Elsevier, Amsterdam: 237pp.

Dregne, E. E. (ed.) (1977). Managing saline water for irrigation. *Proceedings of an International Salinity Conference*. Texas Tech University, Lubbook, Texas: 618pp.

Dregne, H. E. (1997). Land degradation control in drylands: Establishing priorities. In: Barakat, H. N. and Hegazy, A. K. (eds.), *Reviews in Ecology: Desert Conservation and Development. A festschrift for Prof. Kassas on the Occasion of His 75th Birthday*. IDRC, CRD, UNESCO, South Valley University, Cairo, Egypt: pp. 73–88.

Dubief, J. (1963). Contribution au probléme de changements du climat survenus au cours de la périod couverte par le observations meteologiques faites dans Le Nord de l'Afrique. *Arid Zone Res.*, 20: 75–78.

Duvdevani, S. (1953). Dew gradients in relation to climate, soil and topography. *Desert Research Proceedings of Special Publication No. 2 of the Research Council*, Israel: 17pp.

Ebrahim, M. E. (1999). *Vegetation and Flora of a Sector Along Mediterranean Coast of Libya from Tobruk to Egyptian Border*. M.Sc. Thesis, Faculty of Science University of Qar Younis: 160pp.

Eckey, E. W. (1954). *Vegetable Fats and Oils*. Reinhold, New York.

Edrawi, M. and El-Naggar, S. (1995). Natural plant life of Libya. *Assiut J. Environ. Stud.*, 9: 149–160 (in Arabic).

Edwards, A. and Head, S. M. (1987). *Red Sea: Key Environments*. Pergamon Press, Oxford: 441pp.

Ehler, W. L., Fink, D. H. and Mitchell, S. T. (1978). Growth and yield of Jojoba in native stands using run-off collecting microcatchments. *Agron. J.*, 740: 1005–1009.

Eid, E. M. (2008). *Population Biology and Nutrient Cycle of Phragmites australis* (Cav.) *Trin. Ex Steud. In Lake Burullus*. Ph.D. Thesis, Faculty of Science, Tanta University, Tanta, Egypt.

Eig, A. (1931–1932). Les elements et les groupes phytogeographiques auxiliaries dans la flore palestininenne. 2 parts. *Feddes Rep. Spec. Nov. Reg. Veg. Beih.*, 63: 1–201, 1–120.

Eig, A. (1932). Revision of the *Erodium* species of Palestine. *Beih. Bot. Centralbl. Abt. II*, 56: 226–240.

Eig, A. (1938). Taxonomic studies on the oriental species of the genus *Anthemis*. *Palest. J. Bot. Jerusalem*, I: 161–224.

Eig, A. (1948). Diagnosis Specierum Novarum Palestinae. *Palest. J. Bot. Jerusalem*, 4: 171–173.

Eig, A. and Zohary, M. (1939). Plants new for Palestine. I. *Palest. J. Bot. Jerusalem*, 2: 97–102.

El-Abyad, M. S. H. (1962). *Studies on the Ecology of Katamiya Desert*, M.Sc. Thesis, Fac. Sci., Cairo University, Egypt.

El-Amin, H. M. (1990). *Trees and Shrubs of the Sudan*. Ithaca Press and Richmond Rou, Exeter, UK: 484pp.

El-Asmar, H. R. and Wood, P. (2000). Quaternary shoreline development: the northwestern coast of Egypt. *Quat. Sci. Rev.*, 19: 1137–1140.

El-Bana, M. I. (2003). *Environmental and Biological Effects on Vegetation Composition and Plant Diversity of Threatened Mediterranean Coastal Desert of Sinai Peninsula*. Ph.D. Thesis, Faculteit Westerschappear University Antwerpen, Antwerpen: 150pp.

El-Bana, M., Shaltout, K., Khalafallah, A. S. and Mosallam, H. (2008). Ecological Status of the Mediterranean *Juniperus phoenicea* L. Relicts in the desert Mountains of North Sinai, Egypt. *Proceedings of the 5th International Conference on Biological Society*, Tanta University, Tanta, Egypt: pp. 1–9.

El-Demerdash, M. A. (1996). The vegetation of the Farasan Islands, Red Sea, Saudi Arabia. *J. Veg. Sci.*, 7: 81–88.

El-Demerdash, M. A., Henidu, S. Z. and El-Kadu, H. F. (1996). Vegetation and conservation measures in Ras Muhammed sector of the protected areas in southern Sinai, Egypt. *J. Union Arab Biol.* Cairo: 3 (B) Botany: pp. 23–47.

El-Dingawi, A. A. (1990). *Contribution to the Studies on Kochia Plants and their Potentialities in Fodder Production*. Ph.D. Thesis, Faculty of Science, Mansoura University, Mansoura, Egypt.

El-Gadi, A. (1989). *Flora of Libya*, Vol. 145. Department of Botany, Al-Faateh University, Tripoli, Libya.

El-Gazzar, A., El-Demerdash, M., El-Kady, H. F. and Henedy, S. Z. (1995). *Plant Life in the Gulf of Aqaba (Sinai)*. Terminal Report, Department of Protectorates, EEAA, Cairo: 128pp.

El-Ghonemy, A. A. (1973). Phytosociological and ecological studies of the maritime sand dune communities in Egypt. 1. Zonation of vegetation and soil along a dune side. *Bull. Inst. Desert Egypte*, 23(2): 463–473.

El-Ghonemy, A. A. and Tadros, T. M. (1970). Socio-ecological studies of the natural plant communities along a transect between Alexandria and Cairo. *Bull. Fac. Sci. Alexandria Univ. Egypt*, 10: 329–407.

El-Ghonemy, A. A., Shalout, K, Valentine, W. and Wallace, A. (1977). Distributional pattern of *Thymelaea hirsuta* (L.) end and associated species along the Mediterranean coast of Egypt. *Bot. Gaz.*, 138(4): 479–489.

El-Hadeedy, M. E. A. (1984). *Some Studies on Jujoba Plant*. M.Sc. Thesis, Faculty of Agriculture, Ain Shams University, Cairo, Egypt.

El-Hadidi, M. N. (1969). Observations on the Flora of the Sinai mountainous region. *Bull. Soc. Geogr. Egypt*, 40: 124–155.

El-Hadidi, M. N. and Ayyad, M. A. (1975). Floristic and ecological features of Wadi Habis (Egypt). In: *La Flore du Bassin Mediterrane en: Essaide Systematique Synthetique Colloquues Internationaux du C.N.R.S.*, 235: 247–258.

El-Hadidi, M. N. and Hosny, H. (2000). *Flora Egyptiaca*, Vol. 1. The Palm Press, Cairo: 151pp.

El-Housini, A. A., Khalifa, S. A. and Nassar, Z. M. (2004). Growth and Forage yield of *Atriplex canescens* as affected by different soil amendments under saline conditions of Wadi Sudr. *Ann. Agri. Sci. (Moshtohor)*, 42(2): 415–425.

El-Kabalawy, A. (2004). Salinity effect on seed germination of the common desert range grass *Panicum turgidum*. *Seed Sci. Technol.*, 32: 873–878.

El-Kady, H. F. (1980). *Effect of Grazing Pressures and Certain Ecological Parameters on Some Fodder Plants of the Mediterranean Coast of Egypt*. M.Sc. Thesis, Faculty of Science, Tank University, Tanta, Egypt.

El-Kady, H. F. (1987). *A Study of Range Ecosystems of the Western Mediterranean Coastal Desert of Egypt*. Ph.D. Thesis, Faculty of Landscape Developments, Technical University, Berlin, Germany: 136pp.

El-Kady, H. F. (2000). Vegetation analysis along Slouk-Msus road. South eastern Benghazi, Libya. *El-Minia Sci. Bull.*, 13: 61–71.

Ellison, L. (1960). Influence of grazing on plant succession of rangelands. *Bot. Rev.*, 26: 1–78.

El-Masry, M. G. (1961). *Sociological and Ecological Studies on the Vegetation of Lake Idku*. M.Sc. Thesis, Faculty of Science, Alexandria University, Egypt.

El-Mogy, N. S. (2008). *Propagation of Hoh-Hoh-Bah: An Egyptian Experiment*. The Egyptian Company for Natural Oil Production, Cairo (non-published Report): 3pp (in Arabic).

El-Mogy, N. S. (2009). *Hoh-Hoh-Bah Tree: "The Green Gold", A Coloured Pamphlet*. The Egyptian Company for Natural Oil Production, Cairo: 6pp (in Arabic).

El-Morsy, E. S. R. (2008). *Ecological and Taxonomical Studies on the Vegetation of the Western Section of Libyan Mediterranean Coast*. Ph.D. Thesis, Faculty of Science, Ain Shams University, Cairo.

El-Qattamy, M. A. (1975). Transformation of sun energy into electric energy. *Proceedings of an International Conference on Energy*, Sheraz, Iran: pp. 32–38.

El-Saiegh, A. M. (1976). Sun-energy of the desert. *Proceedings of an International Symposium on the Deserts and Utilization*, Saudi Biological Society, Riyadh, Saudi Arabia: pp. 85–120 (in Arabic).

El-Shazly, M. and Shata, A. A. (1969). Geomorphology and Pedology of Mersa Matruah area, western Mediterranean litteral zone. *Bull. Desert Inst.*, 19: 1–29.

El-Sharkawi, H. M. (1961). Phytosociological studies on the vegetation of Bagoush area. *Bull. Inst. Deserte Egypte*, 11(1), 1–18.

Embabi, N. S. (2004). *The Geomorphology of Egypt, Land Forms and Evolution*. The Egyptian Geographical Society, Cairo: 447pp.

Emberger, L. (1930). La Vegetation Méditerranéenne Essai de classification des groupements vegetaux. *Rev. Gen. De Bot.*, 42: 611–622, 705–721.

Emberger, L. (1951). *Report sur les regions arideset semi-arides de l'Afrique du Nord*. Union Internationale des Sciences Biologiques, Serie B, Colloques, Paris: pp. 50–61.

Emberger, L. (1955). Afrique du nord-esert, Ecologie Vegetale. *Comptes Rendus de Recherches Plant Ecology, Reviews of Research*, UNESCO, Paris: pp. 219–249.

Erhilich, P. R. and Erhilich, A. H. (1981). *Extinction: The Causes and Consequences of the Disappearance of Species*. Random House, Inc., New York.

Erinc, S. (1949). The climates of Turkey according to Thornthwartes classification. *Ann. Ass. Amer. Geogr.*, 39(1): 26–46.

Erinc, S. (1950). Climatic types and the variation of moisture regions in Turkey. *Geogr. Rev.*, 40(2): 224–234.

Evenari, M. (1985). The desert environment chapter. In: Evenary, M., Noy-Meir, I. and Goodal, D. W. (eds.), *Hot Deserts and Shrublands*, Vol. 12A. Elsevier, Amsterdam: pp. 1–21.

Evenari, M., Sharon, L. and Tadmor, N. (1971–1982). *The Negev: The Challenge of a Desert.* Harvard University Press, Cambridge, MA: 345pp, 437pp.

Evenary, M., Noy Meer, I. and Goodal, D. W. (eds.) (1985). *Ecosystems of the World*, Vol. 12A. Hot Deserts and Arid Shrublands. Elsevier, Amsterdam: 365pp.

Fanshaw, D. B. (1966). The doum palm-*Hyphaene thebaica,* (Del.) Mart. E. *Afr. Agric. J.*, 32: 108–116.

FAO (1980). *Range and Fodder Crop Development.* Synai Arab Republic, Natural Range Management and Fodder Cop Production, FAO, Rome: 95pp.

FAO (1982). *Date Production and Protection.* FAO Plant Production and Protection Paper 35, FAO, Rome.

FAO (1994). *Mangrove Forest Management Guidelines.* FAO Forestry, Paper No. 11, FAO, Rome.

Fathi, A. A., Abdelhazer, H. M. A., Flower, R. J., Ramadani, M. and Karaiem, M. M. (2001). Phytoplankton communities of North African wetland lakes: the CASSAINA A project. *Aquatic Ecol.*, 35: 303–318.

Feinburn, M. and Zohary, M. (1955). A geobotanical survey of Transjordan. *Bull. Res. Counc. Israel*, 50: 5–35.

Ferrar, H. T. (1914). Notes on a mangrove swamp at the mouth of the Gulf of Suez. *Cairo Sci. J.*, CIII(88): 23–24.

Feth, J. E. (1973). *Water Facts and Figures for Planners and Managers.* U.S. Geological Survey Circular 601: 1.

Fink, D. H. and Ehrler, W. L. (1979). Runoff Farming for Jojoba. *Proceedings of an International Arid Lands Conference on Plant Resources*, Texas Tech University, Texas: pp. 212–224.

Fisher, W. B. (1978). *The Middle East*, 5th ed. Methuen, London: 615pp.

Flower, R. J. (2001). Change, stress, sustainability and aquatic ecosystem resilience in North African wetland lakes during the 20th century: an introduction to integrated biodiversity studies within the CASSARINA Project. *Aquatic Ecol.*, 35: 261–280.

Flower, R. J. et al. (2001). Recent environmental change in North African wetland lakes: diatoms and other stratigraphic evidence from the CASSARINA sites. *Aquatic Ecol.*, 35: 369–388.

Forsskal, P. (1775). *Flora Aegyptiaca-Arabica.* In: Niebuhr, C. (ed.), Hauniaa, Type Moller, 32, CXXVI, 219pp + 1 map.

Forti, M. (1970). *Grazing Trials on Fodder Bushes in the Migda Farm (Hebrew).* Negev Institute of Arid Zone Research Publication, Beer Sheva, Israel.

Franshawe, D. B. (1966). *The Doum Palm.* Harvard University Press, Cambridge, MA.

Fresenius, G. (1834). *Beitrage Zur Flora von Aegypten un Arablen.* Museum Sencken Bergianum, Frankfurth. a. Main: pp. 9–94, 165–188.

Garrod, D. A. E. and Bate, D. M. A. (1937). *The Stone Age of Mount Carmel*, Vol. I. Clarendon Press, Oxford: 246pp.

Gentry, H. (1958). The natural history of Jojoba (*Simmondsia chinensis*) and its cultural aspects. *Econ. Bot.*, 12: 261–295.

Gibali, M. A. (1988). *Studies on the Flora of North Sinai.* M.Sc. Thesis, Faculty of Science, Cairo University, Cairo.

Girgis, W. A. (1973). Phytosociological studies on the vegetation of Ras El-Hekma – Marsa Matruh Coastal plain. *Egypt. J. Bot.*, 16: 393–409.

Gomez-Campo, C. (1985). *Plant Conservation in the Mediterranean Area.* Dr W. Junk Publisher, Dordrecht, Boston and Lancaster.

Goodman, S. M. and Meininger, P. L. (1989). *The Birds of Egypt.* Oxford University Press, Oxford, UK: 489pp.

Goudie, J. W. (ed.) (1990). *Technique for Desert Reclamation.* John Wiley & Sons, New York: 271pp.

Greco, J. (1966). *L'Ersian La Defence of La Resauration De Sols le Roboisement En Algerie*. Publ. Au Ministere De L' Agric. et de La reforme agraire, Alger: 393pp.

Green, R. A. and Foster, E. O. (1933). The liquid wax of seeds of *Simmondsia chenensis*, California. *Bot. Gaz.*, 94: 826–828.

Greuter, W. (1991). Botanical diversity endemism and extinction in the Mediterranean area: an analysis based on the published volumes of Mediterranean checklist. *Bot. Chron.*, 10: 63–79.

Greuter, W., Burdet, H. and Long, G. (eds.) (1984–1986, 1989). *Med-Checklist: A Critical Inventory of Vascular Plants of the Circum-Mediterranean Countries*. Conservatoire et Jardin Botanique de la Ville de Geneve, Geneve.

Grove, A. T. (1985). The arid environment. Chapter 2. In: Wicken, G. E., Gooden, J. R. and Field, D. V. (eds.), *Plants for Arid Lands*. George Allen & Unwin, London: pp. 9–18.

Guest, E. R. (1966). *Introduction to the Flora of Iraq I*. Ministry of Agriculture, Baghdad: 213pp.

Haas, W. and Mittelbach, M. (2000). Detoxification experiment with the seed oil from *Jatropha curcas* L. *Ind. Crops Prod.*, 12: 111–118.

Hall, D. O. (1985). Plant hydrocarbon resources in arid and semi-arid lands. Chapter 27. In: Wickens, G. E., Gooden, T. R. and Field, D. V. (eds.), *Plants for Arid Lands*. George Allen & Unwin, London: pp. 364–384.

Halwagy, R. (1962). The incidence of biotic factors in northern Sudan. *Oikos*, 13: 97–107.

Hamed, A. I. (2005). *Solenostemma arghel* (Del.) Hayne. In: Batanoumy, K. H. (ed.), *Encyclopaedia of Wild Medicinal Plants in Egypt Part I. Conservation and Sustainable Use of Medicinal Plants in Arid and Semi-Arid Ecosystem in Egypt*. EEAA, Egypt: pp. 59–81.

Hammond, P. M. (1995). The current magnitude of biodiversity. In: Heywood, V. H. (ed.), *Global Biodiversity Assessment*. Cambridge University Pres, Cambridge, UK: pp. 113–128.

Hammouda, F. M. (2005). *Conservation and Sustainable Use of Medicinal Plants: National Survey. Report 3, The Western Desert and Oases*. EEAA, UNDP & GEF, Egypt: 154pp.

Hammouda, F. M. (2006). *Origanum syriacum* L. subsp. *sinaicum*. In: Batanouny, K. H. (ed.), *Encyclopedia of Wild Medicinal Plants in Egypt*, Vol. 2. EEAA, MPCP, GEF, UNDP, Egypt: pp. 64–83.

Hart, H. C. (1891). *Some Account of the Fauna and Flora of Sinai, Petra and Wady Arabah*. A. P. Watt for the Committee of the Palestine Exploration Fund, London: 255pp.

Hassan, H. M. (1974). *An Illustrated Guide to the Plants of Erkwit*. Herbarium Publication Botany Department, University of Khartoum, The Sudan: 106pp.

Hassan, M. M. (1975). *Geography of Libya and Arab Countries*. Benghazi University Press, Benghazi, Libya: 437pp.

Hasselequist, F. (1757). *Iter Palaestinum*. Typ. L Salvii, Stockholm: 619pp.

Hassib, M. (1951). Distribution of plant communities in Egypt. *Bull. Fac. Sci. Univ. Fouad I*, Cairo, Egypt, 29: 59–61.

Hefny, M. B. (1953). Two climatic maps of the Nile Basin and Vicinity. *Bull. Soc. Geogr. Egypte*, 26: 183–192.

Hepkins, S. T. and Jones, D. (1983). *Research Guide to the Arid Lands of the World*. Oryx Press, Phoenix, Arizona.

Hemming, C. F. (1961). The ecology of the coastal area of Northern Eriterea. *J. Ecol.*, 49: 55–78.

Higginis, L. G. and Riley, N. D. (1988). *A Field Guide to the Butterflies of Britain and Europe*. Collins, London.

Hills, E. S. (1966). *Arid Lands: A Geographical Appraisal*. Metheun, London: 461pp.

Hume, W. F. (1925). *Geology of Egypt*, Vol. 1. The Surface Features of Egypt, Their Determinating Causes and Relation to Geological Structure. Egyptian Survey Department, Cairo: 418pp.

Hus, K. J. (1971). Origin of the alps and western Mediterranean. *Nature*, 233: 44–48.

Ireland, A. W. (1948). The climate of the Sudan. In: Tothill, J. D. (ed.), *Agriculture in the Sudan*. Oxford University Press, Oxford, UK: pp. 62–82.

Isman, M. B. and Rodriguez, E. (1983). Larval growth inhibitors from species of *Parthenium* (Asteraceae). *Phytochemistry*, 22: 2709–2713.

IUCN (1986). *Marine Conservation Survey of the Yemen Republic, Red Sea Coast:* 84pp.

IUCN (2008). *Mediterranean Mosaic Centero de Ediciones de la Diputacion de Ma'laga (CEDMA)*. IUCN, Spain: 188pp.

IUCN/PERSGA (1987). *Preliminary Coastal Zone Management Recommendations for Y.A.R.* IUCN/PERSGA: 54pp.

Jafri, S. M. and El-Gadi, A. (1978). *Flora of Libya*, Vols. 25–144. Department of Botany, El-Faateh University, Tripoli.

Jeftic, L. Keckes, S. and Pernetta, J. C. (1996). Implications of future climatic changes for the Mediterranean coastal region. Chapter 1. In: Jeftic, L., Keckes, S. and Pernetta, J. C. (eds.), *Climatic Change in the Mediterranean*, Vol. 2. Arnold, London: pp. 1–26.

Jewitt, Y. N. (1966). Soil of arid lands. Chapter VI. In: Hills, E. S. (ed.), *Arid Lands: A Geographical Appraisal*. Methuen & Co. Ltd., London: pp. 103–125.

Johnson, D. V. (1985). Present and potential economic use of palms in arid and semi-arid areas. Chapter 14. In: Wickens, G. E., Goaden, J. R. and Field, D. V. (eds.). *Plants for Arid Lands*. George Allen & Unwin, London: pp. 189–202.

Jones, P. D., Paper, S. C. B., Bradely, R. S., Diaz, H. F., Kelly, P. M. and Wigley, T. M. L. (1986a). Northern hemisphere surface air temperature variations 1851–1984. *J. Climatol. Appl. Meteorol.*, 25: 161–179.

Jones, P. D., Paper, S. C. B. and Wigley, T. M. L. (1986b). Southern hemisphere surface air temperature variations: 1851–1984. *J. Climatol. Appl. Meteorol.*, 25: 1213–1230.

Jones, P. D., Wigly, T. M. L. and Wright, P. B. (1986c). Global temperature variations between 1861–1984. *Nature*, 32: 430–434.

Jurgens, K. D., Fons, R., Peters, T. and Sender, S. (1996). Heart and respiratory rates and their significance for connective oxygen transport rates in the smallest mammal, the Eruscan shrew *Suncus etruscus*. *J. Exp. Biol.*, 199: 2579–2584.

Karaiem, M. M. and Ben Hamza, C. (2000). Sites description, water chemistry and vegetation transects of Tunisian Lakes. Fish population study of the nine North African investigated lakes. CASSARINA project, Tunisian Final Report, University of Tunis (Unpublished report).

Kassas, M. (1952). Habitats and plant communities in the Egyptian desert. I. Introduction. *J. Ecol.*, 40: 342–351.

Kassas, M. (1955). Rainfall and vegetation in arid N.E. Africa. *Plant Ecology Proceedings of Mantpelier Symposium*, UNESCO, Paris: pp. 49–77.

Kassas, M. (1956). The mist oasis of Erkwit, Sudan. *J. Ecol.*, 44: 180–194.

Kassas, M. (1957). On the ecology of the Red Sea coastal land. *J. Ecol.*, 45: 187–203.

Kassas, M. (1960). Plant life is the deserts. Chapter III. In: Hills, E. S. (ed.), *Arid Lands. A Geographical Appraisal*. Methuen & Co. Ltd., UNESCO, Paris: pp. 145–179.

Kassas, M. (1970). Desertificaiton versus potential for recovery in circum-saharan territories. In: Dregne, H. (ed.), *Arid Lands in Transition*, Vol. 13. American Association for Advanced Science, Washington, DC: pp. 123–142.

Kassas, M. (1972a). Ecological consequences of water development project, Keynote paper. In: Polumin, N. (ed.), *The Environmental Future*, 7 Major Water ETC Development Projects. Macmillan, London: pp. 215–246.

Kassas, M. (1972b). A brief history of land use in Mareotis region. *Minerva Biol.*, 1: 167–174.

Kassas, M. (1996). Two elements of an agenda 21 for North Africa. In: Batanouny, K. and Ghabbour, S. I. (eds.), *Proceedings of Workshop on Arid Lands Biodiversity in North Africa*. IUCN, SDC, Cairo: pp. 7–11.

Kassas, M. (1998). Fragile ecosystems in the near east countries: problems and management. In: Gopal, B., Pathak, P. S. and Saxena, K. G. (eds.). *Ecology Today: An Anthology of Contemporary*. Ecological Research International Scientific Publication, New Delhi, India: pp. 307–332.

Kassas, M. (2004). *The Nile in Danger*, Vol. 705. Dar El-Maaref, Cairo: 185pp (in Arabic).

Kassas, M. and Imam, M. (1954). Habitat and plant communities in the Egyptian desert. III. The Wadi bed ecosystem. *J. Ecol.*, 42: 424–441.

Kassas, M. and Imam, M. (1959). Habitat and plant communities in the Egyptian desert. IV The gravel desert. *J. Ecol.*, 47: 289–310.

Kassas, M. and Zahran, M. A. (1962). Studies on the ecology of the Red Sea coastal land I. The district of Gebel Ataqa and El-Galala El-Bahariya. *Bull. Soc. Geogr. Egypte*, 35: 129–175.

Kassas, M. and Zahran, M. (1965). Studies on the ecology of the Red Sea coastal land II. The district from El-Galala El-Qibliya to Hurghada. *Bull. Soc. Geogr. Egypte*, 38: 185–193.

Kassas, M. and Zahran, M. A. (1967). On the ecology of the Red Sea littoral salt marsh, Egypt. *Ecol. Manogr.*, 37(4): 297–315.

Kassas, M. and Zahran, M. A. (1971). Plant life on the coastal mountains of the Red Sea, Egypt. *J. Ind. Bot. Soc. Gold. Jubilee*, 50A: 571–589.

Keay, R. W. I. (1959). *Vegetation Map of Africa*. Oxford University Press, Oxford, UK: 24pp.

Keith, L. B. (1958). Some effects of increasing soil salinity on plant communities. *Can. J. Bot.*, 36: 79–89.

Keith, H. G. (1965). *A Prelimiary Check List of Libyan Flora*, 2 Vols. The Government of Libyan Arab Republic, Ministry of Agriculture and Agrarian Reform, Tripoli.

Kendrew, W. G. (1961). *The Climates of the Contents*, 5th ed. Oxford University Press, Oxford, UK: 608pp.

Khalifah, S. F. (2000). *Energy Crops: Biodiesels and Oils*. Publications of Horticulture Institute, Agriculture Research Center, Cairo: 32pp.

Khalil, M. T. and Shaltout, K. H. (2006). *Lake Bardawil and Zaranik Protected Area*, Vol. 15. Publication of Biodiversity Unity, EEAA, Cairo: 594pp.

Khedr, A. A. (1989). *Ecological Studies on Lake Manzala, Egypt*. M.Sc. Thesis, Faculty of Science, Mansoura University, Egypt.

Khedr, A. A. (1999). Floristic composition and phytogeograph in a Mediterranean deltaic lake (Lake Burollos), Egypt. *Ecol. Mediterr.*, 25(1): 1–11.

Khedr, A. A. and Zahran, M. A. (1999). Comparative study on the plant life of two Mediterranean deltaic lakes in Egypt. *Assuit Univ. Bull. Environ. Res.*, 2(1): 1–14.

Khedr, A. A. and Lovett-Doust, J. (2000). Determination of floristic diversity and vegetation composition in the islands of Burollos Lake, Egypt. *Appl. Veg. Sci.*, 3: 147–156.

Khedr, A. A. and El-Gazzar, A. (2006). *Phytoecology of Zaranik Lagoon, Lake Bardawil, North Sinai*. First Progress Report, Mediterranean – Wet Coast Project, EEAA, Cairo: 27pp.

Khoshoo, T. N. and Subrahmanyam, G. V. (1985). Ecodevelopment of arid laids in India with non-agricultural economic plants – a holistic approach. Chapter 18. In: Wickens, G. E., Goodin, J. R. and Field, D. V. (eds.), *Plants for Arid Lands*. George Allen & Mnwia, London: pp. 243–265.

Konig, P. (1986). *Zonation of Vegetation in the Mountainous Region of South West Saudi Arabia (Asir, Tihama),* Ph.D. Thesis, University of Berlin, Berlin, Germany.

Koppen, W. (1931). *Grundriss der Klimakunde*. W. De Gruyter, Berlin, Germany.

Kotschy, T. (1861). *Umrisse Von Sud-Palastina in Kleide der Frühlingsflora*. Wien: 16pp.

Kumar, R., Manning, E. W. and Murck, B. (1993). *Challenge of Sustainability*. Foundation for International Training, Don Mills, Ontarie, Garade: 275pp.

Lange, O. L., Schulze, E. D. and Koch, W. (1970). Experimental – Okhologische Untersuchurgen an Flachten der Negev – Wuste. II $CO_2$ Gaswechsel and Wasserhaushalt Von *Ramalian maciformis* am naturlichen Standert Wahrend der sommerlichen Trochenperiod. *Flora*, 159: 38–62.

Lieth, H. V. Al-Masoom, A. A. (eds.) (1993a). *Towards the Rational Use of High Salinity Tolerant Plants*, Vol. 1. Deliberation About High Salinity Tolerant Plants and Ecosystem. Kluwer Academic Publishers, London: 521pp.

Lieth, H. and Al-Masoom, A. A. (1993b). *Towards the Rational Use of High Salinity Tolerant. Plants*, Vol. 2. *Agriculture and Forest Under Marginal Soil Water Conditions*. Kluwer Academic Publishers, London: 447pp.

Le Houerou, H. N. (1959). *Recherches Ecologiques et Floristiques sur la Vegetation de la Tunisie Meridionale*. Inst. Des Res. Sahariennes, University de Algiers, Algiers: 510pp.

Le Houerou, H. N. (1969a). Le vegetation de la Tunisie Steppique (avec reference aux vegetations analogues d'Algerie, de Libya et du Maroc). *Ann. Inst. Natl. Rech. Agron. Tunisie*, 42(5): 624pp.

Le Houerou, H. N. (1969b). *Princepes, Methods et Techniques de Amelioration Fourragere et Pastrodle Tunisie*. FAC, Rome: 291pp.

Le Houerou, H. N. (1970). North Africa: past, present and future. In: Dregue, H. (ed.), *Arid Lands in Transition*. American Association for the Advancement of Science, Washington, DC: pp. 227–278.

Le Houerou, H. N. (1972). An assessment of the primary and secondary productions of the arid grazing lands of North Africa. In: Rodin, L. E. (ed.), *The Ecophysiological Foundation of Ecosystem Productivity in Arid Zones*. Nauka, Leningrad: pp. 168–178.

Le Houerou, H. M. (1973). Problemes et potentialities de lerres arid de l'Afrique du nord. *Opt. Mediterr.*, 26: 17–36.

Le Houerou, H. N. (1975). Edue preliminaire sur la compatibilitie des Flores Nord African et Palestinienne. In: *La Flora du Basin Med. Essai de Systematique Synthetique NRS*, Paris: pp. 345–350.

Le Houerou, H. N. (1981). Impact of man and his animals on Mediterranean vegetation Chapter 25. In: Di Castri, F., Goodal, D. W. and Specht, R. L. (eds.), *Mediterranean Type Shrublands Ecosystems of the World*, Vol. 11. Elsevier, Amsterdam: pp. 479–521.

Le Houerou, H. N. (1985). The desert and arid zones of North Africa. Chapter 4. In: Evenari, M., Noy-Meir, I. and Goodal, D. W. (eds.), *Hot Desert and Arid Shrulands Ecosystem of the World*, Vol. 12A. Elsevier, Amsterdam: pp. 101–147.

Le Houerou, H. N. (1992). Vegetation and land use in the Mediterranean basin by the year 2050: a prospective study. In: Jeftic, L., Milliman, J. D. and Sestini, G. (eds.), *Climatic Changes and the Mediterranean Environmental and Societal Impacts of Climatic Changes and Sea-Level Rise in the Mediterranean Region*. Arnold, London: pp. 175–231.

Le Houerou, H. N. (2001). Bioclimatology and phytogeography of the Red Sea basin and Aden Gulf. *J. Arid Environ.*, 1: 128:

Le Houerou, H. N. and Hoste, C. (1977). Rangeland production and annual rainfall relations in the Mediterranean basin in the African Sahilo–Sudanian Zone. *J. Range Manage.*, 3: 181–184.

Lister, N.-M. E. (2008). Bridging science and values. The challenge of biodiversity conservation. Chapter 6. In: Waltner-Toews, D., et al. (eds.), *The Ecosystem Approach Complexity, Uncertainty and Managing for Sustainability*. Columbia University Press, New York: pp. 83–107.

Long, G. A. (1957). *The Bioclimatology and Vegetation of Eastern Jordan*. FAO, Rome: 97pp.

Mabbutt, J. A. (1978). Problems of development and technology transfer in the world's dryland. In: Mann, H. S. (ed.). *Arid Zone Research and Development. Proceedings of an International Symposium on Arid Zone Research and Development*. Arid Zone Research Association, Jodhpur, India: pp. 459–468.

Mainguet, M. (1972). Le modele de gres problems generaux. *Inst. Geogr. Natl.*, Paris: 657pp.

Maire, R. (1933, 1940). Etude sur la flore et la vegetation du sahara central. *Mém. Soc. Hist. Nat.*, 3. Mission du Hoggar 2, 433pp + 36 pls.

Maire, R. (1952–1977). *Flore de l'Afrique du Nord*, Vols. 1–14. Encyclopaedia Biologique, Lechevalier, Paris.

Malcolm, C. V. (1972). *Establishing Shrubs in Saline Environment. Technical Bull.* 14, Department of Agriculture, Western Australia: 37pp.

Mandura, A. S., Saifullah, S. M. and Khaffagii, A. K. (1987). Mangrove ecosystem of southern Red Sea coast of Saudi Arabia. *Proc. Saudi Biol. Soc.*, 10: 165–193.

Mann, H. S. (ed.) (1978). *Arid Zone Research and Development. Proceedings of an International Symposium on Arid Zone Research and Development*. Arid Zone Research Association, Jodhpur, India: 531pp.

Marks, M., Lapin, B. and Randall, J. (1994). *Phragmites australis* (*P. communis*) threats, Management and monitoring. *Natl. Areas J.*, 14: 285–294.

Mc Kell, C. M. (1993). Salinity tolerance of *Atriplex* species fodder shrubs in arid lands. In: Pessarakli, M. (ed.), *Handbook of Plant and Crop Stress*, Marcel Dekker Inc., New York: pp. 497–503.

Mclillo, J. and Salama, O. (2008). Threats to ecosystem services. Chapter 3. In: Chiven, E. and Bernsteins, A. (eds.), *Sustaining Life: How Human Depends on Biodiversity*. Oxford University Press, Oxford, UK: pp. 107–114.

Medial, F and Verlaque, V. (1997). Ecological characteristics and rarity of endemic plants from S. E. France and Corsica. Implications for biodiversity conservation. *Biol. Conserv.*, 80: 269–281.

Medial, F. and Quezel, P. (1997). Hot-spots analysis for conservation of plant biodiversity in the Mediterranean basin. *Ann. Mo. Bot. Gard.*, 84: 112–127.

Meigs, P. (1953). World distribution of arid and semi-arid homo-climates. In: *Reviews of Research in Arid Zone Hydrology*, Arid Zone Program, 1: 202–210, UNESCO, Paris.

Meigs, P. (1962). Classification and occurrence of Mediterranean type dry climates. *Proceedings of the Land Use in Semi-Arid Mediterranean Climates. International Geographical Union Symposium,* UNESCO, Iraklion (Greece): pp. 17–21.

Meigs, P. (1966). *Geography of the Coastal Deserts*. Arid Zone Research No. 28, UNESCO, Paris: 140pp.

Meigs, P. (1973). World distribution of coastal deserts. In: Amiran, D. H. K. and Wilson, A. W. (eds.), *Coastal Deserts Their Natural and Human Environment*. The University of Arizona Press, Tucson, AZ: pp. 3–12.

Meteorological Office (1951). *Weather in the Indian Ocean to Latitude 30°S and Longitude 95°E Including the Red Sea and Persian Gulf,* Vol. II. Local Information Part I. Red Sea. M.O. 451(I), London.

Migahid, M. A., Abdel Rahman, A. A., El-Shafei, A. M. and Hammouda, M. A. (1955). Types of habitats and vegetation at Ras El-Hikma. *Bull. Inst. Desert Egypte*, 2: 107–190.

Migahid, A. M., El-Shafei, A. M. and Abdel Rahman, A. A. (1959). Ecological observations in the western and southern Sinai. *Bull. Soc. Geogr. Egypte*, 32: 165–205.

Migahid, M. A. and Hammouda, M. A. (1978). *Flora of Saudi Arabia*, 2 Vols. Riyadh University Publication, Riyadh: 939pp.

Miller, T. J. R. (1997). *Environmental Science Sustaining the Earth*, 4th ed. Wadsowrth Publishing Company, California: 470pp.

Monod, Th. (1937). *Meharees*, Paris: 300pp.

Monod, Th. (1958). Paris perspectives de l'homme et des phénoménes naturels dans la degradation des paysages et le declin des civilizations a travers le monde Mediterranean L. s. avec les deserts et semi-deserts adjacent, au Reun. *Technology Archives*: 38pp.

Montasir, A. H. (1937). On the ecology of Lake Manzla. *Bull. Fac. Sci. Egypt. Univ. Cairo*, 12: 50.

Morcos, S. A. (1990). Physical and chemical oceanography. In: Morcos, S. A. and Varely, A. (eds.). *Red Sea, Gulf of Aden and Suez Canal: A Bibliography on Oceanographic and Marine Environmental Research*. ALECSO, PERSGA, UNESCO, Belgium: pp. xvi–xxii.

Murray, G. W. (1951). The Egyptian climate. An historical outline. *Geogr. J.*, 117(4): 422–434.

Muschler, R. (1912). *A Manual Flora of Egypt*, Vol. II. R. Friedlander, Berlin: 1312pp.

Myers, N. (1990). The biodiversity challenge expanded hot spots analysis. *Environmentalists*, 10: 243–256.

Nabhan, G. D. and Felger, R. S. (1989). Wild desert relatives of crops. Their direct uses as food. Chapter 3. In: Wicken, G. E., Gooden, T. R. and Field, D. V. (eds.), *Plants for Arid Lands*. George Allen & Unwin, London: pp. 19–33.

Nadji, H, Diouf, P. N., Banaboura, A., Bedard, Y., Riedi, B. and Stevanovic, T. (2009). Comparative study of lignins isolated from Alfa Grass (*Stipa tenacissima* L.). *Bioresour. Technol.*, 100(14): 3585.

Nahal, I. (1997). Renewable natural resources: degradation and desertification in Syria. In: Barakat, H. N. and Hegazy, A. K. (eds.), *Reviews in Ecology: Desert Conservation and Development. A*

*festschrift for Prof. M. Kassas on the Occasion of His 75th Birthday*. IDRC/CRDI, UNESCO, South Valley University, Cairo, Egypt: pp. 99–117.
Newbigen, M. I. (1924). *The Mediterranean Lands: An Introductory Study in Human & Historical Geography*. Christophers, London: 222pp.
Newton, L. E. (1980). Phytogeographical associations of the succulent plant flora of south-west Arabia and the Horn of Africa. *Nat. Cact. Succ. J.*, 35: 83–88.
Oldfield, Sera (2004). *Deserts: The Living Drylands*. The MIT Press, Cambridge: 160pp.
Oliver, F. W. (1938). The flowers of Mareotis: an impression. Part I. *Trans. Norfolk Norwich Nat. Soc.*, 14: 397–437.
Oliver, F. W. (1945). The flowers of Mareotis: An impression. Part II. *Trans. Norfolk Norwich Nat. Soc.*, 14: 397–437.
Orshan, G. (1985). The desert of the Middle East. Chapter I. In: Evenari, M, Noy-Meir, I. and Goodal, D. W. (eds.), *Ecosystems of the World*, Vol. 12A. Hot Deserts and Arid Shrublands. Elsevier, Amsterdam: pp. 1–28.
O'Toole, C. and Raw, A. (1991). *Bees of the World*. Blanford, London.
Ozenda, P. (1958). *Flora du Sahara*. Centre National de le Recherche Scientifique Alger, Paris: 486pp.
Pabot, H. (1954). Lu Vegetation Naturelle de la Syrie Apercu Floristique et ecologique. *Proceedings of the Symposium on the Protection and Conservation of Nature in the Near East*, Beirut: pp. 80–84.
Pearce, D., Barbier, E. and Markandya, A. (1990). *Sustainable Development: Economics and Environment in the Third World*. Edward Elgar, London: 217pp.
Pessarakli, M. (1993). *Hand Book of Plant and Crop Stress*. Marcel Dekker Inc., New York: 692pp.
Peters, A. J., et al. (2001). Recent environmental change in North African wetland lakes: a baseline study of organochlorine contaminant residues in sediments from nine sites in the CASSARINA Project. *Aquatic Ecol.*, 35: 449–459.
Petts, G. G. (1984). *Impounded Rivers Perspective for Ecological Management*. John Wiley, New York: 326pp.
Poiret, A. (1789). *Voyage en Barbarie*, 2 Vols. in 8 Parts; 470 species from NE Algeria and NW Tunise.
Post, G. E. (1896). *Flora of Syria, Palestine and Sinai*. The American Press, Beirut: 919pp.
Post, G. E. and Autran, E. (1899). Plantae Postianae Fasc. X. *Bull. Herb. Boiss*, 7: 146–161.
Post, G. E. and Dinsmore, J. E. (1932–1933). *Flora of Syria, Palestine and Sinai*, 2nd ed., 2 Vols. American University of Beirut, Beirut, Publ. Fac. Arts & Sci., Nat. Sci. Ser. I 639pp, II 928pp.
Pottier-Alapetite, G. (1981). *Flora la Tunisie: Angiospermes-Dicotyledone, Apetales-Dialypetales*. Publications Scientifique Tunisiennes Programme Flore et Vegetation Tunisienne, Tunis: 1190pp.
Pratov, V. and El-Gadi, A. (1980). On Flora composition of the pasture zone of the Libyan Jamahirya. *Libyan J. Sci.*, 10B: 19–43.
Qiser, M. and El-Gadi, A. (1984). A critical analysis of the Flora of Libya. *Libyan J. Sci.*, 13: 31–40.
Quezel, P. (1965). *La Vegetation du Sahara Du Tchad a le Mauritanie*. Fischer, Stuttgart: 333pp.
Quezel, P. (1985). Définition of the Mediterranean region and origin of its Flora. In: Gomez-Campo, C. (ed.), *Plant Conservation in the Mediterranean Area*. Dr. W. Junk Publisher, Dordrecht: pp. 9–24.
Raheja, P. C. (1966). Aridity and salinity: a survey of soils and land use. In: Boyko, H. (ed.), *Salinity and Aridity New Approaches to Old Problems*. Dr. W. Junk Publishers, The Hague, Netherlands: pp. 43–113.
Ramadani, M., et al. (2001). North African wetland lakes: characterization of nine sites included in the CASSARINA project. *Aquatic Ecol.*, 35: 281–302.
Rashed, M. A (1998). Calcrete an Pleistocene coastal ridges West Alexandria, Egypt. Sedementary nature and applications. *Sedementol. Egypt*, 6: 113–128.

Ratallack, G. and Dilcher, D. L. (1981). A coastal hypothesis for the dispersal and rise to dominance of flowering plants. In: *Palaeobotany, Palaeoecology and Evolution*, Vol. 2. Praeger Publishers, New York: pp. 27–77.

Raunkiaer, C. (1934). *The Life Forms of Plants and Statistical Plant Geography*. Oxford Clarendon Press, London: 631pp.

Rauwolf, L. (1583). Aigentliche beschreibung der Raiss so er vor diser zeit gegen auffgang inn die Morgenlander fürnemlich syriamp Judaeam, Arabian, Mesoptamian, Babyloniam, Assyrian, Armenian, etc. Laugigen G. Willer: 487pp + 44 pls.

Raven, P. H. (1964). Catastrophic selections and edaphic endemism. *Am. Nat.*, 98: 336–338.

Raymond, A. and Phillips, T. S. (1983). Evidence for an upper carboniferous mangrove community. Chapter 2. In: Teas, H. H. (ed.), *Biology and Ecology of Mangrove*. Tives Dr. W. Junk Publishers, Netherlands: pp. 19–30.

Reyadh, M. (2006). *The Cultivation of* Jatropha curcas *in Egypt*. FAO Cooperative Document Repositry, Cairo: 5pp.

Rodriguez, E. (1977). The ecogeographical distribution of secondary constituents in *Parthenium*. *Biochem. Syst. Ecol.*, 5: 207–218.

Rodriguez, E. (1985). Rubber & Phytochemical specialities from desert plants of North America. Chapter 29. In: Wickens, G. E., Gooden, T. R. and Field, G. V. (eds.), *Plants for Arid Lands*. George Allen & Unwin, London: pp. 399–412.

Sadek, J. (1926). *The Geography and Geology of the District Between Gebels: Ataqa and El-Galala El-Bahariya (Gulf of Suez)*. Geological Survey, Egypt, Cairo: 120pp.

Saenger, P. (2001). *Mangrove Ecology: Silviculture and Conservation*. Kluwer Academic Publishers, the Netherlands: 372pp.

Said, R. (1962). *The Geology of Egypt*. Elsevier, Amsterdam: 377pp.

Salama, F. M., Abdel Ghani, M. M., El-Naggar, S. M. and Baayo, K. A. (2005). Vegetation structure and environmental gradient in the Sallum area, Western Mediterranean coast, Egypt. *Ecol. Med.*, 31(1): 1–13.

Samaan, A. A. (1974). Primary production of the Edku Lake, Egypt. *Bull. Inst. Ocean Fish ARE*, 4: 260–317.

Sample, L. (2003). Jojoba oil could fuel cars and trucks. *New Scientist* (abstract).

Sauvage, C. (1954). Le Hammadas du Sud Marocain. *Trav. Inst. Sci. Cherifien, Ser. Gen.*

Schechter, J. (1978). Some new directions in arid zone research. In: Mann, H. S. (ed.), *Arid Zone Research and Development. Proceedings and Selected Papers International Symposium on Arid Zone Research and Development*. Arid Zone Research Association, Jodhpur, India: pp. 483–489.

Schechter, Y. and Galai, C. (1980). The Negev: a desert reclaimed. In: Biswas, R. M. and Biswas, A. K. (eds.), *Desertification*. Pergamon Press, Oxford: pp. 225–308.

Schishor, L. L. (1980). *Soil Studies in the Eastern Zone of the Socialist People's Libyan Arab Jamahiriya*, Vol. 10. Soil Expedition, Selkozpromexport, Tripoli: 392pp.

Schousboe, P. K. A. (1801a). Iakttaggelser Över Vaxtriget I Marokko. K. *Dan Vidensk. Selsk Skr*, 1800, Bd, 1: 204pp.

Schousboe, P. K. A. (1801b). *Beobachtuegen Uber das Gewachsreich in Marokko*, Leipzig. (Latin–Danish edition b German Translation).

Schweinfurth, G. A. (1865). Flora der Soturba an der nubischen Küste. *Verh. Zool Bot. Ges. Wien.*, 15: 537–560.

Semple, E. C. (1932). *The Geography of the Mediterranean Region: Its Relation to Ancient History*. Constable & Co., London: 737pp.

Sen, D. N. and Rajpurohit, K. S. (eds.) (1982). *Contributions to the Ecology of Halophytes*. Tasks to Vegetation Sciences, Dr. W. Junk Publisher, The Hague, Netherlands: 227pp.

Shachak, M., Chapman, E. A. V. and Orv, Y. (1976). Some aspects of the ecology of the desert snail (*Sphineterochila boissieri*) in relation to water and energy flow. *Israel J. Med. Sci.*, 12: 887–891.

Shaltout, K. H. (1983). *An Ecological Study of Thymelaea hirsuta (L.) Endl. In Egypt*. Ph.D. Thesis, Tanta University, Tanta, Egypt.

Shaltout, K. H. and Khalil, M. T. (2005). *Lake Burullus (Burullus Protected Area)*. Publication of National Biodiversity Unit No. 13, EEAA, Cairo: 578pp.

Shaltout, K. H. and Al-Sodany, Y. M. (2008). Vegetation analysis of Burullus wetland: a RAMSAR site in Egypt. *Wetlands Ecol. Monogr.*, 16: 421–439.

Sharon, D. (1972). The spottiness of rainfall in a desert area. *J. Hydrol.*, 17: 161–175.

Sharaf El-Din, A. and Shaltout, K. H. (1985). On the phytosociology of Wadi Araba in the eastern desert of Egypt. *Proc. Egypt. Bot. Soc. Jsmailia Conf.*, 4: 1311–1317.

Shaw, T. (1738). *Specimen Phytographiae Africanae or a Catalogue of Some of the Rarer Plants of Barbary, Egypt and Arabia* (632 species collected in the regencies of Algier, Tunis and in Arabia).

Shawar, A. (1989). Geographical framework of the Red Sea and its group of islands. In: *The Islands of the Red Sea*. The Scientific Royal Society of Jordan, Arab Research Center and Research Institute of the Arabian Studies, ALESCO, Jeddah: pp. 160–195 (in Arabic).

Shay, G. (ed.) (1990). *Saline Agriculture: Salt Tolerant Plants for Develping Countries*. National Academic Press, Washington, DC: pp. IIV–IV.

Shmida, A., Barbour, M. (1982). *A comparison of two types of Mediterranean scrub in Israel and Califonia*. Gen Tech Rep PSW-58, Pacific Southwest Forest and Range Experiment Station, Forest Services, USA, Department of Agriculture, Berkeley, CA: pp. 100–106.

Siegfried, A. (1948). *The Mediterranean*. Translated from French by Doris Hemming, Jona Than Cape, London: 221pp.

Shreve, F. (1942). The desert vegetation of North America. *Bot. Rev.*, 8: 195–2046.

Smith, R. L. (1996). *Elements of Ecology*, 3rd ed. Harper Collins Publishers, New York: 617pp.

Spottswood, J. (1696). Phytologia tingitana. *Philos. Trans.*, 19: 239–249 (600 Species of plants collected around tanger in 1673).

Springuel, I. (2006). *The Desert Garden: A. Practical Guide*. The American University in Cairo Press, Cairo, Egypt: 156pp.

Steinen, H. (1985). *Prosopis tamarugo* in the Chilean Atacama – ecophysiological and reforestation aspects. Chapter 9. In: Wicken, G. E., Goodin, J. R. and Field, D. V. (eds.), *Plants for Arid Lands*. George Allen & Unwia, London: pp. 103–116.

Strand, B. J. (1756). "*Flora Palaestina*". C. Linneaus Amoen. Acad., Upsaliae, 4: 443–467.

Sutton, L. J. (1947). *Rainfall in Egypt*. Phys. Dept., Paper No. 53, Gov. Press, Cairo: 129pp.

Täckholm, V. (1932). Some new plants from Sinai and Egypt. *Svensk Botanisk Tidskrift*, 26(1–2): 370–380.

Täckholm, V. (1956). *Student's Flora of Egypt*, 1st ed. Anglo-Egyptian Bookshop, Cairo: 649pp.

Täckholm, V. (1974). *Student's Flora of Egypt*, 2nd ed. Cairo University Publishing Cooperative Printing Company, Beirut: 888pp.

Tadros, T. M. (1949). Geobotany in Egypt. A historical review. *Vegetatio*, 2: 38–42.

Tadros, T. M. (1953). A phytosociological study of halophlous communities from Mareotis, Egypt. *Vegetatio*, 4: 102–124.

Tadros, T. M. (1956). An ecological survey of the semi-arid coastal strip of the western desert of Egypt. *Bull. Inst. Desert Egypte*, 6(2): 28–56.

Tadros, T. M. and Atta, B. A. M. (1958). The plant communities of barely fields and uncultivated desert areas of Mareotis, Egypt. *Vegetatio*, 8: 161–175

Thornthwait, C. W. (1948). An approach towards a rational classification of climate. *Geogr. Rev.*, 39(1): 55–94.

Tolba, M. K. (1997). Preface. In: Barakat, H. and Hegazy, A. K. (eds.), *Reviews in Ecology: Desert Conservation and Development. A Festschrift for Prof. M. Kassas on the Occasion of His 75th Birth Day*. IDRC, UNESCO & South Valley University, Egypt: pp. 4–18.

Tothill, J. D. (1948). *Agriculture in the Sudan*. Oxford University Press, Oxford, UK.

Trewartha, G. T. (1954). *An Introduction to Climate*. McGraw Hill, New York: 377pp.

Tristram, H. B. (1884). *The Survey of Western Palestine. The Fauna and Flora of Palestine*. London Committee of the Palestine Exploration Fund, London: 455pp.

UNEP (1992). *World Atlas of Desertification*. Edwards Arnolds, London: 1 × +69pp.

UNESCO/FAO (1963). Bioclimatic map of the Mediterranean zone. Explanatory notes. *Arid Zone Res.*, 22: 17.
UNESCO (1977). *Map of the World Distribution of Arid Regions*. MAB Technical Notes, UNESCO, Paris: 7pp.
Vahl, M. (1790–1794). Symbolae Botanica I. Vol. in 3 parts including plants collected by him in 1783 in northern Tunisia.
Vernet, A. (1955). Climate and vegetation. In: *Plant Ecology Proceedings of the Montpellier Symposium*: pp. 75–101.
Vesey-FitzGerald, D. F. (1955). Vegetation of the Red Sea south of Jeddah, Saudi Arabia. *J. Ecol.*, 43: 477–489.
Vesey-FitzGerald, D. F. (1957). Vegetation of the Red Sea north of Jeddah, Saudi Arabia. *J. Ecol.*, 45: 547–562.
Von Chi-Bonnardel, R. (1973). *The Atlas of Africa*. Jeune Afrique, Paris: 335pp.
Waisel, Y. (1971). Pattern of distribution of some xerophytic species on the Negev, Israel. *Israel J. Bot.*, 20: 101–110.
Walsh, G., Snedaker, S. and Teas, H. (eds.) (1975). *Proceedings of an International Symposium on Biology and Management of Mangroves*. Honolulu, Hawai: Vol. 1: 401pp, Vol. 2: pp. 402–846.
Warner, T. T. (2004). *Desert Meteorology*. Cambridge University Press, Cambridge, UK: 595pp.
Watson, O. M. (1990). *Atriplex* spp. irrigated forage crops. *Agric. Ecosyst. Environ.*, 32: 107–118.
Welch, D. C. (1989). *Nutritive Value of Shrubs. Shrub Biology and Utilization*. Academic Press, New York: pp. 405–422.
Wigley, T. M. L. (1992). Future climate of the Mediterranean basin with particular emphasis on changes in precipitation. In: Jeftic, L., Milliman, T. D. and Sestini, G. (eds.), *Climatic Changes in the Mediterranean*. Arnold, London: pp. 15–43.
Wilson, J. M. and Witcombe, J. R. (1985). Crops for arid lands. Chapter 4. In: Wickens, G. E., Gooden, J. R. and Field, D. V. (eds.), *Plant for Arid Lands*. George Allen & Unwin, London: pp. 35–52.
Wood, J. R. I. (1997). *A Handbook of the Yemen Flora*. Royal Botanic Garden, Kew, London: 434pp + 40 Coloured Plates.
Yermanos, D. M. and Gonzales, R. (1976). Mechanical harvesting of Jojoba. *Calif. Agric.*, 30(1): 8–9.
Younes, H. A., Zahran, M. A. and El Qurashy, M. E. (1983). Vegetation–soil relationships of a sea-landword transect, Red Sea coast, Saudi Arabia. *J. Arid Environ.*, 6: 349–356.
Zahran, M. A. (1962). *Studies on the Ecology of the Red Sea Coastal Land*. M.Sc. Thesis, Faculty of Science, Cairo University, Egypt.
Zahran, M. A. (1964). *Contributions to the Study on the Ecology of the Red Sea Coast*. Ph.D. Thesis, Faculty of Science, Cairo University, Egypt.
Zahran, M. A. (1965). Distribution of the mangrove vegetation in Egypt. *Bull. Inst. Desert Egypte*, 15(2): 7–12.
Zahran, M. A. (1967). On the ecology of the eastern coast of the Gulf of Suez. I. Littoral salt marsh. *Bull. Inst. Desert Egypte*, 17(2): 225–252.
Zahran, M. A. (1977). Africa: Wet formation of the African Red Sea coast. Chapter II. In: Chapman, V. J. (ed.), *Ecosystems of the World. Wet Coastal Ecosystems*. Elsevier Science Publications, Amsterdam: pp. 215–231.
Zahran, H. A. (1982a). Ecology of the halophytic vegetation of Egypt Chapter I. In: Sen, D. N. and Rajpurohit, K. S. (eds.), *Contributions to the Ecology of Halophytes*, Vol. 2. Tasks for Vegetation Science, Dr. W. Junk Publisher, The Hague, Netherlands: pp. 3–20.
Zahran, M. A. (1982b). *Vegetation Types of Saudi Arabia*. King Abdul Aziz University, Jeddah, Saudi Arabia: 61pp.
Zahran, M. A. (1986). Forage potentialities of *Kochia Indicia* and *K. scoparia* in arid lands with particular reference to Saudi Arabia. *Arab Gulf J. Sci. Res.*, 4(1): 53–68.

Zahran, M. A. (1993a). Dry coastal ecosystems of the Asian Red Sea coast. In van der Maarel, E. (ed.), *Ecosystems of the World 2B. Dry Coastal Ecosystems: Africa, America, Asia and Oceania*. Elsevier, Amsterdam: pp. 17–29.

Zahran, M. A. (1993b). *Juncus* and *Kochia*: Fiber and Fodder Producing Halophytes Under Salinity and Aridity Stress. Chapter 26. In: Pessarakli, M. (ed.), *Handbook of Plant and Crop Stress*. Marcel Dekker Inc., New York: pp. 505–530.

Zahran, M. A. (2002). Phytogeography of the Red Sea littorals of Egypt and Saudi Arabia. *Bull. Soc. Geogr. Egypte*, LXXV(75): 149–158.

Zahran, M. A. (2004). *The Natural Vegetation: Renewable Resource for the Sustainable Development of the Deserts in the Arab World*. Environmental Series, Zayed International Prize, Dubai: 489pp (in Arabic).

Zahran, M. A. (2007). *Mangrove Ecosystem of the Coastal Belts of the Red Sea and Arabian Peninsula (in Arabic)*, Vol. 7. The World of Environmental Series, Zayed International Prize, Dubai: 335pp.

Zahran, M. A., Kamal El-Din, H. and Boulos, S. T. (1972). Potentialities of the fibrous plants of the Egyptian Flora in National Economy. I. *Juncus rigidus* Desf. and paper industry. *Bull. Inst. Desert Egypte*, 22(1): 193–203.

Zahran, M. A. and Negm, S. A. (1973). Ecological and pharmacological studies on *Salsola tetrandra* Forssk. *Bull. Fac. Sci. Mansoura Univ. Egypt*, 1: 67–75.

Zahran, M. A. and Boulos, S. T. (1974). Potentialities of the fiber plants of the Egyptian Flora in National Economy. II. *Thymelaea hirsuta* (L.) Endl. *Bull. Fac. Sci. Mansoura Univ. Egypt*, 1: 77–87.

Zahran, M. A. and Abdel Wahid, A. A. (1982). Halophytes and human welfare. Part 3. In: Sen, D. N. and Rajpurohit, K. S. (eds.), *Contributions to the Ecology of Halophytes*. Dr W. Junk Publishers, The Hague, Netherlands: pp. 235–257.

Zahran, M. A., Younes, H. A. and Hajrah, H. H. (1983). On the ecology of mangal vegetation of the Saudi Arabian Red Sea Coast. *J. Univ. Kuwait (Sci.)*, 10(1): 87–99.

Zahran, M. A., El-Demerdash, M. A. and Mashali, I. A. (1985). On the ecology of the deltaic Mediterranean Coast, Egypt. I. General Survey. *Proc. Egypt. Bot. Soc. Ismailia Conf.*, 4: 1392–1407.

Zahran, M. A. El-Demerdash, M. A. and Mashali, I. A. (1990). Vegetation types of the deltaic Mediterranean coast of Egypt and their environment. *J. Veg. Sci.*, 1: 305–310.

Zahran, M. A. and Willis, A. J. (1992). *The Vegetation of Egypt*, 1st ed. Chapmen & Hall, London: 424pp.

Zahran, M. A., Muhammed, B. K. and El-Dingawi, A. A. (1992). Establishment of *Kochia* forage halophytes in the salt affected lands of the Arab countries. *J. Environ. Sci. Mansoura Univ. Egypt*, 4: 93–119.

Zahran, M. A. and Al-Kaf, H. F. (1996). Introduction to the ecology of the littoral halophytes of Yemen. *Arab Gulf J. Sci. Res.*, 14(3): 691–703.

Zahran, M. A. and Willis, A. J. (2008). *The Vegetation of Egypt*, 2nd ed. Springer Publishers, The Netherlands: 437pp.

Zahran, M. A., Willis, A. J., Mosallam, H. M. and Bazaid, S. (2009). *Ecology and Sustainable Development of the Red Sea Coastal Deserts*. Publishing Al-Taif University, Saudi Arabia: 455pp.

Zohary, M. (1932). Neue Beitrage Zur Kenntnis de Flora Palastine. *Beih. Bot. Centralbl. Abt. II*, 50: 44–53.

Zohary, M. (1935). Die phytogeographische Gliederung dir Flora der Halbinsel Sinai. *Beih. Bot. Centralbl. Abt. II*, 52: 249–621.

Zohary, M. (1940). Geobotanical analysis of the Syrian desert. *Palest. J. Bot.*, 2: 46–96.

Zohary, M. (1941). Taxonomical studies I the Flora of Palestine and neighbouring countries. *Palest. J. Bot. Jerusalem*, 2: 151–184.

Zohary, M. (1944). Vegetation transects through the desert of Sinai. *Palest. J. Bot.*, III(2): 57–78.

Zohary, M. (1962). *Plant Life of Palestine, Israel and Jordan*. The Ronald Press, New York: 262pp.

Zohary, M. (1973). *Geobotanical Foundations of the Middle East*. Gustar Fisher Verlag, Stuttgart: 737pp.

Zohary, M. (1978). *Flora Palaestina Part III*. The Israel Academy of Science and Humanities, Jerusalem, Israel: 481pp.

Zohary, M. and Orshan, G. (1949). Structure and ecology of the vegetation in the Dead Sea region of Palestine. *Palest. J. Bot. Jerusalem*, 4: 177–206.

Zanoni, G. (1675). *Historia Botanice Bologna*: 224pp. (Plants collected around Tanger by Balaam).

# List of Genera

In this book, scientific names are currently accepted (valid) and widely used in literature; although some names are outdated. Recommended references to confirm and update these names are: Post and Dinsmore (1932–1933), Andrews (1950–1956), Maire (1952–1977), Zohary (1962), Guest (1966), Danin (1972), Täckholm (1974), Boulos (1975, 1977, 1979a, b, 1983, 1989, 1995, 1997, 1999, 2000, 2002, 2005, 2009), Ali and Jafri (1976), Migahid and Hammouda (1978), Jafri and El-Gadi (1978), Pottier-Alapetite (1981), Le Houerou (1985, 2001), Auder et al. (1987–1989), Ayyad and Ghabbour (1993), Al-Eisawi (1996), Wood (1997), Dallman (1998), and Blondel and Aronson (1999).

For ease of reference, genera of the scientific names are listed below:

*Abies*
*Abutilon*
*Acacia*
*Acalypha*
*Acantholimon*
*Acanthophyllum*
*Acanthus*
*Acer*
*Achillea*
*Achyranthus*
*Aciynonyx*
*Acomys*
*Actiniopteris*
*Addax*
*Adenostoma*
*Adenium*
*Adiantum*
*Adonis*
*Aegilops*
*Aeluropus*
*Aerva*
*Agathopora*
*Alcephalus*
*Alectoris*
*Alhagi*
*Alisma*
*Alkanna*
*Allium*
*Alnus*
*Aloe*
*Alternanthera*
*Alternaria*
*Althaea*
*Amaranthus*
*Ammi*
*Ammodaucus*
*Ammophila*
*Ammosperma*
*Amygdalus*
*Anabasis*
*Anacyclus*
*Anagallis*
*Anarrhinum*
*Anastatica*
*Anemone*
*Anisotes*
*Anthemis*
*anthyllis*
*Antirrhinum*
*Anvillea*
*Apium*
*Apolonia*
*Arbutus*
*Arctostaphylos*
*Arenaria*
*Argania*
*Argemone*
*Argyrolobium*
*Arisarum*
*Aristida*
*Aristolochia*
*Arnebia*
*Arrhenatherum*
*Artemisia*
*Aspergo*
*Arthrocnemum*
*Asplenium*
*Aster*
*Asteriscus*
*Asthenatherum*
*Astomaea*
*Astragalus*
*Atractylis*
*Atraphaxis*
*Atriplex*
*Avena*
*Avicennia*
*Azolla*
*Bacopa*
*Balanites*
*Balanus*
*Ballota*
*Barbeya*
*Barbula*
*Barleria*
*Bassia*
*Battendiera*
*Bellevalia*

*Caralluma*
*Cardopatium*
*Carduncellus*
*Carduus*
*Carex*
*Carissa*
*Carlina*
*Carpinus*
*Carrichtera*
*Carthamus*
*Cassia*
*Castanea*
*Casuarina*
*Colchicum*
*Coleus*
*Combretum*
*Cometes*
*Commicarpus*
*Commiphora*
*Companula*
*Conium*
*Convolvulus*
*Conyza*
*Corchorus*
*Corchus*
*Corciduru*
*Danthiopsis*
*Danthonia*
*Daphne*
*Datura*
*Daucus*
*Delonix*
*Delosperma*
*Delphinium*
*Desmostachya*
*Deverra*
*Dianthus*
*Dichanthium*
*Dichrocephala*
*Eremopoa*
*Eremopogon*
*Eremostachys*
*Erigeron*
*Erodium*
*Erophila*
*Eruca*
*Erucaria*
*Eryngium*
*Erysimum*
*Escherichea*
*Eucalyptus*
*Euclea*
*Eupatorium*
*Euphorbia*
*Exoacantha*
*Fagonia*
*Farsetia*
*Felicia*
*Ferula*
*Festuca*
*Ficus*
*Filago*
*Fimbrystylis*
*Flaveria*
*Foleyola*
*Forsskaolea*
*Francoeuria*
*Frankenia*
*Fredolia*
*Fuirena*
*Fumana*
*Fumaria*
*Galium*
*Gagae*
*Gaillonia*
*Gazella*
*Genetta*
*Genista*
*Geranium*
*Gladiolus*
*Glaucium*
*Glebiones*
*Glinus*
*Globularia*
*Haplophyllum*
*Hedysarum*
*Helianthemum*
*Helichrysum*
*Helicophyllum*
*Heliotropium*
*Helosciadium*
*Herniaria*
*Herpestes*
*Hevea*
*Hibiscus*
*Hippocrepis*
*Hippolais*
*Hippomarathrum*
*Hordeum*
*Hyacinth*
*Hymenocarpus*
*Hyoscyamus*
*Hyoseris*
*Hyparrhenia*
*Hypericum*
*Hyphaene*
*Hypoestes*
*Ifloga*
*Imperata*
*Indigofera*
*Inula*
*Iphiona*
*Ipomaea*
*Iris*
*Isatis*
*Jasminum*
*Lancophora*
*Lappula*
*Larix*
*Lasiurus*
*Lathyrus*
*Launaea*
*Laurus*
*Lavandula*
*Lavatera*
*Lawsonia*
*Leersia*
*Lemna*
*Lens*
*Leontica*
*Leontodon*
*Leontopodium*
*Leopoldia*
*Leptadenia*
*Leptochloa*
*Leucas*
*Ligustrum*
*Lilium*
*Limoniastrum*
*Limonium*
*Linaria*
*Lindenbergia*
*Linum*
*Lipidium*
*Lippia*
*Lobularia*
*Lolium*
*Lonicera*
*Marselea*
*Mastomys*
*Matricaria*
*Matthiola*
*Maytenus*
*Mecomishus*
*Medicago*
*Melanocorypha*
*Melhanea*
*Melilotus*
*Mellivora*
*Mentha*
*Mesembryanthemum*
*Michauxia*
*Micromeria*
*Mimusops*
*Minuartia*
*Mirabilis*
*Mollugo*
*Moltkia*
*Moltkiopsis*
*Molucella*
*Momordica*
*Monerma*
*Monodiella*
*Monsonia*
*Morettia*
*Moricandia*
*Moringa*
*Morus*
*Murica*
*Muricarea*

*Glossonema*
*Glycyrrhiza*
*Gnaphalium*
*Gomphocarpus*
*Gossybium*
*Grewia*
*Grimmia*
*Gundelia*
*Gusima*
*Gymnarrhena*
*Gymnocarpos*
*Gypsophila*
*Haemanthus*
*Halimione*
*Halocnemum*
*Halodule*
*Halogeton*
*Halopeplis*
*Halophila*
*Halopyrum*
*Haloxylon*
*Hammada*
*Jasonia*
*Jatropha*
*Juncus*
*Juniperus*
*Jurinea*
*Jussiaea*
*Kalanchoe*
*Kanahia*
*Kermococcus*
*Kickxia*
*Kleinia*
*Kochia*
*Koeleria*
*Koelpinia*
*Koniga*
*Krascheninnkova*
*Kyllinga*
*Lacnophyllum*
*Lactuca*
*Lagurus*
*Lamarckia*
*Lamium*
*Loptochloa*
*Loranthus*
*Lotononis*
*Lotus*
*Ludwigia*
*Lunularia*
*Lupinus*
*Luzula*
*Lyceon*
*Lycium*
*Lygeum*
*Lygos*
*Lythrum*
*Macowania*
*Maerua*
*Magastoma*
*Majorana*
*Malva*
*Mandragora*
*Mania*
*Maresia*
*Marrubium*
*Muscari*
*Myriophyllum*
*Myrsine*
*Myrtus*
*Najas*
*Narcissus*
*Nasturtium*
*Nepeta*
*Nerium*
*Neslia*
*Neurada*
*Nicotiana*
*Nigella*
*Nitella*
*Nitraria*
*Noea*
*Notholaena*
*Notobasis*
*Notoceras*
*Nucularia*
*Nuphar*
*Nymphaea*
*Nymphoides*
*Nypa*
*Obione*
*Ochna*
*Ochradenus*
*Octea*
*Octodium*
*Oenanthe*
*Oenothera*
*Olea*
*Oncoba*
*Onobrychis*
*Ononis*
*Onopordum*
*Onosma*
*Onychium*
*Ophioglossum*
*Ophrys*
*Opohyllum*
*Opophytum*
*Opuntia*
*Orchis*
*Origanum*
*Pegolettia*
*Pennisetum*
*Pentatropis*
*Pentzia*
*Pergularia*
*Periploca*
*Perraldiria*
*Persea*
*Persicaria*
*Phagnalon*
*Phalaris*
*Phelypaea*
*Phillyrea*
*Phleum*
*Phlomis*
*Phoenix*
*Pholiurus*
*Phollyres*
*Phragmites*
*Phyla*
*Phyllanthus*
*Phytophthora*
*Picris*
*Pteranthus*
*Pterocephalus*
*Pterogaillonia*
*Pterolobium*
*Pulicaria*
*Putaria*
*Pyrethrum*
*Pyrus*
*Pythrum*
*Quercus*
*Randonia*
*Ranunculus*
*Ranunculus*
*Raphanus*
*Rapistrum*
*Rapitianus*
*Reaumuria*
*Reichardia*
*Reseda*
*Retama*
*Rhamnus*
*Rhanterium*
*Rhazya*
*Salvadora*
*Salvia*
*Samolus*
*Sanseviera*
*Sarcocornia*
*Sarcopoterium*
*Sarothamnus*
*Satureja*
*Savignya*
*Saxifrage*
*Scabiosa*
*Scadoxus*
*Scandix*
*Schangenia*
*Schimpera*
*Schismus*
*Schoenefeldia*
*Schoenus*
*Schouwia*
*Scilla*
*Scirpus*
*Sclerocephalus*
*Scleropae*

*Orlaya*
*Ormocarpum*
*Ornithogalum*
*Ornithopsis*
*Orobanche*
*Oropetium*
*Oryx*
*Oryza*
*Oryzopsis*
*Ostrya*
*Osyris*
*Otostegia*
*Oudneya*
*Oxalis*
*Oxytropis*
*Paeonia*
*Pallenis*
*Palmoxylon*
*Pancratium*
*Panicum*
*Papver*
*Papyrus*
*Paracaryum*
*Parapholis*
*Parietaria*
*Paronychia*
*Parralderia*
*Parthenium*
*Paspalidium*
*Paspalum*
*Peganum*
*Sorghum*
*Sparganium*
*Spartium*
*Specularia*
*Spergula*
*Spergularia*
*Sphacranthus*
*Sphaerocoma*
*Sphenopus*
*Spirodela*
*Spitzelia*
*Sporobolus*
*Stachys*
*Pimpinella*
*Pinus*
*Pistacia*
*Pistia*
*Pisum*
*Pituranthos*
*Pitymus*
*Plantago*
*Platanus*
*Plectranthus*
*Plumbago*
*Poa*
*Podonosma*
*Polycarpaea*
*Polycarpon*
*Polygonum*
*Polypogon*
*Populus*
*Portulaca*
*Posidonia*
*Potamogeton*
*Poterium*
*Prasium*
*Premna*
*Primula*
*Prosopis*
*Prunus*
*Pseuderucaria*
*Pseudorlaya*
*Psilurus*
*Psoralea*
*Tamarix*
*Tanacetum*
*Taphopzous*
*Taraxacum*
*Targiona*
*Taverniera*
*Taxus*
*Teline*
*Tephrosia*
*Terminalia*
*Tetradicilis*
*Tetragonolobus*
*Tetrapogon*
*Rhetinolepis*
*Rheum*
*Rhizophora*
*Rhus*
*Rhynchosia*
*Ribes*
*Ricia*
*Ricinus*
*Ridolfia*
*Robbairea*
*Roemeria*
*Romula*
*Roripa*
*Rosa*
*Rosmarinus*
*Rosularia*
*Rousettus*
*Rubia*
*Rubus*
*Ruellia*
*Rumex*
*Ruppia*
*Ruscus*
*Ruta*
*Saccharum*
*Saccocalyx*
*Sageratia*
*Saldonella*
*Salicornia*
*Salix*
*Salsola*
*Tragus*
*Trianthema*
*Tribulus*
*Trichilia*
*Trichodesma*
*Tricholaena*
*Trichoneura*
*Trifolium*
*Trigonella*
*Tripteris*
*Trisetaria*
*Trisetum*
*Triticum*
*Scolymus*
*Scorpiurus*
*Scorzonera*
*Scrophularia*
*Scutellaria*
*Sebestena*
*Securigera*
*Seddera*
*Sedum*
*Seidlitzia*
*Selaginella*
*Senecio*
*Senna*
*Serratula*
*Sesbania*
*Seselifolium*
*Setaria*
*Sevada*
*Sida*
*Silene*
*Silybum*
*Simmondsia*
*Sinapis*
*Sisymbrium*
*Smilax*
*Solanum*
*Soldanella*
*Solenostemma*
*Sonchus*
*Sonneratia*
Sorbus
*Vallisnera*
*Varthemia*
*Verbascum*
*Verbena*
*Veronica*
*Viburnum*
*Vicia*
*Vigna*
*Viola*
*Viscum*
*Visnea*
*Vitex*
*Vitis*

*Statice*
*Steinheilia*
*Stellaria*
*Stenbergia*
*Stephanochilus*
*Sterculia*
*Sternbergia*
*Stipa*
*Stipagrostis*
*Striga*
*Styphyllococcus*
*Styrax*
*Suaeda*
*Sus*
*Sylvia*
*Tamarindus*
*Teucrium*
*Thapsia*
*Themeda*
*Thesium*
*Thrincia*
*Thymbra*
*Thymelaea*
*Thymus*
*Tolpis*
*Tolypella*
*Tordylium*
*Tortula*
*Tourneuxia*
*Toxicodendron*
*Traganopsis*
*Traganum*
*Tuberaria*
*Tulipa*
*Typha*
*Ulex*
*Ulmus*
*Umbilicus*
*Urginea*
*Urochondra*
*Urospermum*
*Ursus*
*Urtica*
*Utricularia*
*Vaccaria*
*Vaillantia*
*Valantia*
*Valeriana*
*Volutaria*
*Warionia*
*Withania*
*Wolffia*
*Xanthium*
*Ximenia*
*Zannichellia*
*Zea*
*Zekova*
*Zilla*
*Ziziphus*
*Zosima*
*Zostera*
*Zygophyllum*

MIX
Papier aus verantwortungsvollen Quellen
Paper from responsible sources
FSC® C105338

If you have any concerns about our products,
you can contact us on
**ProductSafety@springernature.com**

In case Publisher is established outside the EU,
the EU authorized representative is:
**Springer Nature Customer Service Center GmbH**
**Europaplatz 3, 69115 Heidelberg, Germany**

Printed by Libri Plureos GmbH
in Hamburg, Germany